Advanced Vehicle Dynamics

Reza N. Jazar

Advanced Vehicle Dynamics

Springer

Reza N. Jazar
Aerospace, Mechanical and Manufacturing
Engineering
RMIT University
Melbourne
VIC, Australia

ISBN 978-3-030-13060-2 ISBN 978-3-030-13062-6 (eBook)
https://doi.org/10.1007/978-3-030-13062-6

Library of Congress Control Number: 2019933099

This Springer imprint is published by the registered company Springer Nature Switzerland AG
The registered company address is: Gewerbestrasse 11, 6330 Cham, Switzerland

We live twice.
The first round is serious.
The second round is funny.
Interestingly, we are mixed together on Earth.

Dedicated to
Kavosh,
Vazan,
Mojgan.

Preface

In this book, we revise the vehicle dynamics based on a new mathematical model for combined tire forces. The force system at the tireprint of a loaded, rolling, steered, cambered tire includes forward, lateral, and vertical forces, as well as aligning, roll, and pitch moment. The forward and lateral forces are the most significant forces in vehicle maneuvering. Tire force modeling has been introduced more than a century ago and improved in several steps through experiments and empirical modeling.

In dynamic modeling of tire forces and vehicles motion, it is traditional to introduce the longitudinal force by slip ratio and lateral force by sideslip angle. Our experiments as well as other available experiments conducted by other investigators during the past decades show that there are decreasing interactions between the lateral and longitudinal tire forces. We introduce a set of equations to model interaction of the lateral and longitudinal tire forces and develop the equations of motion of vehicles based on a new model. Such realistic tire force model has the potential to improve the control strategies and increase the safety of vehicles at critical conditions.

In summary, the vehicle dynamic models that are developed for combined tire forces modeling, four-wheel planar, bicycle roll, and four-wheel roll are new theories introduced in this book. Several examples have been included to show the effectiveness of the mathematical equations as well as good results to compare other analysis and projects. The newly developed equations of motion and the mathematical modeling are perfect for investigation, study, predication, and development of control strategy specially for vehicle drift, sliding, and skidding on slippery, icy, snowy, wet, dirt pavements. The equations are also supporting the new vehicle designs equipped with in-wheel electrical motors, and steering by wires, as well as traditional vehicle designs.

I deeply appreciate the extensive helps from my colleagues, Hormoz Marzbani, Sina Milani, Nguyen Dang Quy, and Amir Salemi, for their valuable comments and reviews, simulation, and tests. This book would have not been prepared without their contributions.

Level of the Book

This book has been developed from nearly a decade of research and experiments in nonlinear vehicle dynamics and teaching courses in vehicle dynamics. It is addressed primarily to the graduate student in engineering. Hence, it is an advanced releaser book that may also be used as a textbook. It provides fundamental and advanced topics needed in computerizing vehicle handling. The whole book can be covered in one course in 12–16 weeks. Students are required to know the fundamentals of vehicle dynamics, kinematics and dynamics, as well as have an acceptable knowledge of numerical methods and differential equations.

The contents of the book have been kept at a fairly theoretical–practical level. All concepts are deeply explained and their application emphasized, and most of the related theories and formal proofs have been explained. The book places a strong emphasis on the physical meaning and applications of the concepts. Topics that have been selected are of high interest in the field. An attempt has been made to expose students and researchers to the most important topics and applications.

Organization of the Book

The book is organized so it can be used for teaching or for self-study. Chapter 1, "Tire Dynamics," contains kinematics and coordinate frame transformation between different frames in tire, wheel, and vehicle body. It also includes the main theory behind combined tire force equations and their interactions. It also covers the vehicle load transfer by forward and lateral acceleration. Chapter 2, "Vehicle Planar Dynamics," develops the equations of motion of a planar rigid vehicle, both bicycle and four-wheel models. Several examples for normal and critical maneuvers are presented. Chapter 3, "Vehicle Roll Dynamics," follows the same method as the Chap. 2 to present the equations of motion of a roll rigid vehicle, also for both bicycle and four-wheel models. There are several new phenomena that appear only in the roll model. Several examples for normal and critical maneuvers are presented. Chapter 4, "Road Dynamics," deals with the main concept of road design to help vehicles move safer and smoother.

Method of Presentation

This book uses a "*fact–reason–application*" structure. The "fact" is the main subject we introduce in each section. Then the reason is given as a "proof." The application of the fact is examined in some "examples." The "examples" are a very important part of the book as they show how to implement the "facts." They also cover some other facts that are needed to expand the "fact."

Prerequisites

Since the book is written for researchers and advanced graduate level students of engineering, the assumption is that users are familiar with matrix algebra, numerical analysis, differential equations, as well as principles of kinematics and dynamics. Therefore, the prerequisites are the fundamentals of kinematics, dynamics, vector analysis, matrix theory, numerical methods, and differential equations.

Unit System

The system of units adopted in this book is, unless otherwise stated, the international system of units (SI). The units of degree (deg) or radian (rad) are utilized for variables representing angular quantities.

Symbols

- Lowercase bold letters indicate a vector. Vectors may be expressed in an n-dimensional Euclidian space. Example:

$$\begin{array}{cccccc} \mathbf{r}, & \mathbf{s}, & \mathbf{d}, & \mathbf{a}, & \mathbf{b}, & \mathbf{c} \\ \mathbf{p}, & \mathbf{q}, & \mathbf{v}, & \mathbf{w}, & \mathbf{y}, & \mathbf{z} \\ \boldsymbol{\omega}, & \boldsymbol{\alpha}, & \boldsymbol{\epsilon}, & \boldsymbol{\theta}, & \boldsymbol{\delta}, & \boldsymbol{\phi} \end{array}$$

- Uppercase bold letters indicate a dynamic vector or a dynamic matrix, such as force and moment. Example:

$$\mathbf{F} \quad \mathbf{M}$$

- Lowercase letters with a hat indicate a unit vector. Unit vectors are not bold. Example:

$$\begin{array}{cccccc} \hat{\imath}, & \hat{\jmath}, & \hat{k}, & \hat{u}, & \hat{u}, & \hat{n} \\ \hat{I}, & \hat{J}, & \hat{K}, & \hat{u}_{\theta}, & \hat{u}_{\varphi}, & \hat{u}_{\psi} \end{array}$$

- Lowercase letters with a tilde indicate a 3×3 skew symmetric matrix associated with a vector. Example:

$$\tilde{a} = \begin{bmatrix} 0 & -a_3 & a_2 \\ a_3 & 0 & -a_1 \\ -a_2 & a_1 & 0 \end{bmatrix} \qquad \mathbf{a} = \begin{bmatrix} a_1 \\ a_2 \\ a_3 \end{bmatrix}$$

- An arrow above two uppercase letters indicates the start and end points of a position vector. Example:

$$\overrightarrow{ON} = \text{a position vector from point } O \text{ to point } N$$

- The length of a vector is indicated by a non-bold lowercase letter. Example:

$$r = |\mathbf{r}| \qquad a = |\mathbf{a}| \qquad b = |\mathbf{b}| \qquad s = |\mathbf{s}|$$

- Capital letter B is utilized to denote a body coordinate frame. Example:

$$B(oxyz) \qquad B(Oxyz) \qquad B_1(o_1x_1y_1z_1)$$

- Capital letter G is utilized to denote a global, inertial, or fixed coordinate frame. Example:

$$G \qquad G(XYZ) \qquad G(OXYZ)$$

- Right subscript on a transformation matrix indicates the *departure* frames. Example:

$$R_B = \text{transformation matrix from frame } B(oxyz)$$

- Left superscript on a transformation matrix indicates the *destination* frame. Example:

$$\begin{aligned} {}^{G}R_B = &\text{ transformation matrix from frame } B(oxyz) \\ &\text{to frame } G(OXYZ) \end{aligned}$$

- Capital letter R indicates rotation or a transformation matrix, if it shows the beginning and destination coordinate frames. Example:

$${}^{G}R_B = \begin{bmatrix} \cos\alpha & -\sin\alpha & 0 \\ \sin\alpha & \cos\alpha & 0 \\ 0 & 0 & 1 \end{bmatrix}$$

- Whenever there is no subscript or superscript, the matrices are shown in a bracket. Example:

$$[T] = \begin{bmatrix} \cos\alpha & -\sin\alpha & 0 \\ \sin\alpha & \cos\alpha & 0 \\ 0 & 0 & 1 \end{bmatrix}$$

- Left superscript on a vector denotes the frame in which the vector is expressed. That superscript indicates the frame that the vector belongs to, so the vector is expressed using the unit vectors of that frame. Example:

$$ {}^{G}\mathbf{r} = \text{position vector expressed in frame } G(OXYZ) $$

- Right subscript on a vector denotes the tip point that the vector is referred to. Example:

$$ \begin{aligned} {}^{G}\mathbf{r}_P &= \text{position vector of point } P \\ &\quad \text{expressed in coordinate frame } G(OXYZ) \end{aligned} $$

- Right subscript on an angular velocity vector indicates the frame that the angular vector is referred to. Example:

$$ \boldsymbol{\omega}_B = \text{angular velocity of the body coordinate frame } B(oxyz) $$

- Left subscript on an angular velocity vector indicates the frame that the angular vector is measured with respect to. Example:

$$ \begin{aligned} {}_{G}\boldsymbol{\omega}_B &= \text{angular velocity of the body coordinate frame } B(oxyz) \\ &\quad \text{with respect to the global coordinate frame } G(OXYZ) \end{aligned} $$

- Left superscript on an angular velocity vector denotes the frame in which the angular velocity is expressed. Example:

$$ \begin{aligned} {}_{G}^{B_2}\boldsymbol{\omega}_{B_1} &= \text{angular velocity of the body coordinate frame } B_1 \\ &\quad \text{with respect to the global coordinate frame } G, \\ &\quad \text{and expressed in body coordinate frame } B_2 \end{aligned} $$

Whenever the subscript and superscript of an angular velocity are the same, we usually drop the left superscript. Example:

$$ {}_{G}\boldsymbol{\omega}_B \equiv {}_{G}^{G}\boldsymbol{\omega}_B $$

Also for position, velocity, and acceleration vectors, we drop the left subscripts if it is the same as the left superscript. Example:

$$ {}_{B}^{B}\mathbf{v}_P \equiv {}^{B}\mathbf{v}_P $$

- Left superscript on derivative operators indicates the frame in which the derivative of a variable is taken. Example:

$$\frac{{}^{G}d}{dt}x \qquad \frac{{}^{G}d}{dt}{}^{B}\mathbf{r}_P \qquad \frac{{}^{B}d}{dt}{}^{G}_{B}\mathbf{r}_P$$

If the variable is a vector function, and also the frame in which the vector is defined is the same frame in which a time derivative is taken, we may use the following short notation,

$$\frac{{}^{G}d}{dt}{}^{G}\mathbf{r}_P = {}^{G}\dot{\mathbf{r}}_P \qquad \frac{{}^{B}d}{dt}{}^{B}_{o}\mathbf{r}_P = {}^{B}_{o}\dot{\mathbf{r}}_P$$

and write equations simpler. Example:

$${}^{G}\mathbf{v} = \frac{{}^{G}d}{dt}{}^{G}\mathbf{r}(t) = {}^{G}\dot{\mathbf{r}}$$

- If followed by angles, lowercase c and s denote cos and sin functions in mathematical equations. Example:

$$c\alpha = \cos\alpha \qquad s\varphi = \sin\varphi$$

- Capital bold letter $\mathbf{I}$ indicates a unit matrix, which, depending on the dimension of the matrix equation, could be a 3×3 or a 4×4 unit matrix. $\mathbf{I}_3$ or $\mathbf{I}_4$ are also being used to clarify the dimension of $\mathbf{I}$. Example:

$$\mathbf{I} = \mathbf{I}_3 = \begin{bmatrix} 1 & 0 & 0 \\ 0 & 1 & 0 \\ 0 & 0 & 1 \end{bmatrix}$$

- An asterisk ★ indicates a more advanced subject or example that is not designed for undergraduate teaching and can be dropped in the first reading.

Melbourne, VIC, Australia Reza N. Jazar

Contents

Chapter 1
Tire Dynamics

The dynamic performance of a vehicle is mainly determined by the interaction of its tires and road. A vehicle can only move and maneuver by the force systems generated under the tires. In this chapter, we review the generated forces by a tire as well as the kinematics of a tire with respect to the vehicle body. Figure 1.1 illustrates a steered and leaned tire along with the required coordinate frames to analyze its orientation with respect to the vehicle body.

1.1 Wheel and Tire Coordinate Frames

By narrowing a tire, it becomes a flat disk in its mid-plane called the *tire-plane*. We introduce three coordinate frames to express the orientation of such a flat disc with respect to the vehicle. These coordinate frames are: the *wheel frame* W, *wheel-body frame* C, and *tire frame* T. Figure 1.1 illustrates the W and T coordinate frames, and Fig. 1.2 illustrates the W and C coordinate frames (Jazar 2017).

The wheel coordinate frame W (x_w, y_w, z_w) is attached to the center of the wheel and it follows every motion of the wheel except the spin. Therefore, the x_w and z_w axes are always in the tire-plane, and the y_w-axis is always on the spin axis.

When a wheel is straight and upright on the ground, it is at its *neutral* or *rest position*. The W-frame at rest position becomes coincident with the wheel-body coordinate frame C (x_c, y_c, z_c) which is attached to the center of the neutral wheel, parallel to the vehicle coordinate axes, B. The wheel-body frame C is motionless with respect to the vehicle and does not follow any motion of the wheel. The axes x_c, y_c, z_c are always parallel to the axes of the body coordinate frame, B. The vehicle or body coordinate frame B (x, y, z) is attached to the vehicle at its mass center. The x-axis is the longitudinal forward axis of the vehicle parallel to the ground, the y-axis is leftward and parallel to the ground, and the z-axis is the upward axis. The body coordinate frame acts as a reference frame to determine the position and orientation of any component of the vehicle.

R. N. Jazar, *Advanced Vehicle Dynamics*,
https://doi.org/10.1007/978-3-030-13062-6_1

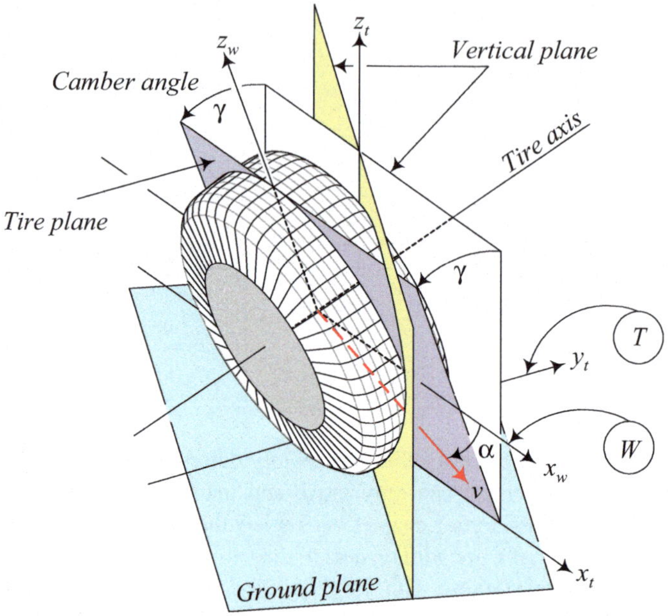

Fig. 1.1 Tire frames T and W, sideslip, and camber angles

The tire coordinate frame T (x_t, y_t, z_t) is set at the center of the *tireprint*, which is assumed to be at the intersection of the tire-plane and the ground. The *tireprint* or *contact patch* is the contact area of the tire and the ground. The z_t-axis is always perpendicular to the ground and upward. The x_t-axis is along the intersection line of the tire-plane and the ground and y_t makes the T coordinate frame right handed. The tire frame does not follow the spin and camber rotations of the tire; however, it follows the steer angle rotation about the z_c-axis (Jazar 2017).

Example 1 Transformation of W to T coordinate frame.

Let us assume that the tire is leaned by an angle γ. The lean of tire is a rotation about the x_t-axis and is called the *camber angle* γ. Camber angle generates some lateral force. Figure 1.2 illustrates the relative configuration of a tire frame T and a wheel frame W.

To determine transformation from W to T, we should begin by assuming W-frame is sitting on the T-frame. Then W needs to go over a rotation and a translation to come to its current position. It should rotate γ about the x_t-axis followed by a displacement ${}^T\mathbf{d}_W$ along the z_w-axis, where R is the radius of the tire. The rotation transformation matrix ${}^TR_W$ and displacement ${}^T\mathbf{d}_W$ are:

$$
{}^TR_W = \begin{bmatrix} 1 & 0 & 0 \\ 0 & \cos\gamma & -\sin\gamma \\ 0 & \sin\gamma & \cos\gamma \end{bmatrix} \tag{1.1}
$$

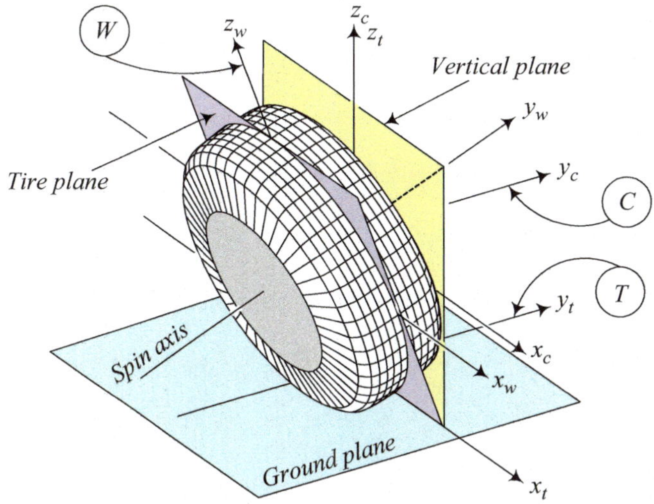

Fig. 1.2 Illustration of tire and wheel coordinate frames

$$ {}^{T}\mathbf{d}_{W} = R\ {}^{T}R_{W}\ {}^{W}\hat{k} \tag{1.2}$$

and the total transformation from W-frame to T-frame is:

$$ {}^{T}\mathbf{r} = {}^{T}R_{W}\ {}^{W}\mathbf{r} + {}^{T}\mathbf{d}_{W} = {}^{T}R_{W}\ {}^{W}\mathbf{r} + R\ {}^{T}R_{W}\ {}^{W}\hat{k} \tag{1.3}$$

where $\mathbf{r}$ is the position vector of a point of the tire in tire-plane.

The displacement ${}^{T}\mathbf{d}_{W}$ indicates the T-expression of the position vector of origin of the wheel frame in the tire frame T. If ${}^{W}\mathbf{r}_{P}$ indicates the position vector of a point P in the wheel frame,

$$ {}^{W}\mathbf{r}_{P} = \begin{bmatrix} x_P & y_P & z_P \end{bmatrix}^{T} \tag{1.4}$$

then the coordinates of the point P in the tire frame ${}^{T}\mathbf{r}_{P}$ are

$$ \begin{aligned} {}^{T}\mathbf{r}_{P} &= {}^{T}R_{W}\ {}^{W}\mathbf{r}_{P} + {}^{T}\mathbf{d}_{W} = {}^{T}R_{W}\ {}^{W}\mathbf{r}_{P} + {}^{T}R_{W}\ {}_{T}^{W}\mathbf{d}_{W} \\ &= \begin{bmatrix} x_P \\ y_P \cos\gamma - R\sin\gamma - z_P \sin\gamma \\ R\cos\gamma + z_P\cos\gamma + y_P \sin\gamma \end{bmatrix} \end{aligned} \tag{1.5}$$

where ${}_{T}^{W}\mathbf{d}_{W}$ is the W-expression of the position vector of the wheel frame in the tire frame.

$$ {}_{T}^{W}\mathbf{d}_{W} = R\ {}^{W}\hat{k} = \begin{bmatrix} 0 & 0 & R \end{bmatrix}^{T} \tag{1.6}$$

As an example, the center of the wheel ${}^W\mathbf{r}_P = {}^W\mathbf{r}_o = \mathbf{0}$ is the origin of the wheel frame W and is always at ${}^W\mathbf{r} = {}^W\mathbf{0} = {}^W\begin{bmatrix} 0 & 0 & 0 \end{bmatrix}$ in the W coordinate frame. Therefore, in the T-frame, the center of the wheel will be at:

$$ {}^T\mathbf{r} = {}^T R_W \, {}^W\mathbf{0} + R \, {}^T R_W \, {}^W\hat{k} = \begin{bmatrix} 0 \\ -R\sin\gamma \\ R\cos\gamma \end{bmatrix} \tag{1.7} $$

The transformation from the wheel W to tire T coordinate frame may also be expressed by a 4×4 homogeneous transformation matrix ${}^T T_W$,

$$ {}^T\mathbf{r}_P = {}^T T_W \, {}^W\mathbf{r}_P = \begin{bmatrix} {}^T R_W & {}^T\mathbf{d}_W \\ 0 & 1 \end{bmatrix} {}^W\mathbf{r}_P \tag{1.8} $$

where (Jazar 2010a, 2011; Mason 2001),

$$ {}^T T_W = \begin{bmatrix} 1 & 0 & 0 & 0 \\ 0 & \cos\gamma & -\sin\gamma & -R\sin\gamma \\ 0 & \sin\gamma & \cos\gamma & R\cos\gamma \\ 0 & 0 & 0 & 1 \end{bmatrix} \tag{1.9} $$

Example 2 Tire T to wheel W frame transformation.

Assume ${}^T\mathbf{r}_P$ to indicate the position vector of a point P in the tire coordinate frame,

$$ {}^T\mathbf{r}_P = \begin{bmatrix} x_P & y_P & z_P \end{bmatrix}^T \tag{1.10} $$

The position vector ${}^W\mathbf{r}_P$ of the point P in the wheel coordinate frame is

$$ \begin{aligned} {}^W\mathbf{r}_P &= {}^W R_T \, {}^T\mathbf{r}_P - {}^W_T\mathbf{d}_W \\ &= \begin{bmatrix} x_P \\ y_P\cos\gamma + z_P\sin\gamma \\ z_P\cos\gamma - R - y_P\sin\gamma \end{bmatrix} \end{aligned} \tag{1.11} $$

because

$$ {}^W R_T = \begin{bmatrix} 1 & 0 & 0 \\ 0 & \cos\gamma & \sin\gamma \\ 0 & -\sin\gamma & \cos\gamma \end{bmatrix} \qquad {}^W_T\mathbf{d}_W = -\,{}^W\mathbf{d}_T = \begin{bmatrix} 0 \\ 0 \\ R \end{bmatrix} \tag{1.12} $$

We may also multiply both sides of Eq. (1.3) by ${}^T R_W^T$ to get the same result.

$$ {}^T R_W^T \, {}^T\mathbf{r}_P = {}^W\mathbf{r}_P + {}^T R_W^T \, {}^T\mathbf{d}_W = {}^W\mathbf{r}_P + {}^W_T\mathbf{d}_W \tag{1.13} $$

$$ {}^{W}\mathbf{r}_P = {}^{W}R_T\, {}^{T}\mathbf{r}_P - {}^{W}_{T}\mathbf{d}_W \tag{1.14} $$

As an example, the center of tireprint in the wheel frame is at

$$ {}^{W}\mathbf{r}_P = \begin{bmatrix} 1 & 0 & 0 \\ 0 & \cos\gamma & -\sin\gamma \\ 0 & \sin\gamma & \cos\gamma \end{bmatrix}^T \begin{bmatrix} 0 \\ 0 \\ 0 \end{bmatrix} - \begin{bmatrix} 0 \\ 0 \\ R \end{bmatrix} = \begin{bmatrix} 0 \\ 0 \\ -R \end{bmatrix} \tag{1.15} $$

The corresponding homogeneous transformation matrix ${}^{W}T_T$ from the tire to wheel frame would be

$$ {}^{W}T_T = \begin{bmatrix} {}^{W}R_T & {}^{W}\mathbf{d}_T \\ 0 & 1 \end{bmatrix} = \begin{bmatrix} 1 & 0 & 0 & 0 \\ 0 & \cos\gamma & \sin\gamma & 0 \\ 0 & -\sin\gamma & \cos\gamma & -R \\ 0 & 0 & 0 & 1 \end{bmatrix} \tag{1.16} $$

We can check that ${}^{W}T_T = {}^{T}T_W^{-1}$, using the inverse of a homogeneous transformation matrix rule.

$$ \begin{aligned} {}^{T}T_W^{-1} &= \begin{bmatrix} {}^{T}R_W & {}^{T}\mathbf{d}_W \\ 0 & 1 \end{bmatrix}^{-1} = \begin{bmatrix} {}^{T}R_W^T & -{}^{T}R_W^T\, {}^{T}\mathbf{d}_W \\ 0 & 1 \end{bmatrix} \\ &= \begin{bmatrix} {}^{W}R_T & -{}^{W}R_T\, {}^{T}\mathbf{d}_W \\ 0 & 1 \end{bmatrix} = \begin{bmatrix} {}^{W}R_T & {}^{W}\mathbf{d}_T \\ 0 & 1 \end{bmatrix} \end{aligned} \tag{1.17} $$

Example 3 Tire T to wheel-body C frame transformation.

Figure 1.3 illustrates the relative configuration of a wheel-body frame C, a tire frame T, and a wheel frame W (Jazar 2017).

To determine the transformation from T-frame to C-frame, we may assume T is on C and find the required maneuvers that T needs to go to its current position. Let us assume that the wheel steers about the z_c-axis by the steer angle δ. The T-frame should go through a rotation of δ about the z_t-axis followed by a displacement of ${}^{C}\mathbf{d}_T$. The rotation matrix ${}^{C}R_T$ and displacement ${}^{C}\mathbf{d}_T$ are:

$$ {}^{C}R_T = \begin{bmatrix} \cos\delta & -\sin\delta & 0 \\ \sin\delta & \cos\delta & 0 \\ 0 & 0 & 1 \end{bmatrix} \tag{1.18} $$

$$ {}^{C}\mathbf{d}_T = -R\, {}^{C}R_T\, {}^{T}\hat{k} \tag{1.19} $$

and therefore, the transformation between the tire T and wheel-body C frames would be:

$$ {}^{C}\mathbf{r} = {}^{C}R_T\, {}^{T}\mathbf{r} + {}^{C}\mathbf{d}_T = {}^{C}R_T\, {}^{T}\mathbf{r} - R\, {}^{C}R_T\, {}^{T}\hat{k} \tag{1.20} $$

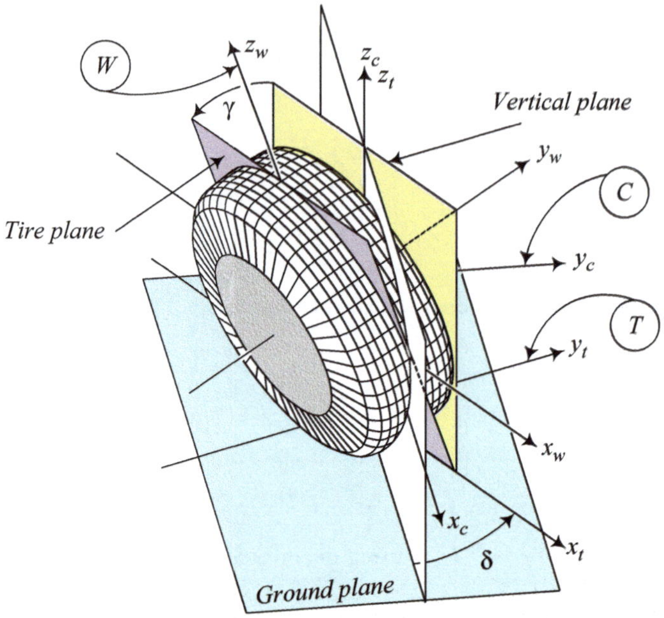

Fig. 1.3 Illustration of tire, wheel, and body coordinate frames

where ${}^T\mathbf{r}$ is the position vector of a point in T-frame and ${}^C\mathbf{r}$ is the same point in the C-frame. As an example, the center of the wheel from (1.7) is always at

$$ {}^T\mathbf{r} = \begin{bmatrix} 0 & -R\sin\gamma & R\cos\gamma \end{bmatrix}^T \tag{1.21} $$

Therefore, in the C-frame, the center of the wheel will be at:

$$ {}^C\mathbf{r} = {}^C R_T \, {}^T\mathbf{r} - R \, {}^C R_T \, {}^T\hat{k} = \begin{bmatrix} R\sin\gamma\sin\delta \\ -R\cos\delta\sin\gamma \\ R\cos\gamma - R \end{bmatrix} \tag{1.22} $$

The origin of the tire frame is at ${}^C\mathbf{d}_T$ in the wheel-body frame.

$$ {}^C\mathbf{d}_T = \begin{bmatrix} 0 & 0 & -R \end{bmatrix}^T \tag{1.23} $$

Therefore, the transformation between the tire and wheel-body frames can be expressed equivalently, by a homogeneous transformation matrix ${}^C T_T$.

$$ {}^C T_T = \begin{bmatrix} {}^C R_T & {}^C\mathbf{d}_T \\ 0 & 1 \end{bmatrix} = \begin{bmatrix} \cos\delta & -\sin\delta & 0 & 0 \\ \sin\delta & \cos\delta & 0 & 0 \\ 0 & 0 & 1 & -R \\ 0 & 0 & 0 & 1 \end{bmatrix} \tag{1.24} $$

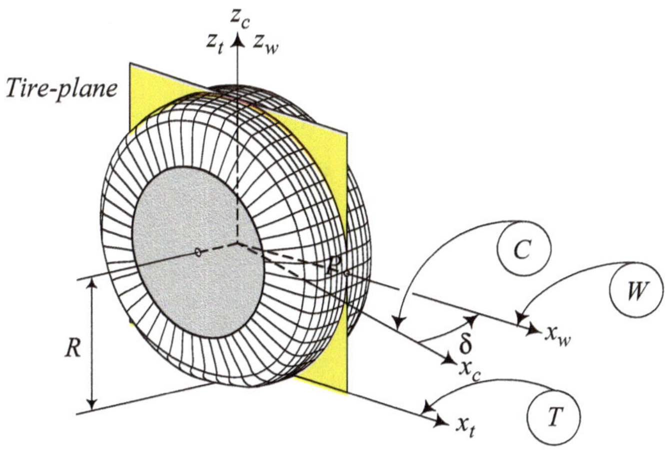

Fig. 1.4 The tire, wheel, and wheel-body frames of a steered wheel

As an example, the wheel-body coordinates of the point P on the tread of a negatively steered tire at the position shown in Fig. 1.4 are:

$$
{}^{C}\mathbf{r} = {}^{C}T_{T}\,{}^{T}\mathbf{r}_{P} = \begin{bmatrix} \cos(-\delta) & -\sin(-\delta) & 0 & 0 \\ \sin(-\delta) & \cos(-\delta) & 0 & 0 \\ 0 & 0 & 1 & -R \\ 0 & 0 & 0 & 1 \end{bmatrix} \begin{bmatrix} R \\ 0 \\ R \\ 1 \end{bmatrix}
$$

$$
= \begin{bmatrix} R\cos\delta \\ -R\sin\delta \\ 0 \\ 1 \end{bmatrix} \tag{1.25}
$$

The homogeneous transformation matrix for wheel-body to tire frame ${}^{T}T_{C}$ is:

$$
{}^{T}T_{C} = {}^{C}T_{T}^{-1} = \begin{bmatrix} {}^{C}R_{T} & {}^{C}\mathbf{d}_{T} \\ 0 & 1 \end{bmatrix}^{-1} = \begin{bmatrix} {}^{C}R_{T}^{T} & -{}^{C}R_{T}^{T}\,{}^{C}\mathbf{d}_{T} \\ 0 & 1 \end{bmatrix}
$$

$$
= \begin{bmatrix} {}^{C}R_{T}^{T} & -{}^{T}_{C}\mathbf{d}_{T} \\ 0 & 1 \end{bmatrix} = \begin{bmatrix} \cos\delta & \sin\delta & 0 & 0 \\ -\sin\delta & \cos\delta & 0 & 0 \\ 0 & 0 & 1 & R \\ 0 & 0 & 0 & 1 \end{bmatrix} \tag{1.26}
$$

Example 4 Wheel W to wheel-body C frame transformation.

The homogeneous transformation matrix ${}^{C}T_{W}$ to go from the wheel frame W to the wheel-body frame C may be found by combined transformations of ${}^{C}T_{T}$ and ${}^{T}T_{W}$.

$$
\begin{aligned}
{}^{C}T_W &= {}^{C}T_T \; {}^{T}T_W \\
&= \begin{bmatrix} c\delta & -s\delta & 0 & 0 \\ s\delta & c\delta & 0 & 0 \\ 0 & 0 & 1 & -R \\ 0 & 0 & 0 & 1 \end{bmatrix} \begin{bmatrix} 1 & 0 & 0 & 0 \\ 0 & c\gamma & -s\gamma & -R\sin\gamma \\ 0 & s\gamma & c\gamma & R\cos\gamma \\ 0 & 0 & 0 & 1 \end{bmatrix} \\
&= \begin{bmatrix} \cos\delta & -\cos\gamma\sin\delta & \sin\gamma\sin\delta & R\sin\gamma\sin\delta \\ \sin\delta & \cos\gamma\cos\delta & -\cos\delta\sin\gamma & -R\cos\delta\sin\gamma \\ 0 & \sin\gamma & \cos\gamma & R\cos\gamma - R \\ 0 & 0 & 0 & 1 \end{bmatrix}
\end{aligned}
\tag{1.27}
$$

If $\mathbf{r}_P$ indicates the position vector of a point P in the wheel coordinate frame,

$$
{}^{W}\mathbf{r}_P = \begin{bmatrix} x_P & y_P & z_P \end{bmatrix}^T \tag{1.28}
$$

then the homogeneous position vector ${}^{C}\mathbf{r}_P$ of the point P in the wheel-body coordinate frame would be:

$$
\begin{aligned}
{}^{C}\mathbf{r}_P &= {}^{C}T_W \; {}^{W}\mathbf{r}_P \\
&= \begin{bmatrix} x_P\cos\delta - y_P\cos\gamma\sin\delta + (R+z_P)\sin\gamma\sin\delta \\ x_P\sin\delta + y_P\cos\gamma\cos\delta - (R+z_P)\cos\delta\sin\gamma \\ -R + (R+z_P)\cos\gamma + y_P\sin\gamma \\ 1 \end{bmatrix}
\end{aligned}
\tag{1.29}
$$

As an example, the position of the wheel center ${}^{W}\mathbf{r} = \mathbf{0}$ for a cambered and steered wheel is at

$$
{}^{C}\mathbf{r} = {}^{C}T_W \; {}^{W}\mathbf{r} = \begin{bmatrix} R\sin\gamma\sin\delta \\ -R\cos\delta\sin\gamma \\ -R(1-\cos\gamma) \\ 1 \end{bmatrix} \tag{1.30}
$$

The $z_c = R(\cos\gamma - 1)$ indicates how far the center of the wheel moves down when the wheel cambers.

If the wheel is not steerable, then $\delta = 0$ and the transformation matrix ${}^{C}T_W$ reduces to

$$
{}^{C}T_W = \begin{bmatrix} 1 & 0 & 0 & 0 \\ 0 & \cos\gamma & -\sin\gamma & -R\sin\gamma \\ 0 & \sin\gamma & \cos\gamma & R(\cos\gamma - 1) \\ 0 & 0 & 0 & 1 \end{bmatrix} \tag{1.31}
$$

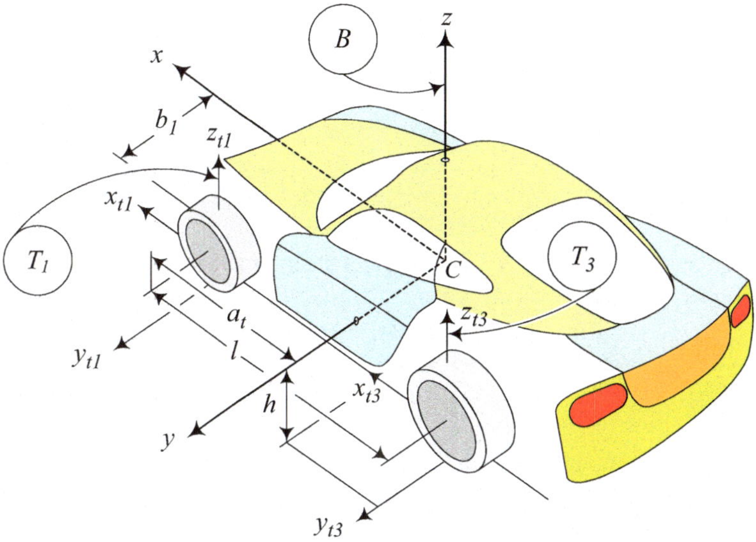

Fig. 1.5 The coordinate frames of the first and fourth tires of a four-wheel vehicle with respect to the body frame

that shows

$$
\begin{aligned}
{}^{C}\mathbf{r}_P &= {}^{C}T_W \, {}^{W}\mathbf{r}_P \\
&= \begin{bmatrix} x_P \\ y_P \cos\gamma - R\sin\gamma - z_P \sin\gamma \\ z_P \cos\gamma + y_P \sin\gamma + R(\cos\gamma - 1) \\ 1 \end{bmatrix}
\end{aligned} \tag{1.32}
$$

Example 5 Tire T to coordinate C frame transformation.

Figure 1.5 illustrates the first and fourth tires of a four-wheel vehicle. There is a body coordinate frame $B\,(x, y, z)$ attached to the mass center C of the vehicle. There are also two tire coordinate frames $T_1\left(x_{t_1}, y_{t_1}, z_{t_1}\right)$ and $T_4\left(x_{t_4}, y_{t_4}, z_{t_4}\right)$ attached to the tires 1 and 4 at the center of their tireprints (Jazar 2017).

The origin of the tire coordinate frame T_1 is at ${}^{B}\mathbf{d}_{T_1}$

$$
{}^{B}\mathbf{d}_{T_1} = \begin{bmatrix} a_1 & b_1 & -h \end{bmatrix}^T \tag{1.33}
$$

where a_1 is the longitudinal distance between C and the front axle, b_1 is the lateral distance between C and the tireprint of the tire 1, and h is the height of C from the ground level. If P is a point in the tire frame at ${}^{T_1}\mathbf{r}_P$

$$
{}^{T_1}\mathbf{r}_P = \begin{bmatrix} x_P & y_P & z_P \end{bmatrix}^T \tag{1.34}
$$

then its coordinates in the body frame are

$$\begin{aligned} {}^{B}\mathbf{r}_P &= {}^{B}R_{T_1}\,{}^{T_1}\mathbf{r}_P + {}^{B}\mathbf{d}_{T_1} \\ &= \begin{bmatrix} x_P\cos\delta_1 - y_P\sin\delta_1 + a_1 \\ y_P\cos\delta_1 + x_P\sin\delta_1 + b_1 \\ z_P - h \end{bmatrix} \end{aligned} \tag{1.35}$$

The transformation matrix ${}^{B}R_{T_1}$ is a result of steering about the z_{t_1}-axis.

$${}^{B}R_{T_1} = \begin{bmatrix} \cos\delta_1 & -\sin\delta_1 & 0 \\ \sin\delta_1 & \cos\delta_1 & 0 \\ 0 & 0 & 1 \end{bmatrix} \tag{1.36}$$

Employing Eq. (1.3), we may examine a wheel point P at ${}^{W}\mathbf{r}_P$

$${}^{W}\mathbf{r}_P = \begin{bmatrix} x_P & y_P & z_P \end{bmatrix}^T \tag{1.37}$$

and find the body coordinates of the point

$$\begin{aligned} {}^{B}\mathbf{r}_P &= {}^{B}R_{T_1}\,{}^{T_1}\mathbf{r}_P + {}^{B}\mathbf{d}_{T_1} \\ &= {}^{B}R_{T_1}\left({}^{T_1}R_W\,{}^{W}\mathbf{r}_P + {}^{T_1}\mathbf{d}_W\right) + {}^{B}\mathbf{d}_{T_1} \\ &= {}^{B}R_{T_1}\,{}^{T_1}R_W\,{}^{W}\mathbf{r}_P + {}^{B}R_{T_1}\,{}^{T_1}\mathbf{d}_W + {}^{B}\mathbf{d}_{T_1} \\ &= {}^{B}R_W\,{}^{W}\mathbf{r}_P + {}^{B}R_{T_1}\,{}^{T_1}\mathbf{d}_W + {}^{B}\mathbf{d}_{T_1} \end{aligned} \tag{1.38}$$

$${}^{B}\mathbf{r}_P = \begin{bmatrix} x_P\cos\delta_1 - y_P\cos\gamma\sin\delta_1 + (R+z_P)\sin\gamma\sin\delta_1 + a_1 \\ x_P\sin\delta_1 + y_P\cos\gamma\cos\delta_1 - (R+z_P)\cos\delta_1\sin\gamma + b_1 \\ (R+z_P)\cos\gamma + y_P\sin\gamma - h \end{bmatrix} \tag{1.39}$$

where

$$\begin{aligned} {}^{B}R_W &= {}^{B}R_{T_1}\,{}^{T_1}R_W \\ &= \begin{bmatrix} \cos\delta_1 & -\cos\gamma\sin\delta_1 & \sin\gamma\sin\delta_1 \\ \sin\delta_1 & \cos\gamma\cos\delta_1 & -\cos\delta_1\sin\gamma \\ 0 & \sin\gamma & \cos\gamma \end{bmatrix} \end{aligned} \tag{1.40}$$

$${}^{T_1}\mathbf{d}_W = \begin{bmatrix} 0 & -R\sin\gamma & R\cos\gamma \end{bmatrix}^T \tag{1.41}$$

As an example let us consider a point P on the z_{t_1}-axis at

$${}^{T_1}\mathbf{r}_P = \begin{bmatrix} 0 & 0 & 2R \end{bmatrix}^T \tag{1.42}$$

The coordinates of P in the body B frame are:

$$\begin{aligned} {}^{B}\mathbf{r}_P &= {}^{B}R_{T_1}\,{}^{T_1}\mathbf{r}_P + {}^{B}\mathbf{d}_{T_1} \\ &= \begin{bmatrix} \cos\delta_1 & -\sin\delta_1 & 0 \\ \sin\delta_1 & \cos\delta_1 & 0 \\ 0 & 0 & 1 \end{bmatrix} \begin{bmatrix} 0 \\ 0 \\ 2R \end{bmatrix} + \begin{bmatrix} a_1 \\ b_1 \\ -h \end{bmatrix} \\ &= \begin{bmatrix} a_1 \\ b_1 \\ 2R - h \end{bmatrix} \end{aligned} \tag{1.43}$$

Example 6 Wheel-body C to vehicle B transformation.

The wheel-body coordinate frame C is always parallel to the vehicle frame B. The origin of the wheel-body coordinate frame of the wheel number 1 is at

$${}^{B}\mathbf{d}_{C_1} = \begin{bmatrix} a_1 & b_1 & -h + R \end{bmatrix}^T \tag{1.44}$$

Hence the transformation between the two frames is only a displacement.

$${}^{B}\mathbf{r} = {}^{B}\mathbf{I}_{C_1}\,{}^{C_1}\mathbf{r} + {}^{B}\mathbf{d}_{C_1} \tag{1.45}$$

As an example consider a point P on the x_{t_1}-axis at a distance R from the center of the tire and assume that the tire is turning about y_{t_1}-axis with angular speed ω. Therefore,

$${}^{C_1}\mathbf{r}_P\,(t) = \begin{bmatrix} R\cos\omega t & 0 & -R\sin\omega t \end{bmatrix}^T \tag{1.46}$$

and the coordinates of the point P in the body coordinate frame B at time t are:

$$\begin{aligned} {}^{B}\mathbf{r}\,(t) &= {}^{B}\mathbf{I}_{C_1}\,{}^{C_1}\mathbf{r}_P\,(t) + {}^{B}\mathbf{d}_{C_1} \\ &= \begin{bmatrix} 1\,0\,0 \\ 0\,1\,0 \\ 0\,0\,1 \end{bmatrix} \begin{bmatrix} R\cos\omega t \\ 0 \\ -R\sin\omega t \end{bmatrix} + \begin{bmatrix} a_1 \\ b_1 \\ -h + R \end{bmatrix} \\ &= \begin{bmatrix} a_1 + R\cos t\omega \\ b_1 \\ R - h - R\sin t\omega \end{bmatrix} \end{aligned} \tag{1.47}$$

Example 7 Difference between tire and wheel frames.

As shown in Fig. 1.6, to express the orientation of a wheel and the force system, three coordinate frames are needed: the wheel frame W, wheel-body frame C, and tire frame T. The wheel coordinate frame W (x_w, y_w, z_w) is attached to the center of the wheel and follows every motion of the wheel except the wheel spin. Therefore, the x_w and z_w axes are always in the tire-plane, while the y_w-axis is always on the spin axis.

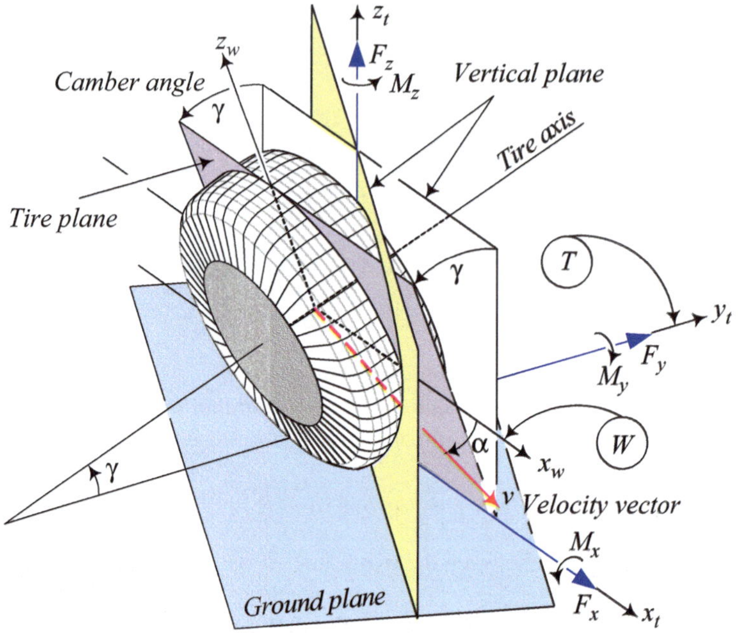

Fig. 1.6 Tire force system is expressed in T-frame

We also attach a wheel-body coordinate frame C (x_c, y_c, z_c) at the center of the wheel parallel and fixed with respect to the vehicle coordinate axes B (x, y, z). The wheel-body frame C is motionless with respect to the vehicle coordinate and does not follow any motion of the wheel. When the wheel is upright straight, the W-frame coincides with C-frame and becomes parallel to the vehicle coordinate frame B. The W-frame makes the steer angle δ and camber angle γ with respect to the C-frame.

The tire coordinate frame T (x_t, y_t, z_t) is set at the center of the tireprint with the z_t-axis perpendicular to the ground and parallel to the z-axis. The x_t-axis is along the intersection line of the tire-plane and the ground. The tire frame follows the steer angle rotation about the z_c-axis but it does not follow the spin and camber rotations of the tire.

To determine the difference between the T and W frames, let us use ${}^T\mathbf{d}_W$ to indicate the T-expression of the position vector of the wheel frame origin relative to the tire frame origin. Having the coordinates of a point P in the wheel frame, we may find its coordinates in the tire frame by:

$$ {}^T\mathbf{r}_P = {}^T R_W \, {}^W\mathbf{r}_P + {}^T\mathbf{d}_W \tag{1.48} $$

If ${}^W\mathbf{r}_P$ indicates the position vector of a point P in the wheel frame,

$$ {}^W\mathbf{r}_P = \begin{bmatrix} x_P & y_P & z_P \end{bmatrix}^T \tag{1.49} $$

then the coordinates of the point P in the tire frame ${}^{T}\mathbf{r}_P$ are

$$
\begin{aligned}
{}^{T}\mathbf{r}_P &= {}^{T}R\,{}^{W}\mathbf{r}_P + {}^{T}\mathbf{d} = {}^{T}R_W\,{}^{W}\mathbf{r}_P + {}^{T}R\,{}^{W}_{T}\mathbf{d}_W \\
&= \begin{bmatrix} x_P \\ y_P \cos\gamma - R\sin\gamma - z_P \sin\gamma \\ R\cos\gamma + z_P\cos\gamma + y_P\sin\gamma \end{bmatrix}
\end{aligned} \tag{1.50}
$$

because

$$
{}^{T}R_W = \begin{bmatrix} 1 & 0 & 0 \\ 0 & \cos\gamma & -\sin\gamma \\ 0 & \sin\gamma & \cos\gamma \end{bmatrix} \qquad {}^{W}_{T}\mathbf{d}_W = \begin{bmatrix} 0 \\ 0 \\ R \end{bmatrix} \tag{1.51}
$$

where ${}^{W}_{T}\mathbf{d}_W$ is the W-expression of the position vector of the wheel frame in the tire frame, R is the radius of the wheel, and ${}^{T}R_W$ is the transformation matrix from the W to T.

The center of the wheel ${}^{W}\mathbf{r}_P = {}^{W}\mathbf{r}_o = \mathbf{0}$ is the origin of the wheel frame W that will be at ${}^{T}\mathbf{r}_o$ in the tire coordinate frame T.

$$
{}^{T}\mathbf{r}_o = {}^{T}\mathbf{d}_W = {}^{T}R\,{}^{W}_{T}\mathbf{d}_W = \begin{bmatrix} 0 \\ -R\sin\gamma \\ R\cos\gamma \end{bmatrix} \tag{1.52}
$$

If the camber angle is zero, $\gamma = 0$, then

$$
{}^{T}\mathbf{r}_o = \begin{bmatrix} 0 \\ 0 \\ R \end{bmatrix} = {}^{W}_{T}\mathbf{d}_W \qquad \gamma = 0 \tag{1.53}
$$

1.2 Tire Force System

The resultant force system that a tire receives from the ground is assumed to be located at the center of the tireprint and can be decomposed along x_t, y_t, and z_t axes of the tire coordinate frame T. Therefore, the interaction of a tire with the road generates a three-dimensional ($3D$) force system including three forces and three moments, as shown in Fig. 1.6. The components of the tire force system are:

1. *Longitudinal force* F_x. It is the force acting along the x-axis. The resultant longitudinal force $F_x > 0$ in accelerating and $F_x < 0$ in braking. Longitudinal force is also called *forward force* or *traction force.*
2. *Lateral force* F_y. It is the force tangent to the ground and orthogonal to both F_x and F_z. The resultant lateral force $F_y > 0$ when it is in the y-direction.

3. *Normal force* F_z. It is the vertical force, normal to the ground plane. The resultant normal force $F_z > 0$ if it is upward. The traditional tires and pavements are unable to provide $F_z < 0$. Normal force is also called *vertical force* or *wheel load*.
4. *Roll moment* M_x. It is the longitudinal moment about the x-axis. The resultant roll moment $M_x > 0$ when it tends to turn the tire about the x-axis. The roll moment is also called the *bank moment*, *tilting torque*, or *overturning moment*.
5. *Pitch moment* M_y. It is the lateral moment about the y-axis. The resultant pitch moment $M_y > 0$ when it tends to turn the tire about the y-axis and move it forward. The pitch moment is also called *rolling resistance torque*.
6. *Yaw moment* M_z. It is the upward moment about the z-axis. The resultant yaw moment $M_z > 0$ when it tends to turn the tire about the z-axis. The yaw moment is also called the *aligning moment*, *self-aligning moment*, or *bore torque*.

This force system is applied on the tire from the ground. All other possible forces on a wheel are assumed to be at the wheel axle. The driving or braking moment applied to the tire from the vehicle about the tire axis is called *wheel torque* T (Jazar 2011, 2017).

1.3 Tire Longitudinal Force

To accelerate or brake a vehicle, a longitudinal force must develop between the tire and the ground. When a moment T is applied to the spin axis of the tire, longitudinal slip ratio s occurs and a longitudinal force F_x is generated at the tireprint proportional to s

$$\mathbf{F}_x = F_x \, {}^T\hat{\imath} \tag{1.54}$$

$$\frac{F_x}{F_z} = C_s \, S\left(s - s_s\right) \tag{1.55}$$

$$C_s = \frac{1}{F_z} \lim_{s \to 0} \frac{\partial F_x}{\partial s} > 0 \tag{1.56}$$

$$s = \frac{R_g \, \omega_w - v_x}{R_g \, \omega_w \, H(R_g \, \omega_w - v_x) + v_x \, H(v_x - R_g \, \omega_w)} \tag{1.57}$$

$$I_w \, \dot{\omega}_w = T - R_w \, F_x \tag{1.58}$$

$$R_h = R_g - \frac{F_z}{k_z} \tag{1.59}$$

where S is the *Saturation* function, H is the *Heaviside* function, I_w is the mass moment of the wheel about its spin axis, R_g is the unloaded geometric radius of tire, R_w is the equivalent tire radius as are shown in Fig. 1.7, and k_z is the vertical stiffness of the tire (Jazar 2011, 2013). The ω_w is tire's angular velocity, and v_x is

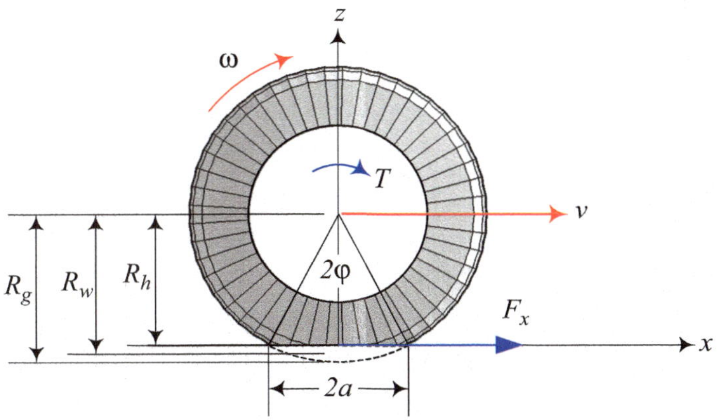

Fig. 1.7 Geometric radius R_g, effective radius R_w, and tire center height R_h of loaded and rolling tire

the tire's forward velocity. Saturation is a linearly proportional function while the variable is within a limit, and constant out of the limit. The *Saturation* function $S(x - x_0)$ is defined as:

$$S(x - x_0) = \begin{cases} x_0 & x_0 < x \\ x & -x_0 < x < x_0 \\ -x_0 & x < -x_0 \end{cases} \tag{1.60}$$

Heaviside is an on-off switching function that is defined as:

$$H(x - x_0) = \begin{cases} 1 & x > x_0 \\ 0 & x < x_0 \end{cases} \tag{1.61}$$

Saturation function may be defined by *Heaviside* function (Andrzejewski and Awrejcewicz 2005).

$$S(x - x_0) = x\, H\,(x_0 - |x|) + x_0\, H\,(|x| - x_0) \tag{1.62}$$

The force F_x is proportional to the normal load F_z, where the coefficient $\mu_x\,(s)$ is called the *longitudinal friction coefficient*. If F_{x_M} is the maximum achievable longitudinal tire force, then

$$F_{x_M} = C_s\, s_s F_z = \mu_x\, F_z \tag{1.63}$$

where s_s is the *saturation slip ratio*.

The *longitudinal slip ratio* of a driving tire is defined by

$$s = \frac{R_g\, \omega_w - v_x}{R_g\, \omega_w} \qquad R_g\, \omega_w > v_x \qquad 0 < s < 1 \tag{1.64}$$

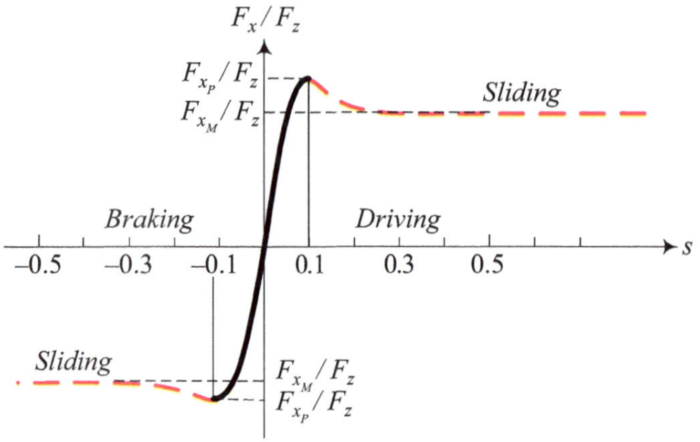

Fig. 1.8 Longitudinal friction coefficient as a function of slip ratio s, in driving and braking

and a braking tire with

$$s = \frac{R_g\,\omega_w - v_x}{v_x} \qquad R_g\,\omega_w < v_x \qquad -1 < s < 0 \tag{1.65}$$

Slip ratio s is positive for driving and is negative for braking. The longitudinal force F_x/F_z is a function of slip ratio s as shown in Fig. 1.8. The longitudinal force reaches a driving peak value F_{xP} at $s = s_s \approx 0.1$, before dropping to an almost constant steady-state saturated driving value F_{xM}. The longitudinal force $F_x\,(s)$ may be assumed proportional to s when s is very small

$$\frac{F_x\,(s)}{F_z} = C_s\,s \qquad |s| < 0.1 \tag{1.66}$$

where C_s is called the *longitudinal slip coefficient.*

The tire will spin when $s \gtrsim 0.1$ and the traction force remains almost constant at F_{xM}. The same phenomena happens in braking and tire slides when $s \lesssim -0.1$ at the values $-F_{xP}$ and $-F_{xM}$. Although the values of F_{xP} and F_{xM} in accelerating and decelerating might be different, we assume they are equal in this book.

A *proportional-saturation* simplification of the longitudinal tire force is exact and safe enough to be used in computer calculation. Figure 1.9 illustrates the model. To include the saturation behavior of the longitudinal force F_x as a function of s for a constant vertical load F_z, we employ this model in which F_x/F_z is proportional to s while s is less than the saturation s_s, and remains constant at the saturated value $F_x = F_{xM}$ equal to the maximum lateral force $F_x = F_{xM}$ for $s > s_s$

$$\frac{F_x}{F_z} = \begin{cases} C_s\,s & -s_s \le s \le s_s \\ \dfrac{F_{xM}}{F_z} = C_s\,s_s = \mu_x & s > s_s \text{ or } s < -s_s \end{cases} \tag{1.67}$$

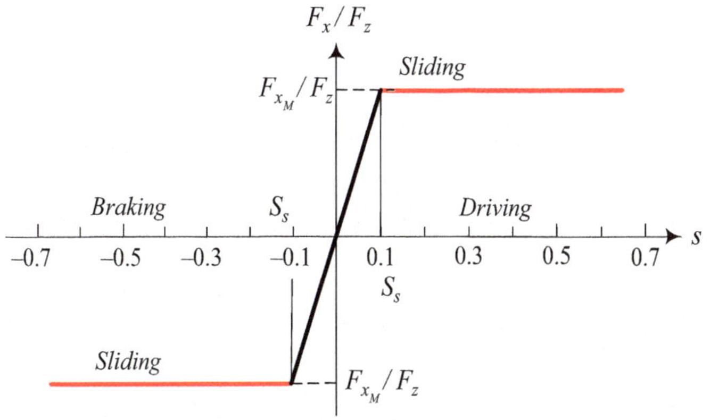

Fig. 1.9 The proportional-saturation model of longitudinal tire force F_x/F_z in vehicle dynamics

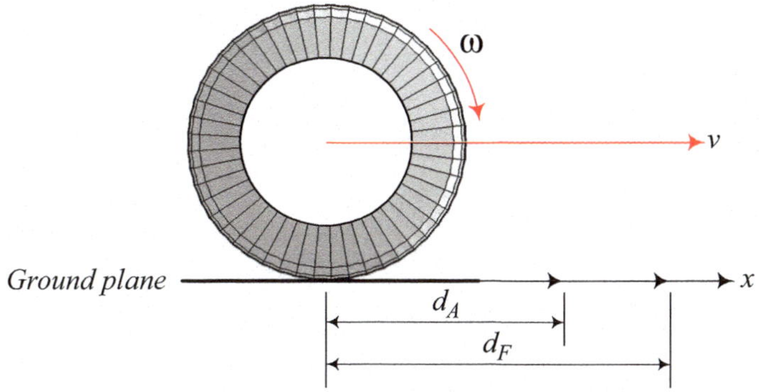

Fig. 1.10 A turning tire on the ground to show the no slip travel distance d_F, and the actual travel distance d_A

where $F_{x_M} = -C_s\, s_s F_z = \mu_x F_z$ is the maximum longitudinal force that the tire can support, which is set by the tire load and the longitudinal friction coefficient μ_x.

Proof The longitudinal slip ratio s, or simply *slip*, is defined as the difference between the actual speed of the tire v_x and the theoretical tire speed $R_g \omega_w$. Figure 1.10 illustrates a rolling tire on the ground. The ideal distance that the tire would freely travel with no slip is denoted by d_F, while the actual distance the tire travels is denoted by d_A. Thus, for a driving tire, $d_F > d_A$, and for a braking tire, $d_F < d_A$.

The difference $d_F - d_A$ is the *tire slip* and therefore, the *tire slip ratio* of the tire is

$$s = \frac{d_F - d_A}{d_F} \qquad d_F > d_A \tag{1.68}$$

$$s = \frac{d_F - d_A}{d_A} \qquad d_F < d_A \tag{1.69}$$

To have the instant value of s, we must measure the travel distances in an infinitesimal time period, and therefore,

$$s \equiv \frac{\dot{d}_F - \dot{d}_A}{\dot{d}_F} \qquad \dot{d}_F > \dot{d}_A \tag{1.70}$$

$$s \equiv \frac{\dot{d}_F - \dot{d}_A}{\dot{d}_A} \qquad \dot{d}_F < \dot{d}_A \tag{1.71}$$

If the angular velocity of the tire is ω_w, then $\dot{d}_F = R_g\,\omega_w$ and $\dot{d}_A = R_w\,\omega_w = v_x$, where R_g is the geometric tire radius and R_w is the effective radius. Therefore, the slip ratio s is defined based on the actual speed $v_x = R_w\,\omega_w$, and the freely rolling speed $R_g\,\omega_w$.

$$\begin{aligned} s &= \frac{R_g\,\omega_w - R_w\,\omega_w}{R_g\,\omega_w} \\ &= \frac{R_g\,\omega_w - v_x}{R_g\,\omega_w} = 1 - \frac{v_x}{R_g\,\omega_w} \qquad R_g\,\omega_w > v_x \end{aligned} \tag{1.72}$$

$$\begin{aligned} s &= \frac{R_g\,\omega_w - R_w\,\omega_w}{R_w\,\omega_w} \\ &= \frac{R_g\,\omega_w - v_x}{v_x} = \frac{R_g\,\omega_w}{v_x} - 1 \qquad v_x > R_g\,\omega_w \end{aligned} \tag{1.73}$$

A tire can exert longitudinal force only if a longitudinal slip is present. During acceleration, the actual velocity v_x is less than the ideal free velocity $R_g\,\omega_w$, and therefore, $s > 0$. During braking, the actual velocity v_x is higher than the free velocity $R_g\,\omega_w$ and therefore, $s < 0$. We may combine the positive and negative slip ratios and define it by a single equation as (1.57) to be used in computer analysis of the equations of motion.

The frictional force F_x between a tire and the road surface is a function of normal load F_z, vehicle speed v_x, and wheel angular speed ω_w. In addition to these variables there are a number of parameters that affect F_x, such as tire pressure, tread design, wear, and road surface conditions (Hartman et al. 2018). It has been determined empirically that a contact friction force of the form

$$F_{x_M} = \mu_x\,F_z \tag{1.74}$$

can effectively model experimental measurements obtained with constant v_x, ω_w and other environmental conditions. Longitudinal slip is also called *circumferential* or *tangential slip*.

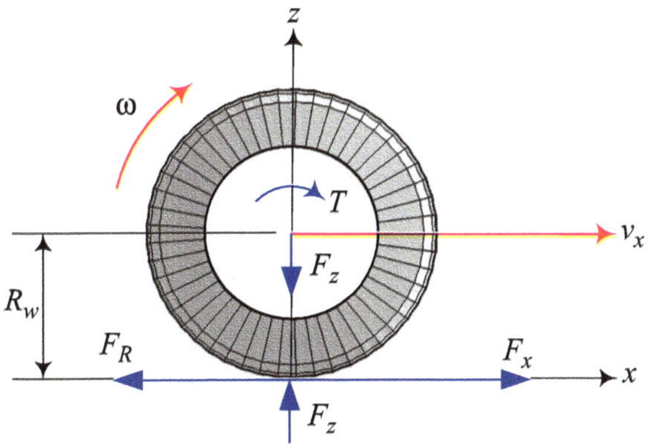

Fig. 1.11 Free-body-diagram of a wheel

Figure 1.11 illustrates free-body-diagram of a tire. In x-direction, there is a traction (or braking) force F_x and a rolling resistance force F_R. In z-direction there is a tire load F_z which will be in balance with the ground reaction as long as the tire is touching the ground. About the y-axis there is a applied torque T. The equations of motion of such tire are:

$$m_w \dot{v}_x = F_x - F_R \tag{1.75}$$

$$I_w \dot{\omega}_w = T - (F_x - F_R)\, R_w \tag{1.76}$$

where m_w is the wheel mass, I_w is the mass moment of the wheel, and R_w is the tire effective radius.

An effective radius $R_w = v_x/\omega_w$ is defined by measuring the tire's angular velocity $\omega = \omega_w$ and forward velocity v_x. When the tire rolls, each part of the circumference is flattened as it passes through the contact area. A practical estimate of the effective radius R_w may be estimated by substituting the arc with the straight length of tireprint as is shown in Fig. 1.7. The tire vertical deflection is

$$R_g - R_h = R_g\,(1 - \cos\varphi) \tag{1.77}$$

and

$$R_h = R_g \cos\varphi \tag{1.78}$$

$$a = R_g \sin\varphi \tag{1.79}$$

If the motion of the tire is compared to the rolling of a rigid disk with radius R_w, then the tire must move a distance $a \approx R_w\,\varphi$ for an angular rotation φ

$$a = R_g \sin\varphi \approx R_w\,\varphi \tag{1.80}$$

hence,

$$R_w = \frac{R_g \sin\varphi}{\varphi} \tag{1.81}$$

Expanding $\frac{\sin\varphi}{\varphi}$ in a Taylor series shows that

$$R_w = R_g \left(1 - \frac{1}{6}\varphi^2 + O\left(\varphi^4\right)\right) \tag{1.82}$$

Using Eq. (1.77) we may approximate

$$\cos\varphi \approx 1 - \frac{1}{2}\varphi^2 \tag{1.83}$$

$$\varphi^2 \approx 2\left(1 - \cos\varphi\right) \approx 2\left(1 - \frac{R_h}{R_g}\right) \tag{1.84}$$

and therefore,

$$R_w \approx R_g \left(1 - \frac{1}{3}\left(1 - \frac{R_h}{R_g}\right)\right) = \frac{2}{3}R_g + \frac{1}{3}R_h \tag{1.85}$$

As R_h is a function of tire load F_z,

$$R_h = R_h\left(F_z\right) = R_g - \frac{F_z}{k_z} \tag{1.86}$$

the effective radius R_w is also a function of the tire load. The angle φ is called *tireprint angle* or *tire contact angle*.

The vertical stiffness of radial tires k_z is less than non-radial tires under the same conditions. So, the loaded height of radial tires, R_h, is less than the non-radials'. However, the effective radius of radial tires R_w is closer to their unloaded radius R_g. As a good estimate, for a non-radial tire, $R_w \approx 0.96 R_g$, and $R_h \approx 0.94 R_g$, while for a radial tire, $R_w \approx 0.98 R_g$, and $R_h \approx 0.92 R_g$.

Generally speaking, the effective radius R_w depends on the type of tire, stiffness, load conditions, inflation pressure, and tire's forward velocity. ■

Example 8 Slip ratio is $0 < s < 1$ in driving.

When we accelerate, a driving moment is applied to the tire axis. The tire is moving slower than a free tire

$$R_w\,\omega_w < R_g\,\omega_w \tag{1.87}$$

and therefore $s > 0$. The equivalent radius for a driving tire is less than the geometric radius

$$R_w < R_g \tag{1.88}$$

Equivalently, we may express the condition using the equivalent angular velocity ω_{eq} and deduce that a driving tire turns faster than a free tire

$$R_g\,\omega_{eq} < R_g\,\omega_w \tag{1.89}$$

The driving moment can be high enough to overcome the friction and turn the tire on pavement while the vehicle is not moving. In this case $v_x = 0$ and therefore, $s = 1$. It shows that the longitudinal slip would be between $0 < s < 1$ when a driving torque is applied.

$$0 < s < 1 \text{ for } T > 0 \tag{1.90}$$

The tire speed $R_w\,\omega_w$ equals vehicle speed v_x only if wheel torque T is zero.

Example 9 Slip ratio is $-1 < s < 0$ in braking.

When we brake, a braking moment is applied to the wheel axis. The tire is moving faster than a free tire

$$R_w\,\omega_w > R_g\,\omega_w \tag{1.91}$$

and therefore, $s < 0$. The equivalent radius for a braking tire is more than the free radius

$$R_w > R_g \tag{1.92}$$

Equivalently, we may express the condition using the equivalent angular velocity ω_{eq} and deduce that a braking tire turns slower than a free tire

$$R_g\,\omega_{eq} > R_g\,\omega_w \tag{1.93}$$

The brake moment can be high enough to lock the tire. In this case $\omega_w = 0$ and therefore, $s = -1$. The longitudinal slip would be between $-1 < s < 0$ when braking.

$$-1 < s < 0 \text{ for } T < 0 \tag{1.94}$$

Example 10 ★Slip ratio based on equivalent angular velocity ω_{eq}.

It is possible to define an effective angular velocity ω_{eq} as an equivalent angular velocity for a tire with radius R_g to proceed with the actual speed $v_x = R_g\,\omega_{eq}$. Using ω_{eq} we have

$$v_x = R_g\,\omega_{eq} = R_w\,\omega_w \tag{1.95}$$

Table 1.1 Average of longitudinal friction coefficients

Road surface	Peak value, μ_x	Sliding value, μ_{x_M}
Asphalt, dry	0.8–0.9	0.75
Concrete, dry	0.8–0.9	0.76
Asphalt, wet	0.5–0.7	0.45–0.6
Concrete, wet	0.8	0.7
Gravel	0.6	0.55
Snow, packed	0.2	0.15
Ice	0.1	0.07

and therefore,

$$s = \frac{R_g\,\omega_w - R_g\,\omega_{eq}}{R_g\,\omega_w} = 1 - \frac{\omega_{eq}}{\omega_w} \qquad \omega_w > \omega_{eq} \tag{1.96}$$

$$s = \frac{R_g\,\omega_w - R_g\,\omega_{eq}}{R_g\,\omega_{eq}} = \frac{\omega_w}{\omega_{eq}} - 1 \qquad \omega_{eq} > \omega_w \tag{1.97}$$

Example 11 Maximum acceleration and longitudinal friction.

Consider a car with mass $m = 1400\,\text{kg}$ that achieves $v_x = 40\,\text{m/s}$ in $t = 10\,\text{s}$. Assuming the car reaches $v_x = 10\,\text{m/s}$ in the first second, then we may assume a maximum acceleration of $a = 10\,\text{m/s}^2$ which needs a traction force of $F_x = ma = 1400 \times 10 = 14{,}000\,\text{N}$. Let us assume that the car is rear-wheel-drive and the rear wheels are driving at the maximum traction under the load 4600 N on each rear tire, then a longitudinal friction coefficient of

$$\mu_x = \frac{F_x}{F_z} = \frac{14{,}000}{2 \times 4600} \approx 1.52 \tag{1.98}$$

is required. Such a large friction needs small grain asphalt and special race car tires.

Table 1.1 shows the average values of longitudinal friction coefficients μ_x for a passenger car tire $215/65R15$.

In this book we accept a linear-saturation model for tire friction. Therefore, the last column of the Table 1.1 depicts the maximum friction coefficients.

Example 12 ★ Alternative slip ratio definitions.

In another alternative definition, the following equation is used for longitudinal slip:

$$s = 1 - \left(\frac{R_g\,\omega_w}{v_x}\right)^n \quad \text{where } n = \begin{cases} +1 & R_g\,\omega_w \le v_x \\ -1 & R_g\,\omega_w > v_x \end{cases} \tag{1.99}$$

$$s \in [0, 1]$$

In this definition s is always between zero and one. When $s = 1$, then the tire is either locked while the car is sliding, or the tire is spinning while the car is not moving.

Example 13 ★ Acceleration-based longitudinal slip.

Employing Eq. (1.70), we may also define s as

$$s \equiv \frac{\ddot{d}_F - \ddot{d}_A}{\ddot{d}_F} \qquad \ddot{d}_F > \ddot{d}_A \tag{1.100}$$

$$s \equiv \frac{\ddot{d}_F - \ddot{d}_A}{\ddot{d}_A} \qquad \ddot{d}_F < \ddot{d}_A \tag{1.101}$$

and therefore,

$$s = \frac{R_g\,\alpha_w - a_x}{R_g\,\alpha_w} = 1 - \frac{a_x}{R_g\,\alpha_w} \qquad R_g\,\alpha_w > a_x \tag{1.102}$$

$$s = \frac{R_g\,\alpha_w - a_x}{a_x} = \frac{R_g\,\alpha_w}{a_x} - 1 \qquad a_x > R_g\,\alpha_w \tag{1.103}$$

where α_w is the tire's angular acceleration, and a_x is the tire's forward acceleration.

Example 14 Velocity-dependent longitudinal force.

Experimental results show that the longitudinal force of a tire drops at higher speeds for the same slip. This fact suggests a correction in the equation of longitudinal force (1.55) as

$$\frac{F_x}{F_z} = C_s\,s - C_{s_1}\,v_x \tag{1.104}$$

or

$$\frac{F_x}{F_z} = C_s\,s - C_{s_2}\,v_x^2 \tag{1.105}$$

or

$$\frac{F_x}{F_z} = C_s\,s - C_{s_1}\,v_x - C_{s_2}\,v_x^2 \tag{1.106}$$

where $C_{s_1}[\,\mathrm{s/m}]$ is the coefficient indication how much F_x/F_z will drop for every 1 m/s of longitudinal velocity of the tire. Similarly, $C_{s_2}[\,\mathrm{s^2/\,m^2}]$ is the coefficient indication how much F_x/F_z will drop for every $1\,[\,\mathrm{m/s}]^2$ of square of longitudinal velocity of the tire.

Example 15 ★Longitudinal slip ratio rate.

Time derivative of s shows that

$$\begin{aligned}\dot{s} &= -\frac{d}{dt}\frac{v_x}{R_g\,\omega_w} = -\frac{a_x\,\omega_w - v_x\,\alpha_w}{R_g\,\omega_w^2} = \frac{v_x}{\omega_w R_g}\frac{\alpha_w}{\omega_w} - \frac{v_x}{\omega_w R_g}\frac{a_x}{v_x} \\ &= (1-s)\left(\frac{\alpha_w}{\omega_w} - \frac{a_x}{v_x}\right) \qquad R_g\,\omega_w > v_x\end{aligned} \tag{1.107}$$

$$\dot{s} = \frac{d}{dt}\frac{R_g\,\omega_w}{v_x} = \frac{R_g\,\alpha_w\,v_x - R_g\,a_x\,\omega_w}{v_x^2} = \frac{R_g\,\omega_w}{v_x}\frac{\alpha_w}{\omega_w} - \frac{R_g\,\omega_w}{v_x}\frac{a_x}{v_x}$$

$$= (s-1)\left(\frac{\alpha_w}{\omega_w} - \frac{a_x}{v_x}\right) \qquad v_x > R_g\,\omega_w \tag{1.108}$$

or

$$\dot{s} = \left(\frac{\alpha_w}{\omega_w} - \frac{a_x}{v_x}\right)(1-s)\;H(R_g\,\omega_w - v_x)$$
$$+\left(\frac{\alpha_w}{\omega_w} - \frac{a_x}{v_x}\right)(s-1)\;H(v_x - R_g\,\omega_w) \tag{1.109}$$

A better model for longitudinal tire force is to include the rate of s such as

$$\frac{F_x}{F_z} = C_s\,s - C_{\dot{s}}\,\dot{s} \tag{1.110}$$

where $C_{\dot{s}}$ is the coefficient indicating how F_x/F_z drops because of the slip rate $\dot{s}$.

Example 16 Activation functions.

Activation functions are utilized to change the level of a command signal from a steady-state value to another steady-state value. The simplest activation function is *Heaviside* function which act as a sudden on/off switch function.

$$H(x - x_0) = \begin{cases} 1 & x > x_0 \\ 0 & x < x_0 \end{cases} \tag{1.111}$$

The *Saturation* function (1.112) does the change with a constant rate between $-x_0$ and x_0.

$$S(x - x_0) = \begin{cases} x_0 & x_0 < x \\ x & -x_0 < x < x_0 \\ -x_0 & x < -x_0 \end{cases} \tag{1.112}$$

$$S(x - x_0) = x\;H\left(x_0 - |x|\right) + x_0\;H\left(|x| - x_0\right) \tag{1.113}$$

In reality, the signal will smoothly gain a rate of change and then smoothly get a negative rate until its level changes to the new value. Therefore, all signals will change their steady-state values on smoother curves. The most practical activation function is the *Logistic* or *Sigmoid* function.

$$f(x) = \text{Sig}(x) = \frac{L}{1 + e^{-k(x-x_0)}} \tag{1.114}$$

The *Sigmoid* function may also be called Soft step. Tangent hyperbolic function is another smooth activation function

$$f(x) = \tanh(x) = \frac{e^x - e^{-x}}{e^x + e^{-x}} \tag{1.115}$$

The *arctan* is another useful activation function.

$$f(x) = \arctan(x) = \tan^{-1}(x) \tag{1.116}$$

The soft sign,

$$f(x) = \frac{x}{1 + |x|} \tag{1.117}$$

inverse square root

$$f(x) = \frac{x}{\sqrt{1 + kx^2}} \tag{1.118}$$

are other practical activation functions.

To have a more realistic change and saturation in longitudinal slip ratio and other saturating characteristics in vehicle dynamics, we may use a continuous activation function instead of the sharp changing *Saturation* function.

1.4 Tire Lateral Force

The capability of generating lateral force is the main advantage of using rubber tires. Only few materials have the characteristics to make a flexible tire while sticking to the ground and generate lateral force when steering. The lateral force is made by sideslip and camber which will be studied in this section (Jazar 2017).

The tire lateral force F_y is the most important tire force in vehicle maneuvering. The lateral force is mainly a function of two angles of the tire: sideslip angle α and camber angle γ as are shown in Fig. 1.6. The lateral force F_y is a linear combination of the angles α and γ assuming the lateral forces generated by α and γ do not affect each other.

$$\frac{F_y}{F_z} = -C_\alpha \, S(\alpha - \alpha_s) - C_\gamma \, S(\gamma - \gamma_s) \tag{1.119}$$

The *Saturation* function $S(x - x_0)$ is defined as:

$$S(x - x_0) = \begin{cases} x_0 & x_0 < x \\ x & -x_0 < x < x_0 \\ -x_0 & x < -x_0 \end{cases} \tag{1.120}$$

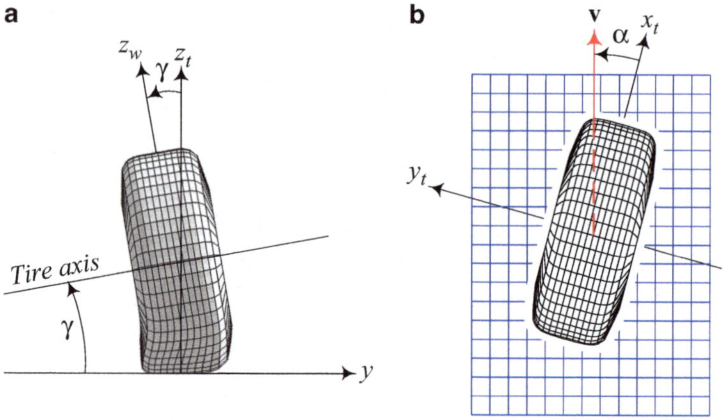

Fig. 1.12 Camber and sideslip angles illustration. (**a**) Front view of a tire and measurement of the camber angle. (**b**) Top view of a tire and measurement of the sideslip angle

The sideslip angle α and camber angle γ are measured in radians $[\text{rad}]$ or degrees $[\text{deg}]$ and therefore, their associated coefficients C_α and C_γ are respectively in $[\text{rad}^{-1}]$ or $[\text{deg}^{-1}]$.

To show the sideslip and camber angles, we assume a flat ground and attach a Cartesian tire coordinate frame T at the center of the tireprint when the tire is vertical and stationary. The x_t-axis is along the intersection line of the tire-plane and the ground. The z_t-axis is upward perpendicular to the ground, and the y_t-axis makes the coordinate system a right-hand triad.

The *camber angle* γ is the angle between the vertical plane and the tire-plane measured from the z_t-axis to the z_w-axis about the x_t-axis. The camber angle can be recognized better in a front view on x-axis as shown in Fig. 1.12a. The *sideslip angle* α is the angle between the x_t-axis and the tire velocity vector $\mathbf{v}$, measured about the z_t-axis. Because the x_t-axis of the T-frame and is always parallel to the x_w-axis of the W-frame, we may also define α as the angle between x_w-axis and $\mathbf{v}$ in a plane parallel to the ground. The sideslip can be recognized better in a top view, as shown in Fig. 1.12b.

Figure 1.13a illustrates a tire, moving along the velocity vector $\mathbf{v}$ at a sideslip angle α. The tire is steered by the steer angle δ. If the angle between the vehicle x-axis and the tire velocity vector $\mathbf{v}$ is shown by β, then

$$\alpha = \beta - \delta \tag{1.121}$$

$$\alpha = \arctan \frac{{}^{T}v_y}{{}^{T}v_x} \tag{1.122}$$

$$\beta = \arctan \frac{{}^{C}v_y}{{}^{C}v_x} \tag{1.123}$$

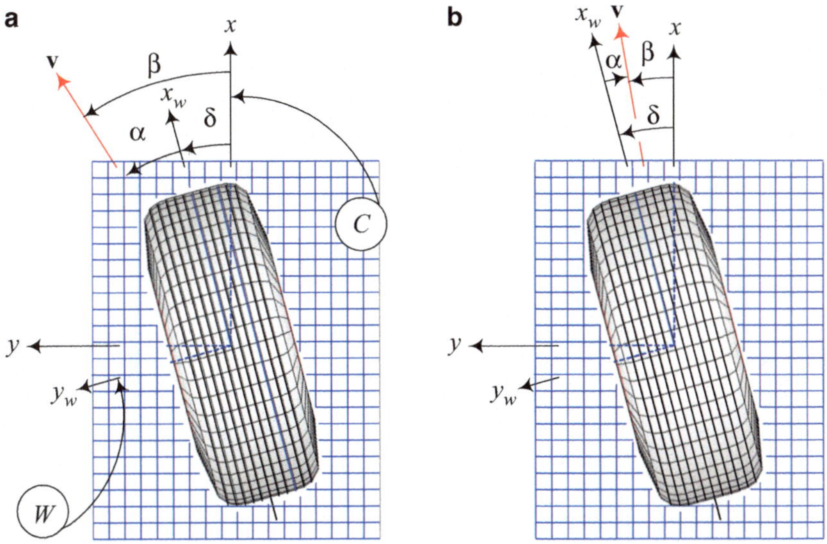

Fig. 1.13 Angular orientation of a moving tire along the velocity vector **v** at a sideslip angle α and a steer angle δ. (**a**) $\alpha > 0$. (**b**) $\alpha < 0$

The angle β is called the *wheel-body sideslip angle* and α is the *tire sideslip angle*. The lateral force F_y, generated by a tire, is proportional to the sideslip angle α for small α.

$$\frac{F_y}{F_z} = -C_\alpha \, \alpha = -C_\alpha \, (\beta - \delta) \qquad |\alpha| < \alpha_s \tag{1.124}$$

Proof A wheel coordinate frame $W(x_w, y_w)$ is attached to the wheel at the center of the wheel as shown in Fig. 1.13a. The orientation of the wheel frame is measured with respect to the wheel-body coordinate frame C (x_c, y_c). The C-frame is always parallel to the vehicle frame $B(x, y)$. The angle between the x_c and x_w axes is the wheel steer angle δ, measured about the z_w-axis. The wheel is moving along the tire velocity vector **v**. The angle between the x_t-axis and **v** is the *tire sideslip angle* α, and the angle between the body x_c-axis and **v** is the *wheel-body sideslip angle* β. The angles α, β, and δ in Fig. 1.13a are all positive. The figure shows that

$$\alpha = \beta - \delta \tag{1.125}$$

Practically, when a steered wheel is moving forward, the relationship between the angles α, β, and δ is such that the velocity vector sits between the x_C and x_w axes. A practical situation is shown in Fig. 1.13b. A steer angle will turn the heading of the wheel by a δ angle. However, because of tire flexibility, the velocity vector of the wheel is lazier than the heading and turns by a β angle, where $\beta < \delta$. As a result, a positive steer angle δ generates a negative sideslip angle α. Figure 1.13b, shows that the relation (1.121) is correct under a practical situation.

Velocity vector of a tire may be decomposed and expressed in the tire frame T or in the wheel-body frame C. The sideslip angle α is used to express $\mathbf{v}$ in T-frame

$$^{T}\mathbf{v} = v\cos\alpha \ {}^{T}\hat{\imath} + v\sin\alpha \ {}^{T}\hat{\jmath} \tag{1.126}$$

and therefore,

$$\alpha = \arctan\frac{^{T}v_y}{^{T}v_x} \tag{1.127}$$

To express $\mathbf{v}$ in C-frame we use the wheel-body sideslip angle β

$$^{C}\mathbf{v} = v\cos\beta \ {}^{C}\hat{\imath} + v\sin\beta \ {}^{C}\hat{\jmath} \tag{1.128}$$

and therefore,

$$\beta = \arctan\frac{^{C}v_y}{^{C}v_x} \tag{1.129}$$

If we indicate the velocity vector $^{T}\mathbf{v}$ of a tire in T-frame with

$$^{T}\mathbf{v} = \begin{bmatrix} v_x & v_y & 0 \end{bmatrix}^{T} \tag{1.130}$$

then velocity vector $\mathbf{v}$ in C-frame will be

$$^{C}\mathbf{v} = {}^{C}R_T \, {}^{T}\mathbf{v} \tag{1.131}$$

$$\begin{bmatrix} v\cos\alpha\cos\delta - v\sin\alpha\sin\delta \\ v\cos\alpha\sin\delta + v\cos\delta\sin\alpha \\ 0 \end{bmatrix} = \begin{bmatrix} \cos\delta & -\sin\delta & 0 \\ \sin\delta & \cos\delta & 0 \\ 0 & 0 & 1 \end{bmatrix} \begin{bmatrix} v\cos\alpha \\ v\sin\alpha \\ 0 \end{bmatrix}$$

Therefore, the wheel-body sideslip angle β is

$$\tan\beta = \frac{^{C}v_y}{^{C}v_x} = \frac{\cos\alpha\sin\delta + \cos\delta\sin\alpha}{\cos\alpha\cos\delta - \sin\alpha\sin\delta} = \tan(\alpha + \delta) \tag{1.132}$$

$$\beta = \alpha + \delta \tag{1.133}$$

Existence of a sideslip angle is sufficient to generate a lateral force F_y, which is proportional to α when the angle is small.

$$\frac{F_y}{F_z} = -C_\alpha\,\alpha = -C_\alpha\,(\beta - \delta) \qquad |\alpha| < \alpha_s \tag{1.134}$$

The sideslip angle α is a tire dynamic parameter and the angles β and δ are vehicle dynamic parameters. Equation (1.121) introduces the way we replace wheel parameter with vehicle parameters in the equations of motion of the vehicle. ■

Example 17 Extreme velocity cases of a wheel.

Consider a wheel as is shown in Fig. 1.13b which has a spinning angular velocity of $\omega \neq 0$ on a frictionless ground. Therefore, the velocity of the wheel center would be zero, $\mathbf{v} = 0$. The sideslip angle of such a wheel would be zero, $\alpha = 0$.

Now consider the wheel which has a zero spinning angular velocity $\omega = 0$ and a nonzero translational velocity $\mathbf{v} \neq 0$. The sideslip angle of such a wheel would be as is shown in Fig. 1.13b. The sideslip angle of a wheel is not a function of the spinning angular velocity of a wheel.

Example 18 Velocity vector from T to wheel-body C-frame.

We have seen the T to C coordinate frame transformation in Example 3 as

$$ {}^{C}\mathbf{r} = {}^{C}R_T \, {}^{T}\mathbf{r} + {}^{C}\mathbf{d}_T = {}^{C}R_T \, {}^{T}\mathbf{r} - R \, {}^{C}R_T \, {}^{T}\hat{k} \tag{1.135} $$

or by a homogeneous transformation matrix ${}^{C}T_T$.

$$ {}^{C}T_T = \begin{bmatrix} {}^{C}R_T & {}^{C}\mathbf{d}_T \\ 0 & 1 \end{bmatrix} = \begin{bmatrix} \cos\delta & -\sin\delta & 0 & 0 \\ \sin\delta & \cos\delta & 0 & 0 \\ 0 & 0 & 1 & -R \\ 0 & 0 & 0 & 1 \end{bmatrix} \tag{1.136} $$

$$ \begin{aligned} {}^{T}T_C &= {}^{C}T_T^{-1} = \begin{bmatrix} {}^{C}R_T & {}^{C}\mathbf{d}_T \\ 0 & 1 \end{bmatrix}^{-1} = \begin{bmatrix} {}^{C}R_T^{T} & -{}^{C}R_T^{T}\,{}^{C}\mathbf{d}_T \\ 0 & 1 \end{bmatrix} \\ &= \begin{bmatrix} {}^{C}R_T^{T} & -{}^{T}_{C}\mathbf{d}_T \\ 0 & 1 \end{bmatrix} = \begin{bmatrix} \cos\delta & \sin\delta & 0 & 0 \\ -\sin\delta & \cos\delta & 0 & 0 \\ 0 & 0 & 1 & R \\ 0 & 0 & 0 & 1 \end{bmatrix} \end{aligned} \tag{1.137} $$

where ${}^{T}\mathbf{r}$ is the position vector of a point in T-frame and ${}^{C}\mathbf{r}$ is the same point in the C-frame.

If we indicate the velocity vector ${}^{T}\mathbf{v}$ of a tire in T-frame with

$$ {}^{T}\mathbf{v} = \begin{bmatrix} v_x & v_y & 0 \end{bmatrix}^{T} \tag{1.138} $$

then velocity vector $\mathbf{v}$ in C-frame will be

$$ {}^{C}\mathbf{v} = {}^{C}R_T \, {}^{T}\mathbf{v} \tag{1.139} $$

$$ \begin{bmatrix} \cos\delta & -\sin\delta & 0 \\ \sin\delta & \cos\delta & 0 \\ 0 & 0 & 1 \end{bmatrix} \begin{bmatrix} v_x \\ v_y \\ 0 \end{bmatrix} = \begin{bmatrix} v_x \cos\delta - v_y \sin\delta \\ v_y \cos\delta + v_x \sin\delta \\ 0 \end{bmatrix} $$

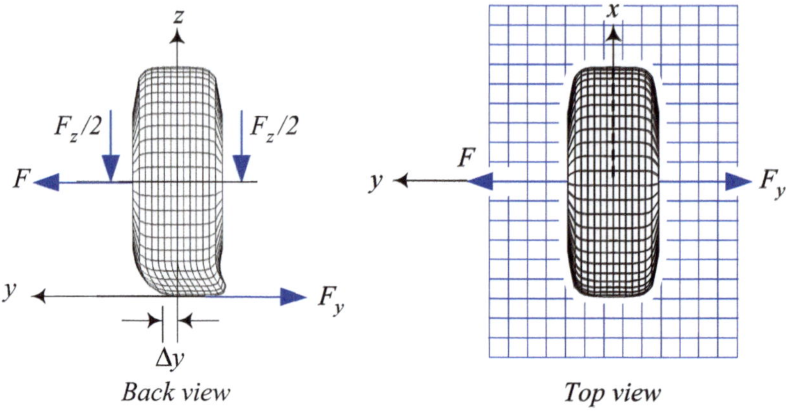

Fig. 1.14 A lateral force F_y generation in tireprint of a rolling tire under vertical and lateral forces

Knowing that

$$ {}^{T}\mathbf{v} = v\cos\alpha\ {}^{T}\hat{\imath} + v\sin\alpha\ {}^{T}\hat{\jmath} \tag{1.140} $$

makes

$$ {}^{C}\mathbf{v} = \begin{bmatrix} v\cos\alpha\cos\delta - v\sin\alpha\sin\delta \\ v\cos\alpha\sin\delta + v\cos\delta\sin\alpha \\ 0 \end{bmatrix} \tag{1.141} $$

Therefore,

$$ \tan\beta = \frac{{}^{C}v_y}{{}^{C}v_x} = \frac{\cos\alpha\sin\delta + \cos\delta\sin\alpha}{\cos\alpha\cos\delta - \sin\alpha\sin\delta} = \tan(\alpha + \delta) \tag{1.142} $$

$$ \beta = \alpha + \delta \tag{1.143} $$

consistent with Eq. (1.121).

1.4.1 *Tire Sideslip*

Consider a rolling tire that is under a vertical load F_z and a side force F at the tire axis, as is shown in Fig. 1.14. Such a rolling tire will move laterally and a reaction lateral force F_y will be generated under the tire. The tire travel velocity $\mathbf{v}$ makes an angle α with respect to the tire longitudinal x-axis as is shown in Fig. 1.15. The

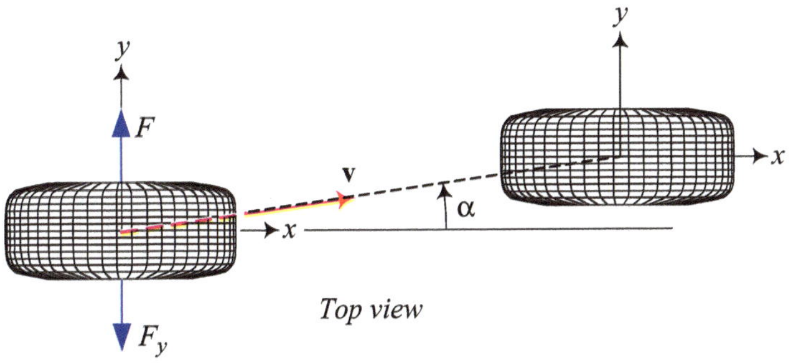

Fig. 1.15 The path of motion of a rolling tire under vertical and lateral forces

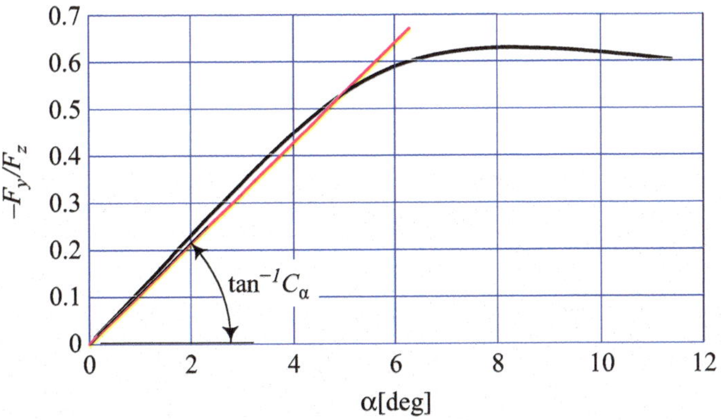

Fig. 1.16 The cornering stiffness C_α is the slope of the curve $F_y = F_y(\alpha)$ at $\alpha = 0$ for a given F_z

angle is called *sideslip* angle which is a reason for generating *lateral force* F_y. The lateral force is proportional with α for small F_y (Jazar 2017).

$$\mathbf{F}_y = F_y \, {}^T\hat{\jmath} \tag{1.144}$$

$$\frac{F_y}{F_z} = -C_\alpha \, S(\alpha - \alpha_s) \tag{1.145}$$

$$\alpha = \arctan \frac{{}^T v_y}{{}^T v_x} \tag{1.146}$$

The coefficient C_α is called the *cornering stiffness* or *sideslip coefficient* of the tire. C_α is the slope of the curve $F_y/F_z = F_y(\alpha)/F_z$ at $\alpha = 0$ at a constant F_z, as is shown in Fig. 1.16.

$$C_\alpha = \lim_{\alpha \to 0} \frac{\partial(-F_y/F_z)}{\partial \alpha} = \frac{1}{F_z} \left| \lim_{\alpha \to 0} \frac{\partial F_y}{\partial \alpha} \right| \tag{1.147}$$

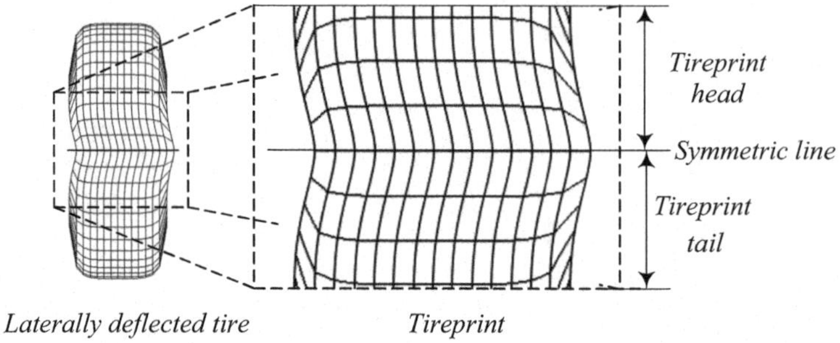

Fig. 1.17 Bottom view of a laterally deflected tire

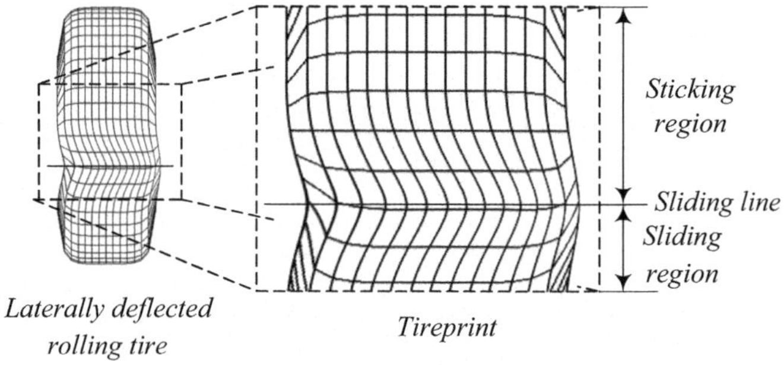

Fig. 1.18 Bottom view of a laterally deflected and turning tire

Proof When a tire is under a constant load F_z and then a side lateral force F is applied on the rim, the tire will deflect laterally while sticking to the ground. A static lateral force F_y is generated at the tireprint as is shown in the back view of Fig. 1.14. The wheel will start sliding laterally if the lateral force F_y reaches a maximum value F_{y_M}. At this point, the lateral force approximately remains constant. The maximum lateral force is proportional to the vertical load

$$F_{y_M} = \mu_y F_z \tag{1.148}$$

where μ_y is the tire friction coefficient in the y-direction. A bottom view of the tireprint of a laterally deflected tire is shown in Fig. 1.17.

If the laterally deflected tire is rolling forward on the road by pushing its spin axis forward, the tireprint will also flex longitudinally. A bottom view of the tireprint for such a laterally deflected and rolling tire is shown in Fig. 1.18. Although the tire-plane of such a tire remains perpendicular to the road, the path of the wheel makes an angle α with tire-plane. The path of the tire is indicated by the velocity vector $\mathbf{v}$ which sits between wheel x_w-axis and body x-axis as is shown in Fig. 1.19.

As the wheel rolls forward, un-deflected treads enter the tireprint region and deflect laterally as well as longitudinally while its normal contact force from the

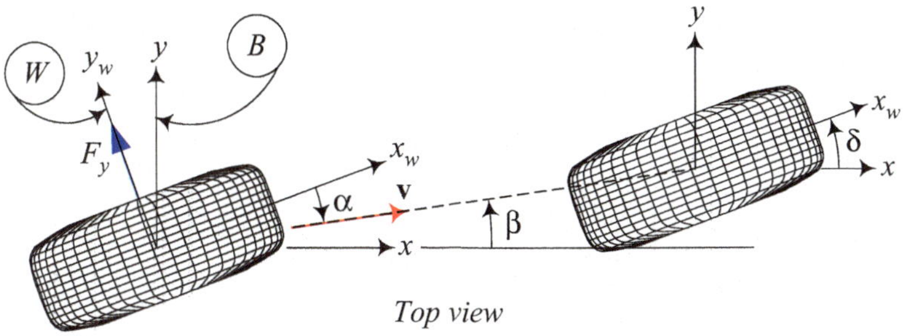

Fig. 1.19 The orientation of the tire velocity vector **v** with respect to the tire and body coordinate frames in a steered wheel

road increases. When a tread moves from the head to the tail of the tireprint, its lateral deflection as well as its normal force increase from zero until it reaches the sliding line. The point where the laterally deflected tread starts sliding back is called *sliding line*. When the tread passes the sliding line and enters the tail region, the deflection and the normal force decrease. They approach zero when tread goes to the tailing edge of the tireprint. As the normal contact force from the road decreases at the tail region of the tireprint, the friction force also decreases and the tread slides back to its original position when leaving the tireprint region. In practical situations, the velocity vector of the center of a steered wheel will be between wheel x_w-axis and the vehicle body x-axis, as is shown in Fig. 1.19. The steering angle δ, tire sideslip angle α, and wheel sideslip angle β are indicated in the figure. The wheel-body sideslip angle β is the angle between the tire velocity **v** and the vehicle longitudinal x-axis, measured from the x-axis about the z-axis.

The slip angle α increases by increasing the lateral force F_y as long as F_y is not saturated, $F_y < F_{y_M}$. It is assumed that the slip angle α and lateral force F_y act as action and reaction. A lateral force generates a slip angle, and a slip angle generates a lateral force. Hence, we can steer the tires of a car to make a slip angle and generate a lateral force to turn the car. Steering causes a slip angle in the tires and creates a lateral force. The slip angle is positive $\alpha > 0$ if the tire should be turned about the z-axis to be aligned with the velocity vector **v**. A positive slip angle α generates a negative lateral force F_y. Hence, steering to the left about the z-axis makes a negative slip angle and produces a positive lateral force to move the tire to the left (Jazar 2017).

Using the velocity vector of the tire in the wheel coordinate frame W, and its components, ${}^W\mathbf{v} = v_x \hat{\imath} + v_y \hat{\jmath}$, we may also define the sideslip angle as

$$\alpha = \arctan \frac{v_y}{v_x} \tag{1.149}$$

A sample of measured lateral force F_y/F_z as a function of slip angle α is plotted in Fig. 1.16. The lateral force F_y/F_z increases linearly for small slip angles;

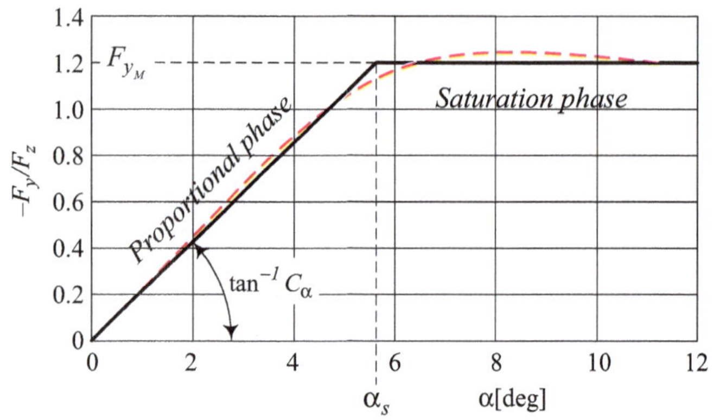

Fig. 1.20 The proportional-saturation model for the lateral force F_y as a function of α, $F_y = F_y(\alpha)$, for a constant vertical load F_z

however, the rate of increasing F_y/F_z decreases for higher α. The lateral force remains constant or drops slightly when α reaches a critical value at which the tire slides on the road. For practical purposes, we may accept different simplified models for the function $F_y/F_z = F_y(\alpha)/F_z$.

1. *Linear model*: We assume the lateral force F_y is proportional to the slip angle α for low values of α.

$$\frac{F_y}{F_z} = -C_\alpha \ S(\alpha - \alpha_s) \tag{1.150}$$

$$C_\alpha = -\frac{1}{F_z} \lim_{\alpha \to 0} \frac{\partial F_y}{\partial \alpha} > 0 \tag{1.151}$$

   Figure 1.16 illustrates the linear model.
2. *Proportional-saturation model*: To include the saturation behavior of the lateral force F_y as a function of α for a constant vertical load F_z, we employ a proportional-saturation model in which F_y/F_z is proportional to α while α is less than the saturation α_s, and remains constant at the saturated value $F_y = F_s$ equal to the maximum lateral force $F_y = F_s = F_{yM}$ for $\alpha > \alpha_s$.

$$\frac{F_y}{F_z} = \begin{cases} -C_\alpha \, \alpha & -\alpha_s \le \alpha \le \alpha_s \\ \dfrac{F_{yM}}{F_z} = -C_\alpha \, \alpha_s = \mu_y & \alpha > \alpha_s \text{ or } \alpha < -\alpha_s \end{cases} \tag{1.152}$$

   The $F_{yM} = -C_\alpha \alpha_s F_z = \mu_y F_z$ is the maximum lateral force that the tire can support. The F_{y_M} is set by the tire load and the lateral friction coefficient μ_y. Figure 1.20 illustrates the *proportional-saturation* model and Eq. (1.152) expresses the *proportional-saturation* function.

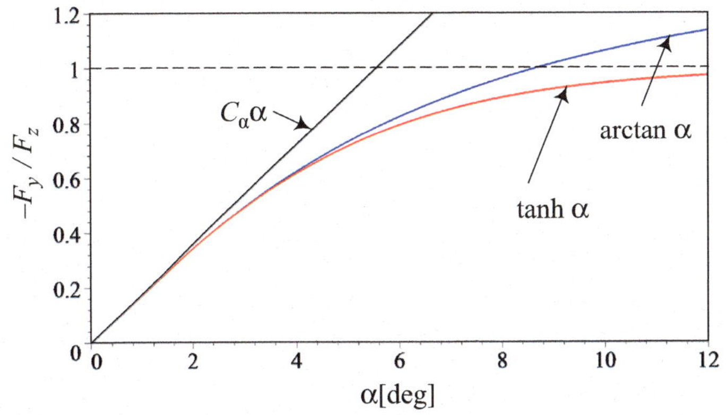

Fig. 1.21 Two nonlinear models for the lateral force F_y/F_z as a function of α, $F_y/F_z = \tanh(C_\alpha \alpha)$ and $F_y/F_z = \arctan(C_\alpha \alpha)$

3. *Nonlinear-saturation model*: A nonlinear saturating function may also be used for better approximation of the behavior of the lateral force F_y as a function of α. Two practical approximate functions are:

$$\frac{F_y}{F_z} = -\tanh(\alpha) \tag{1.153}$$

$$\frac{F_y}{F_z} = -\arctan(\alpha) \tag{1.154}$$

Figure 1.21 illustrates the *nonlinear-saturation* models. ■

Example 19 Lateral force modeling.

There are many mathematical and approximate functions introduced by many researchers to model the lateral force F_y of tires as function of sideslip angle α and F_z. However, none of them can exactly model all tire behaviors in all conditions. This is because F_y is a function of many more parameters, namely: tire size and pattern, speed of the vehicle, tire load, tire air pressure, pavement, ambient temperature and humidity, time history of tire operation, etc. As a result, keeping all parameters constant, only a physical measurement can exactly express $F_y = F_y(\alpha)$. However, such data also cannot be recovered and used for another experiment, as it is practically impossible to keep the conditions the same. Linear approximations are the easiest function for control purposes. Interestingly the linear and proportional-saturation approximations are exact enough for most normal practical driving conditions.

A more exact approximation of the function $F_y = F_y(\alpha)$ is to use a trilinear model as is shown in Fig. 1.22. The first part of the function $F_y/F_z = F_y(\alpha)/F_z = -C_\alpha \alpha$ is applied for all low-speed low-steering maneuvers of vehicles. By increasing speed or higher sideslip angle, the rate of the proportionality decreases and F_y will follow another line, $F_y/F_z = F_y(\alpha)/F_z = -C_1\alpha - C_2$.

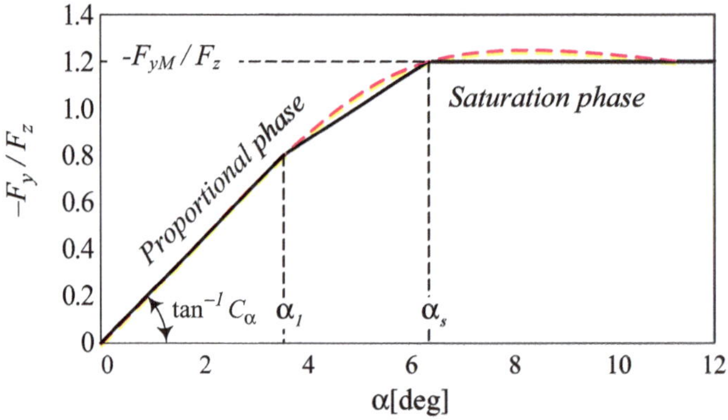

Fig. 1.22 Trilinear model of tire lateral force

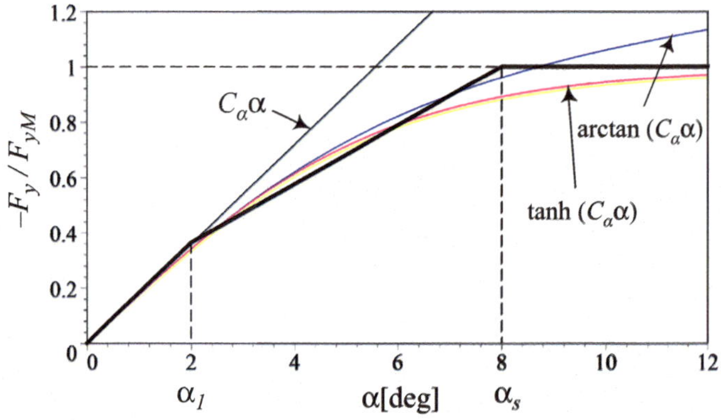

Fig. 1.23 A sample of trilinear model of tire lateral force compared with the linear and nonlinear models

The mathematical expression of the function $F_y = F_y(\alpha)$ is:

$$-\frac{F_y}{F_z} = \begin{cases} C_\alpha \alpha & \alpha \le \alpha_1 \\ \dfrac{C_\alpha \alpha_1 - F_{yM}/F_z}{\alpha_1 - \alpha_s}(\alpha - \alpha_1) + C_\alpha \alpha_1 & \alpha_1 < \alpha \le \alpha_s \\ -F_{yM}/F_z & \alpha_s < \alpha \end{cases} \tag{1.155}$$

Figure 1.23 illustrates a sample of trilinear model of tire lateral force compared with the linear and nonlinear models. Figure 1.23 illustrates a comparison for different models.

Example 20 Cubic model of tire lateral force.

Based on a parabolic normal stress distribution on the tireprint, a third-degree function was presented in the 1950s to calculate the lateral force

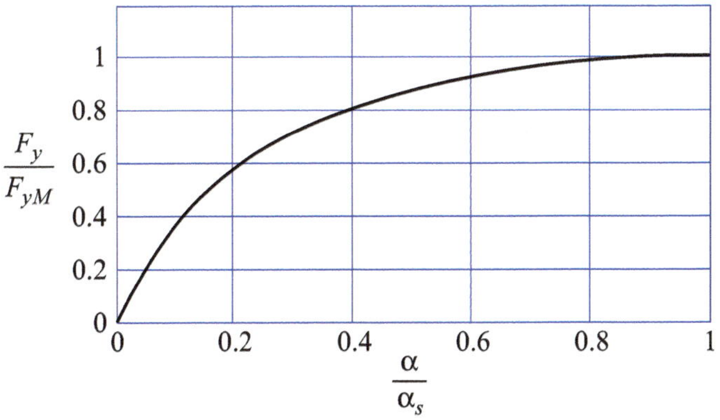

Fig. 1.24 A cubic curve model for lateral force as a function of the sideslip angle

$$\frac{F_y}{F_z} = -C_\alpha \, \alpha \left(1 - \frac{\alpha}{\alpha_s} + \frac{1}{3}\left(\frac{\alpha}{\alpha_s}\right)^2\right) \tag{1.156}$$

Let us show the sideslip angle at which the lateral force F_y reaches its maximum value F_{y_M} by α_s. Equation (1.156) shows that

$$\alpha_s = -\frac{3}{C_\alpha}\frac{F_{y_M}}{F_z} \tag{1.157}$$

and therefore,

$$\frac{F_y}{F_{y_M}} = \frac{3\alpha}{\alpha_s}\left(1 - \frac{\alpha}{\alpha_s} + \frac{1}{3}\left(\frac{\alpha}{\alpha_s}\right)^2\right) \tag{1.158}$$

Figure 1.24 shows the cubic curve model for lateral force as a function of sideslip angle. The equation is applicable only for $0 \le \alpha \le \alpha_M$.

Example 21 Aligning moments and pneumatic trails.

A rolling tire under lateral force and the associated *sideslip angle* α are shown in Fig. 1.25. Lateral distortion of the tire treads is a result of a tangential stress distribution τ_y over the tireprint. Assuming that the tangential stress τ_y is proportional to the distortion, the resultant lateral force F_y over the tireprint area A_P

$$F_y = \int_{A_P} \tau_y \, dA_p \tag{1.159}$$

is at a distance a_{x_α} behind the center line.

$$a_{x_\alpha} = \frac{1}{F_y}\int_{A_P} x \, \tau_y \, dA_p \tag{1.160}$$

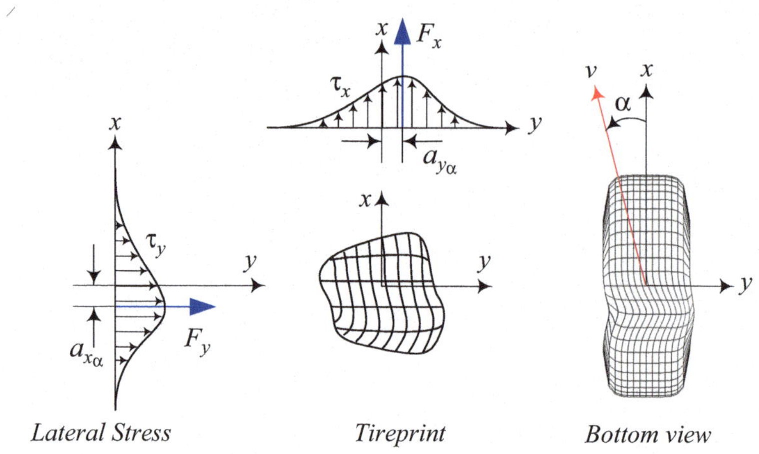

Fig. 1.25 The stress distribution τ_y, the resultant lateral force F_y and longitudinal F_x, and the pneumatic trails a_y and a_x for a turning tire going on a negative slip angle α

The distance $a_{x_\alpha} < 0$ is called the *pneumatic trail*, and the resultant moment $\mathbf{M}_z$ is called the aligning moment, $\mathbf{M}_z = F_y\, a_{x_\alpha}\, \hat{k}$. For a rolling forward tire, the distance $a_{x_\alpha} < 0$ for $\forall\alpha$. The aligning moment tends to turn the tire about the z-axis and make it align with the direction of tire velocity vector $\mathbf{v}$.

A slight shift, a_{y_α}, of the longitudinal force F_x will also contribute in the aligning moment to make the tire parallel to $\mathbf{v}$, $\mathbf{M}_z = -F_x\, a_{y_\alpha}\, \hat{k}$. For a rolling forward tire, the distance $a_{y_\alpha} < 0$ for $\alpha > 0$, and $F_x > 0$ when a positive moment is applied on the wheel to turn it forward. Therefore the total aligning moment of a rolling tire is:

$$\mathbf{M}_z = \left(F_y\, a_{x_\alpha} - F_x\, a_{y_\alpha}\right) \hat{k} \tag{1.161}$$

The stress distributions τ_y and τ_x, the resultant lateral and longitudinal forces F_y and F_x, and the pneumatic trails a_{x_α} and a_{y_α} are illustrated in Fig. 1.25.

There is also a lateral shift in the vertical force F_z under the tire because of slip angle α, which generates a *slip moment* $\mathbf{M}_x$ about the forward x-axis, and *break moment* $\mathbf{M}_y$ about the y-axis.

$$\mathbf{M}_x = -F_z\, a_{y_\alpha}\, \hat{\imath} \tag{1.162}$$

$$\mathbf{M}_y = -F_z\, a_{x_\alpha}\, \hat{\jmath} \tag{1.163}$$

The aligning moments for tires are as illustrated in Fig. 1.26. The pneumatic trails a_{x_α} and a_{y_α} increase for small slip angles up to a maximum value, and decrease to zero and negative values for high slip angles.

The lateral force $F_y/F_z = -C_\alpha\, \alpha$ may be decomposed to $F_y \sin\alpha$, parallel to the path of motion $\mathbf{v}$, and $F_y \cos\alpha$, perpendicular to $\mathbf{v}$ as shown in Fig. 1.27. The component $F_y \cos\alpha$, normal to the path of motion, is called *cornering force*, and

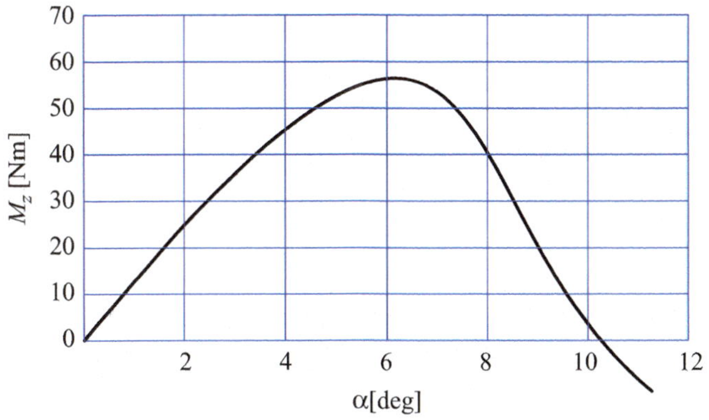

Fig. 1.26 Aligning moment M_z as a function of slip angle α for a constant vertical load

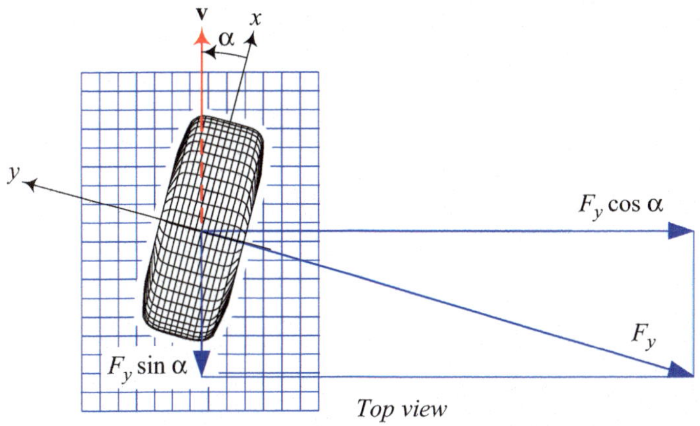

Fig. 1.27 The cornering and drag components of a lateral force F_y

the component $F_y \sin \alpha$, along the path of motion, is called *drag force*. The lateral force F_y is also called *side force* or *grip*. We may combine the lateral forces of all tires of a vehicle and have them acting at the car's mass center C.

Example 22 Effect of velocity.

The curve of lateral force $F_y(\alpha)$ as a function of the sideslip angle α decreases as velocity increases. Hence, we need to increase the sideslip angle at higher velocities to generate the same lateral force. Sideslip angle increases by increasing the steer angle. Figure 1.28 illustrates the effect of velocity on F_y for a radial passenger tire. Because of this behavior, the curvature of trajectory of a one-wheel-car at a fixed steer angle decreases by increasing the driving speed.

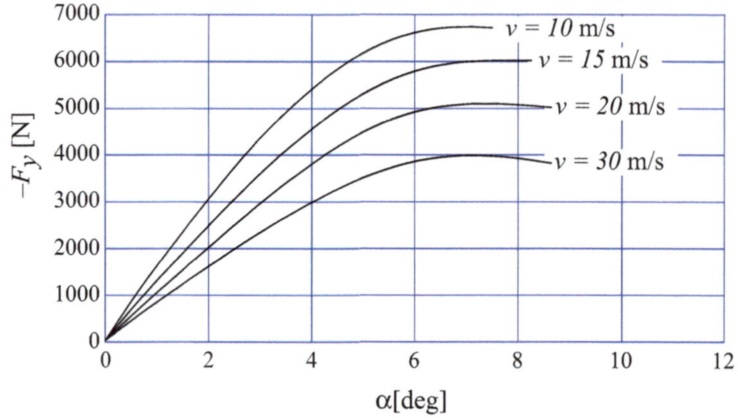

Fig. 1.28 Effect of velocity on F_y for a radial tire

To model the experimental results of lateral force dependency with speed, we suggest a correction in the equation of lateral force (1.145) as

$$\frac{F_y}{F_z} = -C_\alpha \, \alpha + C_{\alpha_1} \, v_x \tag{1.164}$$

or

$$\frac{F_y}{F_z} = -C_\alpha \, \alpha + C_{\alpha_2} \, v_x^2 \tag{1.165}$$

or

$$\frac{F_y}{F_z} = -C_\alpha \, \alpha + C_{\alpha_1} \, v_x + C_{\alpha_2} \, v_x^2 \tag{1.166}$$

where C_{s_1}[s/m] is the coefficient indication how much F_y/F_z will drop for every 1 m/s of longitudinal velocity of the tire. Similarly, $C_{\alpha_2}[\,s^2/\,m^2]$ is the coefficient indication how much F_y/F_z will drop for every 1 $[\,m/s]^2$ of longitudinal velocity of the tire.

Example 23 Effect of tire load on lateral force.

When the tire load F_z increases, the tire treads stick to the road better. Hence, the lateral force increases at a constant slip angle α, and the slippage occurs at a higher slip angles. Figure 1.29 illustrates the lateral force behavior of a sample tire for different normal loads. Increasing the load not only increases the maximum attainable lateral force, but also pushes the maximum of the lateral force to higher slip angles.

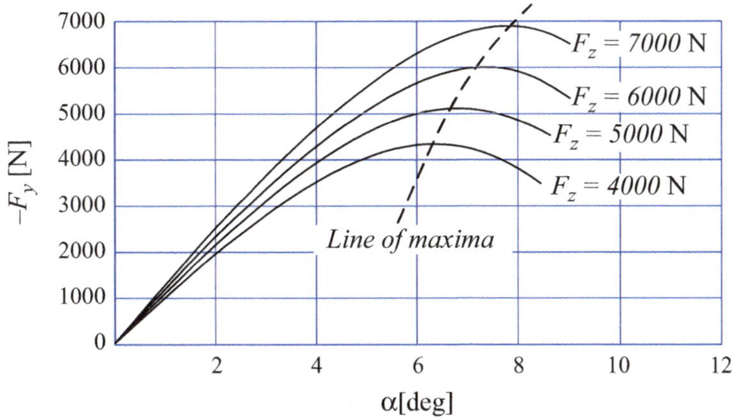

Fig. 1.29 Lateral force behavior of a sample tire for different normal loads as a function of slip angle α

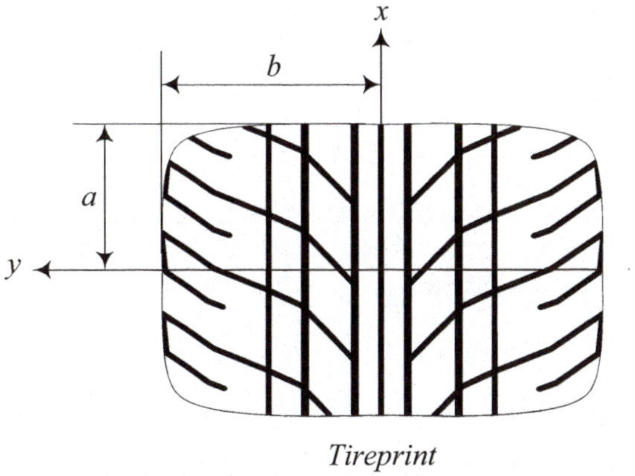

Fig. 1.30 A tireprint

Example 24 Tireprint.

The contact area between a tire and the road is called the *tireprint* and is shown by A. At any point of a tireprint, the normal and friction forces are transmitted between the road and tire. The effect of the contact forces can be described by a resulting force system including force and torque vectors applied at the center of the tireprint. The tireprint is also called *contact patch*, *contact region*, or *tire footprint*. A model of tireprint is shown in Fig. 1.30.

The area of the tireprint is inversely proportional to the tire pressure. Lowering the tire pressure is a technique used for off-road vehicles in sandy, muddy, or snowy areas, and for drag racing. Decreasing the tire pressure causes the tire to slump, so more of the tire is in contact with the surface, giving better traction in low friction

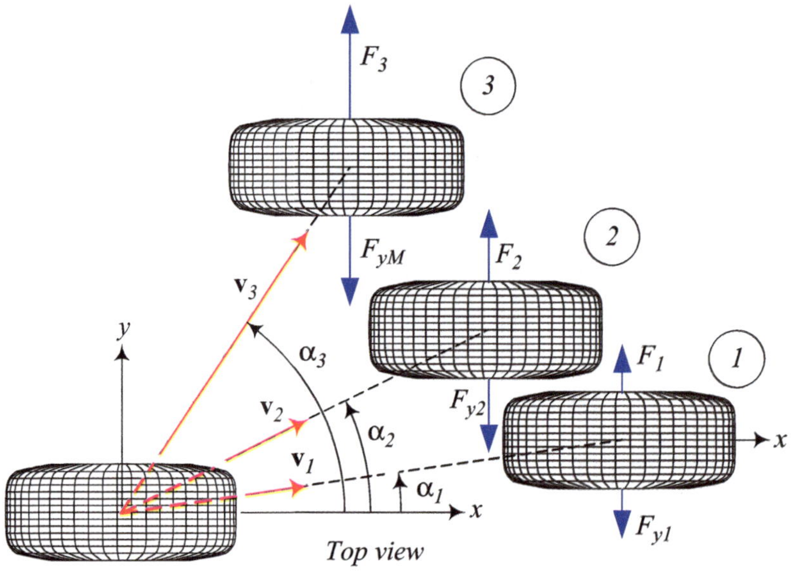

Fig. 1.31 The lateral tire force F_y is proportional to the sideslip angle α assuming the coefficient C_α to remain constant

conditions. It also helps the tire grip small obstacles as the tire conforms more to the shape of the obstacle, and makes contact with the object in more places. The shape of tireprint may be expressed by

$$\frac{x^{2n}}{a^{2n}} + \frac{y^{2n}}{b^{2n}} = 1 \qquad n \in N \tag{1.167}$$

where $n = 2$ or $n = 3$ are showing the tireprint shape for radial tires better.

Example 25 Saturation of lateral force.

Consider a tire that is rolling forward with constant velocity v_x. Applying a constant lateral force F at the wheel center will generate a tire lateral reaction force F_y at the tireprint that causes the tire path to make an angle α with the straightforward x-axis as is shown by label 1 in Fig. 1.31. The tire is in lateral equilibrium force balance. Increasing the applied force F will increase the tire lateral reaction force F_y as well as the angle α. The tire with label 2 in Fig. 1.31 illustrates this situation. The lateral force F_y is proportional to the sideslip angle α and we may assume the coefficient C_α to remain constant.

$$F_{y_1} = -C_\alpha \alpha_1 F_z \tag{1.168}$$

$$F_{y_2} = -C_\alpha \alpha_2 F_z \tag{1.169}$$

The lateral forces are in balance as long as $F \leq F_{y_M}$. If the applied force F is greater than F_{y_M}, then the tire lateral force F_y will saturate and reach its maximum value F_{y_M} and remain at $F_y = F_{y_M}$ for any $F > F_{y_M}$. The lateral forces will not be balanced and the extra force will accelerate the tire laterally at a constant acceleration a_y,

$$F - F_{y_M} = m_t\, a_y \tag{1.170}$$

where m_t is the mass of tire and a_y is the lateral acceleration. This situation is labeled 3 in Fig. 1.31. Assuming a given forward velocity v_x, the tire will be moving along vector $\mathbf{v}$.

$$\mathbf{v}_1 = v_x\ \hat{\imath} + v_x \tan\alpha_1\ \hat{\jmath} = v_1\ \cos\alpha_1\,\hat{\imath} + v_1 \sin\alpha_1\ \hat{\jmath} \tag{1.171}$$

$$\mathbf{v}_2 = v_x\ \hat{\imath} + v_x \tan\alpha_2\ \hat{\jmath} = v_2\ \cos\alpha_2\,\hat{\imath} + v_2 \sin\alpha_2\ \hat{\jmath} \tag{1.172}$$

At the moment where the lateral force is saturated, we have

$$\mathbf{v} = v_x\ \hat{\imath} + v_x \tan\alpha_s\ \hat{\jmath} \tag{1.173}$$

and when $F > F_{y_M}$ the tire will be sliding on the ground along the velocity vector $\mathbf{v}_3$.

$$\mathbf{v}_3 = v_x\,\hat{\imath} + \left(v_x \tan\alpha_s + \int a_y\ dt\right)\ \hat{\jmath} \tag{1.174}$$

When a tire has an acceleration in a direction and constant velocity in the orthogonal direction, the tire force is saturated in the direction of acceleration. In this case, the tire force is saturated in the y direction and the tire force remains at the maximum in the y direction, $F_y = F_{y_M}$.

1.4.2 *Tire Camber*

Camber angle γ is the tilting angle of tire-plane about the longitudinal x-axis. Camber angle generates a lateral force F_y called *camber thrust* or *camber force*. Figure 1.32 illustrates the front view of a positively cambered tire and the generated camber force F_y. Camber angle is considered positive $\gamma > 0$, if it is in the positive direction of the x-axis, measured from the z-axis to the tire-plane. A positive camber angle generates a camber force along the y-axis (Jazar 2017).

The camber force is proportional to γ at low camber angles and depends proportionally on the wheel load F_z. Therefore,

$$\mathbf{F}_y = F_y\,\hat{\jmath} \tag{1.175}$$

$$\frac{F_y}{F_z} = -C_\gamma\ S\left(\gamma - \gamma_s\right) \tag{1.176}$$

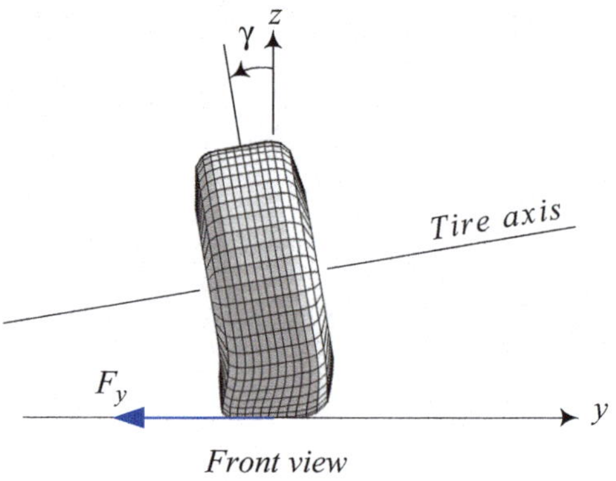

Fig. 1.32 A front view of a cambered tire and the generated camber force

where C_{γ} is called the *camber stiffness* of tire.

$$C_{\gamma} = \frac{1}{F_z} \lim_{\gamma \to 0} \frac{\partial \left(-F_y \right)}{\partial \gamma} \tag{1.177}$$

In the presence of both camber γ and sideslip α, the overall lateral force F_y on a tire is a superposition of the corner force and camber thrust.

$$\frac{F_y}{F_z} = -C_{\gamma}\,\gamma - C_{\alpha}\,\alpha \qquad |\gamma| < \gamma_s \qquad |\alpha| < \alpha_s \tag{1.178}$$

Proof Consider a tire under a constant load while a camber angle is applied on the rim. The tire will deflect laterally such that the tireprint area will be longer in the cambered side. Figure 1.33 compares the tireprint of a straight and a cambered tire on a flat road. As the wheel rolls forward, un-deflected treads enter the tireprint region and deflect laterally as well as longitudinally. The shape of the tireprint depicts that the treads entering the tireprint closer to the cambered side have more time to bestretched laterally. Because the developed lateral stress is proportional to the lateral stretch, the nonuniform tread stretching generates an asymmetric stress distribution and more lateral stress will be developed on the cambered side. The resultant of the nonuniform lateral stress distribution over the tireprint of the cambered tire makes the camber thrust force F_y in the cambered direction.

$$\mathbf{F}_y = F_y\,\hat{\jmath} \qquad F_y = \int_A \tau_y\,dA \tag{1.179}$$

The camber thrust is proportional to the camber angle for small angles.

$$\frac{F_y}{F_z} = -C_{\gamma}\,\gamma \qquad |\gamma| < \gamma_s \tag{1.180}$$

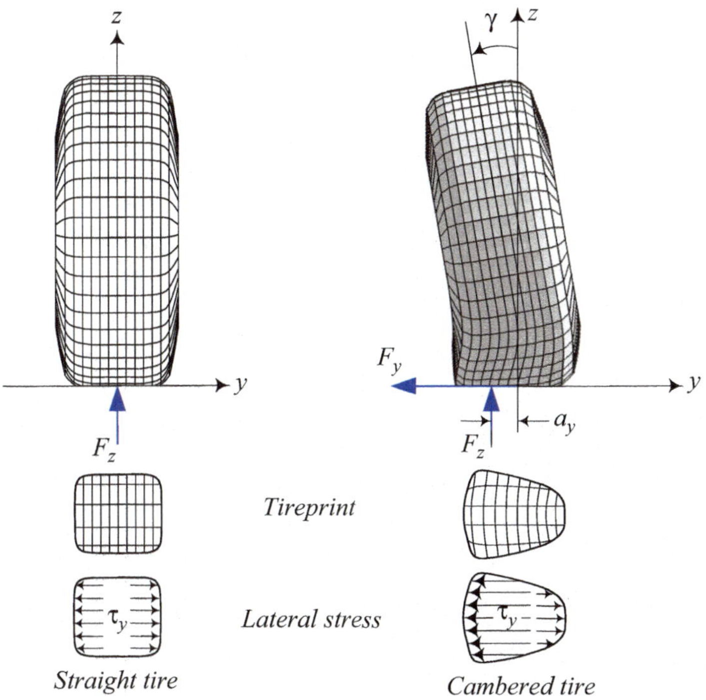

Fig. 1.33 The tireprint of a straight and a cambered tire, turning slowly on a flat road

When the tire is rolling, the resultant camber thrust F_y shifts forward by a distance a_{x_γ}. The resultant moment in the z-direction is called *camber torque*, and the distance a_{x_γ} is called *camber trail.*

$$\mathbf{M}_z = M_z \hat{k} \qquad M_z = F_y \, a_{x_\gamma} \tag{1.181}$$

Camber trail is usually very small, and hence the camber torque is usually ignored in linear analysis of vehicle dynamics.

Because the tireprint of a cambered tire is longer in the cambered side, the resultant vertical force F_z that supports the wheel load

$$F_z = \int_{A_P} \sigma_z \, dA \tag{1.182}$$

shifts laterally by a distance a_{y_γ} from the center of the tireprint.

$$a_{y_\gamma} = \frac{1}{F_z} \int_A y \, \sigma_z \, dA \tag{1.183}$$

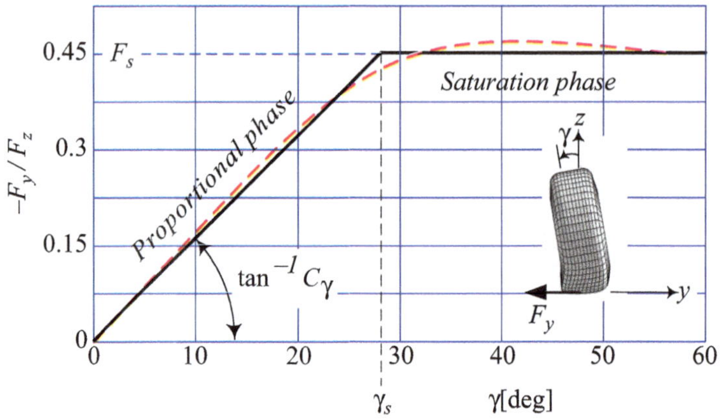

Fig. 1.34 The camber force F_y of a tire as a function of camber angle γ at a constant tire load $F_z = 4500\,\text{N}$

The distance a_{y_γ} is called the *camber arm*, and the resultant moment $\mathbf{M}_x$ is called the *camber moment*.

$$\mathbf{M}_x = M_x\,\hat{k} \qquad M_x = -F_z\,a_{y_\gamma} \tag{1.184}$$

The camber moment tends to turn the tire about the x-axis and tries to make the tire-plane align with the z-axis. The camber arm a_{y_γ} is proportional to the camber angle γ for small angles.

$$a_{y_\gamma} = C_{y_\gamma}\,\gamma \tag{1.185}$$

Figure 1.34 shows the camber force F_y as a function of camber angle γ at a constant tire load $F_z = 4500\,\text{N}$. The lateral force because of camber behaves similar to the lateral force because of sideslip. A linearly increasing and saturating function is a simple and effective model of camber force behavior.

Figure 1.35 depicts the variation of camber force F_y as a function of normal load F_z at different camber angles for a sample radial tire.

The tireprint of a rolling tire will have a longitudinally distortion. Camber will distort the tireprint laterally. Apply a slip angle α to a rolling cambered tire will distort the tireprint laterally in a nonuniform way such that the resultant lateral force would be at a distance a_{x_γ} and a_{y_γ} from the center of the tireprint. Both distances a_{x_γ} and a_{y_γ} are functions of angles α and γ. Camber force due to γ, along with the corner force due to α, gives the total lateral force applied on a tire (Clark 1971). Therefore, the overall lateral force will be calculated as

$$\frac{F_y}{F_z} = -C_\alpha\,\alpha - C_\gamma\,\gamma \tag{1.186}$$

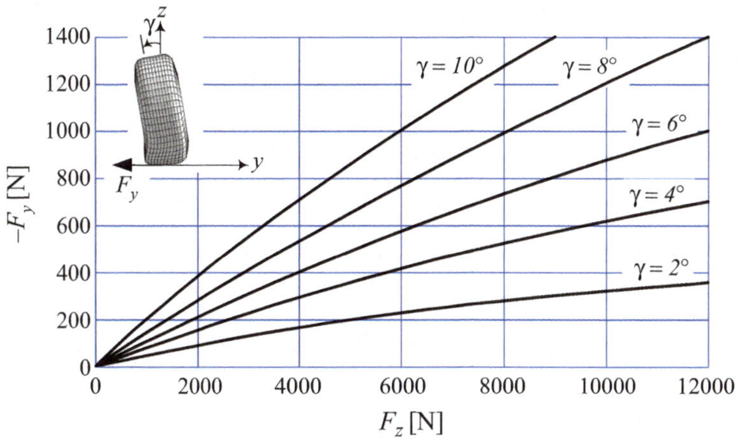

Fig. 1.35 The variation of camber force F_y as a function of normal load F_z at different camber angles for a sample radial tire

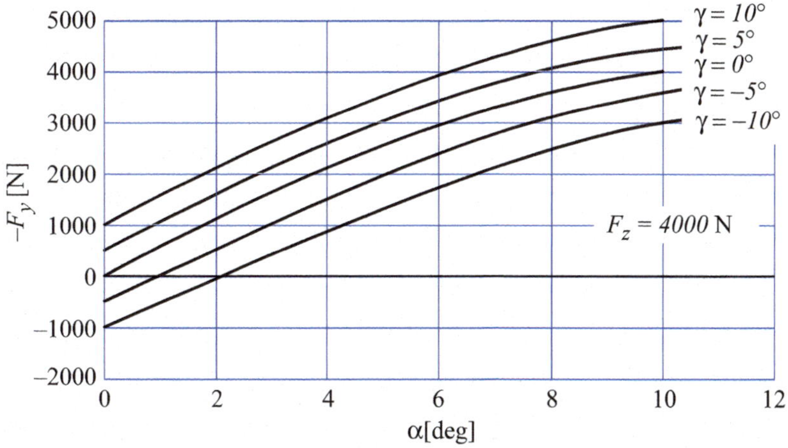

Fig. 1.36 An experimental example for lateral force as a function of γ and α at a constant load $F_z = 4000\,\mathrm{N}$

that is acceptable for $\gamma \lesssim 10\,\mathrm{deg}$ and $\alpha \lesssim 5\,\mathrm{deg}$. Presence of both camber angle γ and slip angle α makes the situation interesting because the total lateral force can be positive or negative according to the directions of γ and α. Figure 1.36 illustrates an example of lateral force as a function of γ and α at a constant load $F_z = 4000\,\mathrm{N}$. Similar to lateral force, the aligning moment M_z can be approximated as a combination of the slip and camber angle effects.

$$M_z = C_{M_\alpha}\,\alpha + C_{M_\gamma}\,\gamma \tag{1.187}$$

The coefficient C_{M_α} is the aligning moment generated by 1 rad sideslip α at a zero γ, and the coefficient C_{M_α} is the aligning moment generated by 1 rad camber γ at zero α. For a radial tire, $C_{M_\alpha} \approx 0.013\,\mathrm{N\,m/\,deg}$ and $C_{M_\gamma} \approx 0.0003\,\mathrm{N\,m/\,deg}$, while for a non-radial tire, $C_{M_\alpha} \approx 0.01\,\mathrm{N\,m/\,deg}$ and $C_{M_\gamma} \approx 0.001\,\mathrm{N\,m/\,deg}$. ■

Example 26 Camber importance and tireprint model.

Cambering of a tire creates a lateral force, even though there is no sideslip. The effects of cambering are particularly important for motorcycles that produce a large part of the lateral force by camber. The following equations are presented to model the lateral deviation of a cambered tireprint from the straight tireprint, and express the lateral stress τ_y due to camber

$$y = -\sin\gamma\left(\sqrt{R_g^2 - x^2} - \sqrt{R_g^2 - a^2}\right) \tag{1.188}$$

$$\tau_y = -\gamma k\left(a^2 - x^2\right) \tag{1.189}$$

where a is half of the longitudinal length of tireprint and the coefficient k is chosen such that the average camber defection is correct in the tireprint.

$$\int_{-a}^{a} \tau_y\,dx = \int_{-a}^{a} y\,dx \tag{1.190}$$

$$-\frac{4}{3}\gamma k a^3 = -\sin\gamma\left(R_g^2 \sin^{-1}\frac{a}{R_g} - a\sqrt{R_g^2 - a^2}\right) \tag{1.191}$$

Therefore,

$$k = \frac{3\sin\gamma}{4a^3\gamma}\left(R_g^2 \sin^{-1}\frac{a}{R_g} - a\sqrt{R_g^2 - a^2}\right) \tag{1.192}$$

$$\approx \frac{3}{4}\frac{R_g\sqrt{R_g^2 - a^2}}{a^2} \tag{1.193}$$

and

$$\tau_y = -\frac{3}{4}\gamma\frac{R_g\sqrt{R_g^2 - a^2}}{a^2}\left(a^2 - x^2\right) \tag{1.194}$$

Example 27 Banked road.

Consider a vehicle moving on a road with a transversal slope β, while its tires remain vertical. There is a downhill component of weight, $F_1 = mg\sin\beta$, that pulls the vehicle down. There is also an uphill camber force due to camber $\gamma \approx \beta$ of tires with respect to the road $F_2 = C_\gamma \gamma F_z$. The resultant lateral force $F_y = C_\gamma \gamma F_z - mg\sin\beta$ depends on camber stiffness C_γ and determines if the vehicle goes uphill or downhill, considering the vehicle to be neutral steering. Since the camber stiffness C_γ is higher for non-radial tires, it is more possible for a non-radial than a radial tire to go uphill.

The effects of cambering are particularly important for motorcycles that produce a large part of their lateral force by cambering. For cars and trucks, the camber angles are much smaller and in many applications their effect can be negligible. However, some suspensions are designed to make the wheels cambered when the axle load varies, or when they steered.

1.5 Tire Combined Force

The force system at the tireprint of a loaded, rolling, steered, cambered tire includes: forward force F_x, lateral force F_y, vertical force F_z, aligning moment M_z, roll moment M_x, and pitch moment M_y. The forward force F_x and lateral force F_y are the most significant forces in vehicle maneuvering. The F_x and F_y take the tire load F_z, sideslip α, longitudinal slip s, and the camber angle γ as input,

$$\frac{F_x}{F_z} = F_x(\alpha, s, \gamma) = C_s\, S(s - s_s) \tag{1.195}$$

$$\frac{F_y}{F_z} = F_y(\alpha, s, \gamma) = -C_\alpha\, S(\alpha - \alpha_s) - C_\gamma\, S\left(\gamma - \gamma_s\right) \tag{1.196}$$

where S is the saturation functions. The saturation function S has been introduced in (1.60) and (1.62).

$$S(x - x_0) = \begin{cases} x_0 & x_0 < x \\ x & -x_0 < x < x_0 \\ -x_0 & x < -x_0 \end{cases} \tag{1.197}$$

$$S(x - x_0) = x\, H(x_0 - |x|) + x_0\, H(|x| - x_0) \tag{1.198}$$

$$H(x - x_0) = \begin{cases} 1 & x > x_0 \\ 0 & x < x_0 \end{cases} \tag{1.199}$$

The *longitudinal slip coefficient* C_s, the *lateral stiffness* C_α, and the *camber stiffness* C_γ are all assumed to be constant. The forward or longitudinal force F_x and the lateral force F_y are in tireprint and tangent to the ground. The tangential forces F_x and F_y may be combined to make a resultant *tangetial force* called the *shear force* (Pacejka 2012; Abe 2009; Dai et al. 2017a,b).

When the tire has a combination of tire inputs, α, s, γ, the tire forces are called *tire combined force.* The most important tire combined force is the shear force because of longitudinal slip and sideslip. However, as long as the angles and slips are very small within the linear range of tire behavior, a superposition can be utilized to estimate the output forces. When the tire is under both longitudinal slip and sideslip, the tire is under *combined slip.*

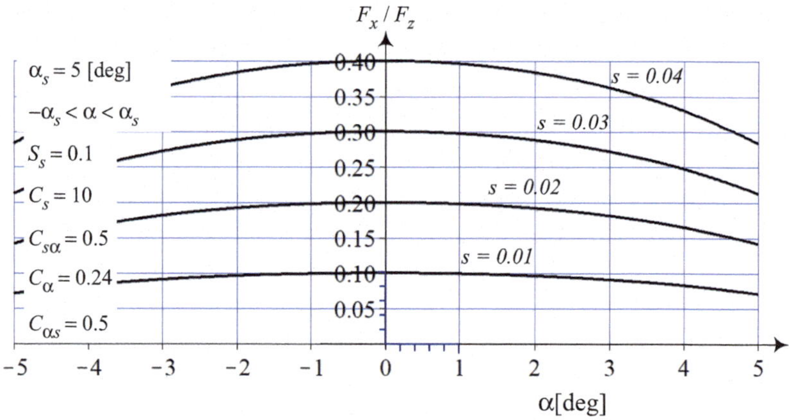

Fig. 1.37 The longitudinal force F_x/F_z drops in the $(\alpha,\ F_x/F_z)$plane, when sideslip α introduces and increases to a tire

It is known that introducing both, longitudinal slip and sideslip together, will reduce their effect and their associated forces. In this section, we study the lateral and longitudinal forces when both of them are present.

1.5.1 Elliptic Model

We adopt the proportional-saturation model for longitudinal and sideslip of tire. When $\alpha = 0$, a small longitudinal slip $s < s_s$ generates the longitudinal force $F_x/F_z = C_s s$, and when $s = 0$, a small sideslip angle $\alpha < \alpha_s$ generates a lateral force of $F_y/F_z = -C_\alpha \alpha$. When there exists a longitudinal slip $s < s_s$ and then we also introduce a sideslip $\alpha < \alpha_s$, the longitudinal force will reduce by

$$\frac{F_x}{F_z} = C_s s \sqrt{1 - C_{s\alpha}\left(\frac{\alpha}{\alpha_s}\right)^2} \qquad |\alpha| < \alpha_s \qquad |s| < s_s \tag{1.200}$$

where α_s and s_s are respectively the *saturation values* of α and s for the tire, and $C_{s\alpha}$ is the *tire longitudinal drop factor* for reduction of F_x in the presence of sideslip α. The tire *longitudinal drop factor* $C_{s\alpha}$ is determined by experiment. Figure 1.37 illustrates how the longitudinal force reduces in the $(\alpha,\ F_x/F_z)$-plane when sideslip α is introduced to a tire with constant longitudinal slip s.

When there exists a longitudinal slip $s < s_s$, the lateral force will drop from its level of $F_y/F_z = -C_\alpha \alpha$ to

$$\frac{F_y}{F_z} = -C_\alpha \alpha \sqrt{1 - C_{\alpha s}\left(\frac{s}{s_s}\right)^2} \qquad |\alpha| < \alpha_s \qquad |s| < s_s \tag{1.201}$$

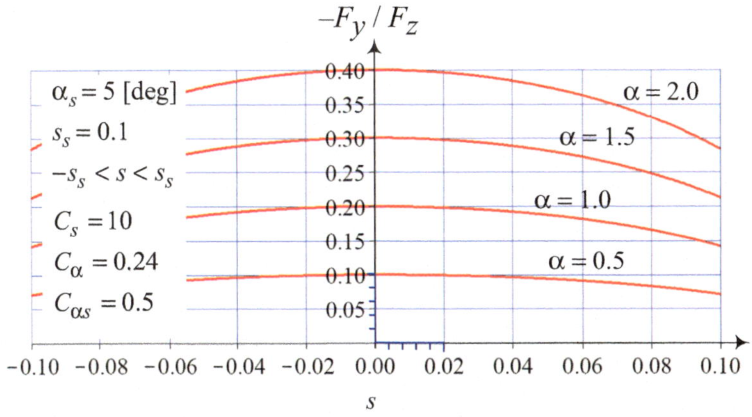

Fig. 1.38 The lateral force drops in the $(\alpha, -F_y/F_z)$-plane when longitudinal slip α introduces to a tire with constant sideslip s

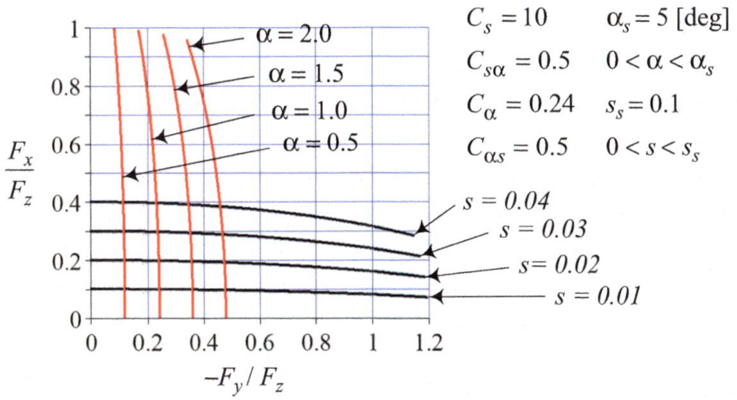

Fig. 1.39 The longitudinal and lateral forces drop in the $(-F_y, F_x)$ plane, when respectively sideslip and longitudinal slip are introduced to a tire

and $C_{\alpha s}$ is the *tire lateral drop factor* for reduction of F_y in the presence of longitudinal slip s. The tire lateral drop factor $C_{\alpha s}$ is determined by experiment. Figure 1.38 illustrates how the lateral force drops in the $(\alpha, -F_y/F_z)$-plane when longitudinal slip is introduced to a tire with constant sideslip and then increases.

A proper way to show the effect of combined slips on tire forces is to plot F_x/F_z and F_y/F_z against each other. Figure 1.39 illustrates both lateral and longitudinal forces when longitudinal slip and sideslip are introduced in the tire dynamics.

Including positive and negative ranges of α and s and plotting the curves for $0 > s > s_s$ and $0 > \alpha > \alpha_s$ for different values of tire drop factors, $C_{\alpha s}$, $C_{s\alpha}$, makes the relationship of F_x and F_y as shown in Fig. 1.40 justifying elliptic model.

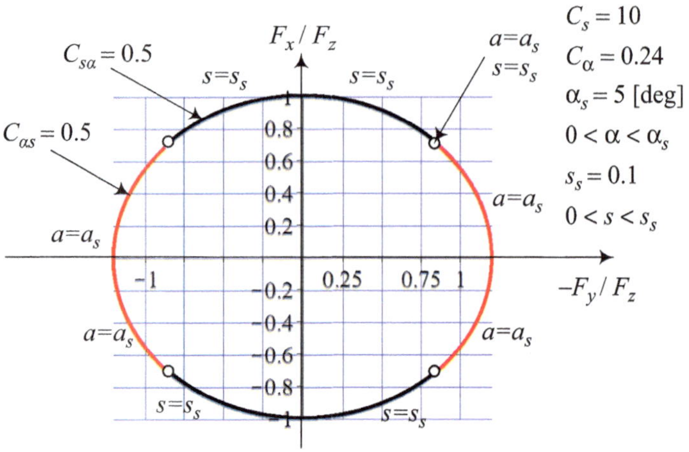

Fig. 1.40 The longitudinal and lateral forces at the limits in the (F_y, F_x) plane, when sideslip and longitudinal slip are introduced to a tire. At any point on limit shape one of the slips, α or s, is saturated. There are four points on the limit shape where both α and s are saturated

Including the saturation in lateral and longitudinal forces, we define F_x and F_y mathematically by the following equations.

$$\frac{F_x}{F_z} = C_s\, S\,(s - s_s)\sqrt{1 - C_{s\alpha}\left(\frac{S\,(\alpha - \alpha_s)}{\alpha_s}\right)^2} \tag{1.202}$$

$$\frac{F_y}{F_z} = -C_\alpha S\,(\alpha - \alpha_s)\sqrt{1 - C_{\alpha s}\left(\frac{S\,(s - s_s)}{s_s}\right)^2} \tag{1.203}$$

$$S(s - s_s) = s\ H\,(s_s - |s|) + s_s\ H\,(|s| - s_s) \tag{1.204}$$

$$S(\alpha - \alpha_s) = \alpha\ H\,(\alpha_s - |\alpha|) + \alpha_s\ H\,(|\alpha| - \alpha_s) \tag{1.205}$$

Figures 1.41 and 1.42 illustrate the elliptic proportional-saturation model for longitudinal and lateral forces indicating the effect of sideslip α on longitudinal force F_x/F_z and the effect of longitudinal slip s on lateral force F_y/F_z.

Proof Assuming $\alpha = 0$, a small longitudinal slip $s < s_s$ generates the longitudinal force $F_x/F_z = C_s s$. When we have longitudinal slip s and then a sideslip α is introduced into the system, the longitudinal force will drop. To have a mathematical model that simulates experimental data, the following nonlinear function is introduced to model the phenomenon

$$\frac{F_x}{F_z} = C_s s\sqrt{1 - C_{s\alpha}\left(\frac{\alpha}{\alpha_s}\right)^2} \qquad |\alpha| < \alpha_s \qquad |s| < s_s \tag{1.206}$$

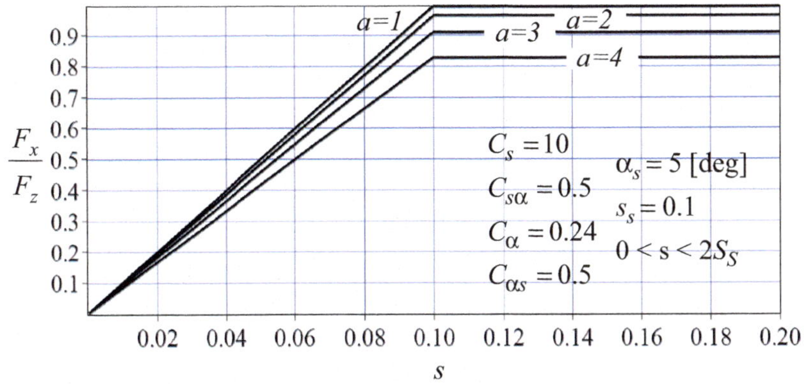

Fig. 1.41 The effect of sideslip angle α on longitudinal force F_x/F_z in saturating model

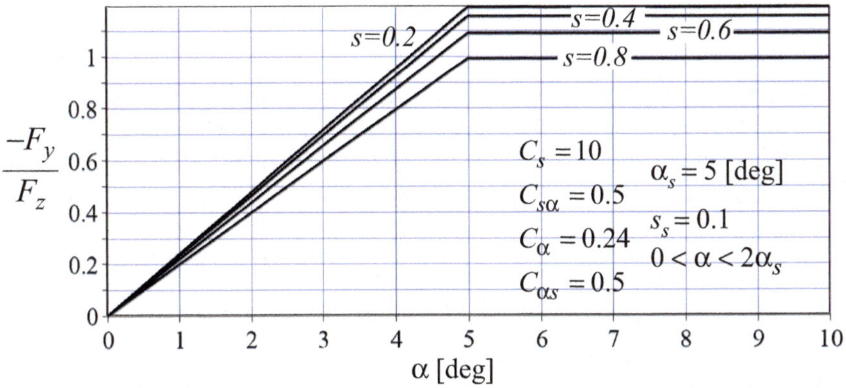

Fig. 1.42 The effect of longitudinal slip ratio s on lateral force F_y/F_z in saturating model

where α_s is the saturation point of α for the tire. The coefficient $C_{s\alpha}$ is the *longitudinal drop factor* that indicates the percentage of reduction in F_x/F_z when α is changing from $\alpha = 0$ to $\alpha = \alpha_s$, for a constant s.

Similarly, when $s = 0$ and a small sideslip angle $\alpha < \alpha_s$ is applied, the tire generates a lateral force of $F_y/F_z = -C_\alpha \alpha$. The lateral force F_y/F_z will drop when longitudinal slip s is added to the dynamics of the tire. The following nonlinear function models this phenomenon

$$\frac{F_y}{F_z} = -C_\alpha \alpha \sqrt{1 - C_{\alpha s} \left(\frac{s}{s_s}\right)^2} \qquad |\alpha| < \alpha_s \qquad |s| < s_s \tag{1.207}$$

where s_s is the saturation point of s for the tire. The coefficient $C_{\alpha s}$ is the *lateral drop factor* that determines the percentage of reduction in F_y/F_z when s is changing from $s = 0$ to $s = s_s$, for a constant α. Figures 1.37 and 1.38 illustrate how the

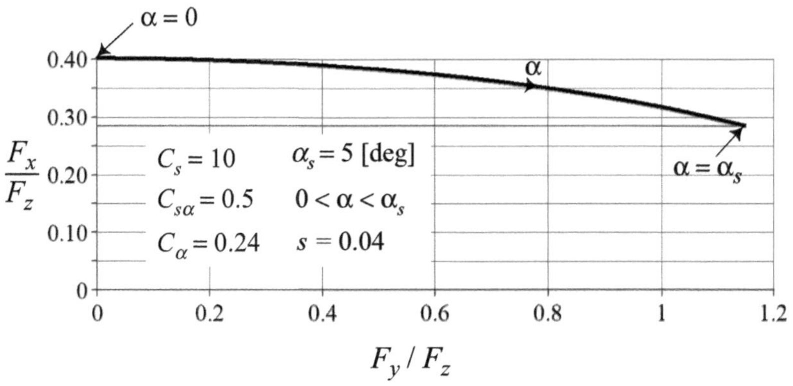

Fig. 1.43 $C_{s\alpha} = 0.5$ means that F_x/F_z drops $1-\sqrt{1-C_{s\alpha}} = 30\%$ when α increases from $\alpha = 0$ to $\alpha = \alpha_s$. If $C_{s\alpha} = 0.3$ then F_x/F_z drops $(1-0.3) \times 100 = 70\%$ when α increases from $\alpha = 0$ to $\alpha = \alpha_s$

functions (1.206) and (1.207) work. Although α and s may go beyond α_s and s_s will saturate at $F_x/F_z = C_s s\sqrt{1-C_{s\alpha}}$ and $F_y/F_z = -C_\alpha \alpha \sqrt{1-C_{\alpha s}}$.

Employing proportional-saturation model for longitudinal and lateral forces, we may use *saturation* and *Heaviside* functions (1.60) and (1.61) to express the forces in a single equation to cover proportional and saturation parts of F_x/F_z and F_y/F_z. Equations (1.202) and (1.203) are expressing the longitudinal and lateral forces of tires using proportional-saturation in elliptic model. ■

Example 28 Measuring of $C_{s\alpha}$.

Figure 1.43 illustrates a sample of the effect of increasing α on F_x/F_z using elliptic model (1.200) from $\alpha = 0$ to $\alpha = \alpha_s$, for a constant $s = 0.04$. When $\alpha = 0$ then, $F_x/F_z = 0.4$ for the given values of $C_s = 10$, $\alpha_s = 5$ deg, $C_\alpha = 0.24$. By increasing α the value of F_x/F_z drops as is shown in the figure. Knowing that when $\alpha = \alpha_s$, then $F_x/F_z = 0.283$ and $F_y/F_z = -1.15$, we have,

$$\frac{F_x}{F_z} = C_s s \sqrt{1 - C_{s\alpha}\left(\frac{\alpha}{\alpha_s}\right)^2} \tag{1.208}$$

$$0.283 = 0.4\sqrt{1 - C_{s\alpha}} \tag{1.209}$$

$$C_{s\alpha} = 0.499 \tag{1.210}$$

which indicates, $C_{s\alpha} = 0.499$, and F_x/F_z drops by $\left(1 - \sqrt{1-0.5}\right) \times 100 \simeq 30\%$ approximately.

In general, when $\alpha = 0$ we have

$$\frac{F_{x_1}}{F_z} = C_s s \tag{1.211}$$

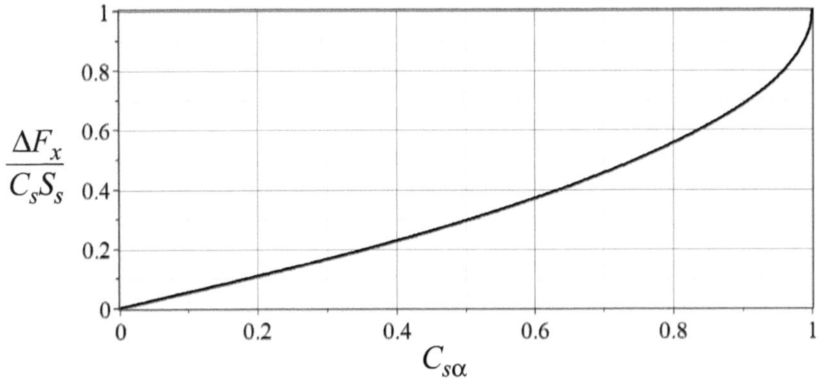

Fig. 1.44 The effect of the longitudinal drop factor $C_{s\alpha}$ on percentage reduction of longitudinal force $\Delta F_x / C_s s_s$

and when we $\alpha = \alpha_s$, then

$$\frac{F_{x2}}{F_z} = C_s s \sqrt{1 - C_{s\alpha}} \tag{1.212}$$

therefore,

$$\frac{F_{x2}}{F_{x_1}} = \sqrt{1 - C_{s\alpha}} \tag{1.213}$$

and therefore, the reduction will be

$$\frac{\Delta F_x}{C_s s_s} = \frac{F_{x_1} - F_{x2}}{F_{x_1}} = 1 - \sqrt{1 - C_{s\alpha}} \tag{1.214}$$

In case of $C_{s\alpha} = 0.3$ then F_x / F_z drops $\left(1 - \sqrt{1 - 0.3}\right) \times 100 = 16\%$ when α increases from $\alpha = 0$ to $\alpha = \alpha_s$.

Figure 1.44 illustrates the effect of the longitudinal drop factor $C_{s\alpha}$ on percentage reduction of longitudinal force $\Delta F_x / C_s s_s$.

Example 29 The limit slip curve.

Let us begin with $\alpha = 0$, $s = s_s$, and

$$\begin{aligned} s_s &= 0.1 \qquad C_s = 10 \qquad C_{s\alpha} = 0.5 \\ \alpha_s &= 5 \deg \qquad C_\alpha = 0.24 \qquad C_{\alpha s} = 0.5 \end{aligned} \tag{1.215}$$

The longitudinal force of the tire is saturated and is at its maximum $F_x / F_z = F_{x_M} / F_z = C_s \, s_s = 1$. This situation is indicated by point A in Fig. 1.45. Increasing sideslip angle α from $\alpha = 0$ to $\alpha = \alpha_s$ will generate a lateral force according to Eq. (1.201) and decreases the magnitude of the longitudinal force according to

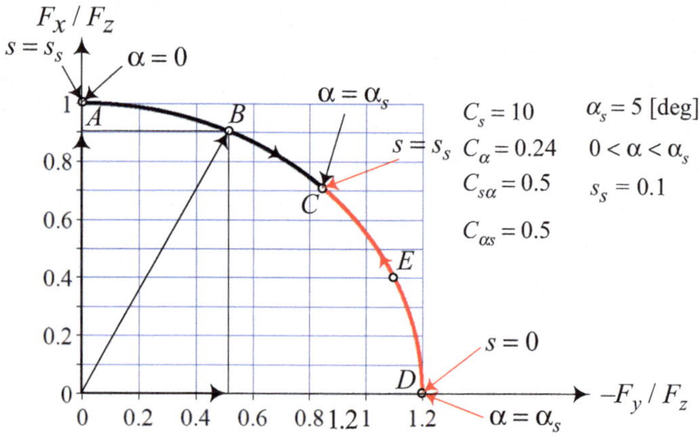

Fig. 1.45 The saturation line. At any point on the saturation line, at least one of s or α is at its saturation level. Only at point C, both s and α are at their saturation levels

Eq. (1.200). The nominal point B is a sample point indicating the tire force vector with a saturated longitudinal component and unsaturated lateral component. The tire at point B is considered to be in longitudinal sliding on the ground and laterally sticking to the ground.

The curve from point A to C indicates how the forces F_x/F_z and F_y/F_z will change when α increases from $\alpha = 0$ at A to $\alpha = \alpha_s$ at C while $s = s_s$. This is true for all points between A and C at which the longitudinal force is saturated and the lateral force is undersaturated and their values are:

$$\frac{F_x}{F_z} = \frac{F_{x_M}}{F_z} = C_s s_s \sqrt{1 - C_{s\alpha}\left(\frac{\alpha}{\alpha_s}\right)^2} \tag{1.216}$$

$$\frac{F_y}{F_z} = -C_\alpha \alpha \sqrt{1 - C_{\alpha s}} < \frac{F_{y_M}}{F_z} \tag{1.217}$$

At a nominal point B between A and C, at which $F_x/F_z = 0.9$, we have

$$\frac{F_{x_M}}{F_z} = C_s s_s \sqrt{1 - C_{s\alpha}\left(\frac{\alpha}{\alpha_s}\right)^2} = \sqrt{1 - 0.5\left(\frac{\alpha}{5}\right)^2} = 0.9 \tag{1.218}$$

$$\alpha = 3.08 \text{ deg} \tag{1.219}$$

and therefore,

$$\begin{aligned}\frac{F_y}{F_z} &= -C_\alpha \alpha \sqrt{1 - C_{\alpha s}\left(\frac{s}{s_s}\right)^2} \\ &= 0.24 \times 3.08\sqrt{1 - 0.5} = 0.522\end{aligned} \tag{1.220}$$

At point C where $\alpha = \alpha_s$, the lateral force will also be saturated and the force conditions become

$$\frac{F_x}{F_z} = \frac{F_{x_M}}{F_z} = C_s s_s \sqrt{1 - C_{s\alpha}} \tag{1.221}$$

$$\frac{F_y}{F_z} = \frac{F_{y_M}}{F_z} = -C_\alpha \alpha_s \sqrt{1 - C_{\alpha s}} \tag{1.222}$$

Having $C_{s\alpha} = 0.5$, the magnitude of F_x/F_z will drop from $F_{x_M}/F_z = 1$ at point A to $F_{x_M}/F_z = 0.707$ at point C.

$$\begin{aligned} \frac{F_x}{F_z} = \frac{F_{x_M}}{F_z} &= C_s s_s \sqrt{1 - C_{s\alpha}\left(\frac{\alpha_s}{\alpha_s}\right)^2} \\ &= 10 \times 0.1\sqrt{1 - 0.5 \times \frac{5}{5}} = 0.707 \end{aligned} \tag{1.223}$$

The curve AC is the limit line of generating force on which the longitudinal force remains saturated. The point C is the absolute terminating point at which both longitudinal and lateral forces are saturated. No matter how much s and α are more than their saturation values of s_s and α_s, the resultant tire tangential force of the tireprint will not move from point C.

Let us now begin with $\alpha = \alpha_s$, $s = 0$, for the same tire (1.215). The lateral force of the tire is saturated and it is at its maximum possible value $F_y/F_z = -F_{y_M}/F_z = C_\alpha s_\alpha = 1.2$. This situation is shown by point D in Fig. 1.45. Let us increase the longitudinal slip of the tire s from $s = 0$ to generate longitudinal force in addition to the lateral force. The lateral force reduces and the tangential tire force vector moves from point D towards C. Point E is a nominal point between D and C at which the lateral force of the tire is saturated while the longitudinal force is unsaturated. The tire at E slides laterally on the ground while remaining stuck to the ground in longitudinally direction. The line from point D to point C indicates how the forces F_y/F_z and F_x/F_z will change when s increases from $s = 0$ to $s = s_s$.

Example 30 Possible tire force zone.

Let us assume a tire that is at conditions that at least one of the longitudinal slips or sideslip α is saturated. Considering both, negative and positive values of s and α and plotting the limit slip curves make a closed elliptic shape that divides the plane of $(F_x/F_z, F_y/F_z)$ into possible and impossible zones of tangential tire force vectors. Figure 1.40 illustrates a sample of the possible force zone. Any resultant tangential force vector at the tireprint of the tire starts at the origin of the coordinate at the center of tireprint and ends to a point inside the friction limit shape or on its boundary.

Let us analytically determine the equation of the closed limit shape of Fig. 1.40.

$$s_s = 0.1 \qquad C_s = 10 \qquad C_{s\alpha} = 0.5$$

$$\alpha_s = 5\,\text{deg} \qquad C_\alpha = 0.24 \qquad C_{\alpha s} = 0.5 \tag{1.224}$$

The curve AC in Fig. 1.45 is a part of the shape on which we have

$$\frac{F_x}{F_z} = C_s s_s \sqrt{1 - C_{s\alpha}\left(\frac{\alpha}{\alpha_s}\right)^2} = \sqrt{1 - 0.02\alpha^2} \tag{1.225}$$

$$\frac{-F_y}{F_z} = C_\alpha \alpha \sqrt{1 - C_{\alpha s}\left(\frac{s_s}{s_s}\right)^2} \simeq 0.17\alpha \tag{1.226}$$

These are parametric equations of the curve AC. The other parts of the shape would have similar parametric equations. Although Fig. 1.40 shows an elliptic shape, the limit shape may not be an actual ellipse.

Example 31 Ellipse condition.

It is common to use the following ellipse as an approximation for the limit of tire force.

$$\left(\frac{F_y/F_z}{C_\alpha \alpha_s}\right)^2 + \left(\frac{F_x/F_z}{C_s s_s}\right)^2 = 1 \tag{1.227}$$

Substituting elliptical model of F_x/F_z and F_y/F_z from (1.200) and (1.201) provide us with

$$\left(\frac{\alpha\sqrt{1 - C_{\alpha s}\left(\frac{s}{s_s}\right)^2}}{\alpha_s}\right)^2 + \left(\frac{s\sqrt{1 - C_{s\alpha}\left(\frac{\alpha}{\alpha_s}\right)^2}}{s_s}\right)^2 = 1 \tag{1.228}$$

and then,

$$\frac{\alpha^2}{\alpha_s^2} + \frac{s^2}{s_s^2} - \frac{\alpha^2 s^2}{\alpha_s^2 s_s^2} C_{s\alpha} - \frac{\alpha^2 s^2}{\alpha_s^2 s_s^2} C_{\alpha s} = 1 \tag{1.229}$$

This is the condition to have the tire force limit to be the ellipse (1.227).

On the partial curve AC of Fig. 1.45, the longitudinal slip is saturated, $s = s_s$. Therefore, the condition (1.229) simplifies to

$$C_{s\alpha} + C_{\alpha s} = 1 \tag{1.230}$$

It means that if we have $C_{s\alpha} + C_{\alpha s} = 1$, then the curve AC of Fig. 1.45 is a part of ellipse (1.227). Equation (1.230) is the condition to have the limit shape as an ellipse.

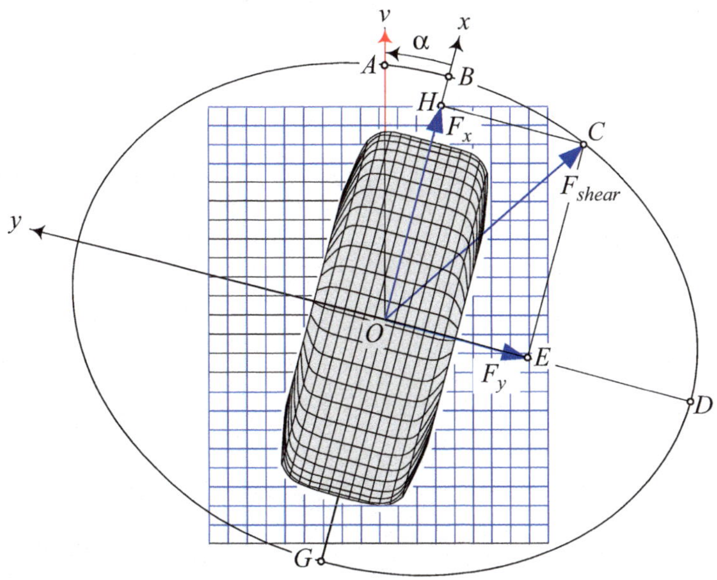

Fig. 1.46 Location and orientation of friction ellipse compared to the tire

Similarly, on the partial curve DC of Fig. 1.45, the sideslip is saturated and therefore, $\alpha = \alpha_s$. The condition (1.229) simplifies to the same *ellipse limit condition* (1.230). Therefore, the ellipse (1.227) matches well with elliptical tire force model if (1.230) is correct.

Figure 1.46 illustrates the friction ellipse compared to the orientation of the tire. The tire friction force or shear force F_{shear} is saturated as it is touching the friction ellipse. This tire is either sliding longitudinally or laterally, or both, and the tire forces are calculated by (1.202) and (1.203). In case the tire force is within the friction ellipse, then the tire sticks to the ground and we have

$$\left(\frac{F_y}{C_\alpha \alpha_s}\right)^2 + \left(\frac{F_x}{C_s s_s}\right)^2 < 1 \tag{1.231}$$

Figure 1.47 illustrates a stuck to the ground tire.

Example 32 Different $C_{s\alpha}$ and $C_{\alpha s}$ with $C_{s\alpha} + C_{\alpha s} = 1$.

Let us assume a tire with different values of $C_{s\alpha}$ and $C_{\alpha s}$ satisfying the ellipse limit condition (1.230).

$$\alpha_s = 5\,\text{deg} \qquad s_s = 0.1 \qquad C_s = 10 \qquad C_\alpha = 0.24 \tag{1.232}$$

Figure 1.48 illustrates the friction ellipse for $C_{s\alpha} = 0.1$, $C_{\alpha s} = 0.9$, and $C_{s\alpha} = 0.3$, $C_{\alpha s} = 0.7$, and $C_{s\alpha} = 0.7$, $C_{\alpha s} = 0.3$, and $C_{s\alpha} = 0.9$, $C_{\alpha s} = 0.1$. The friction

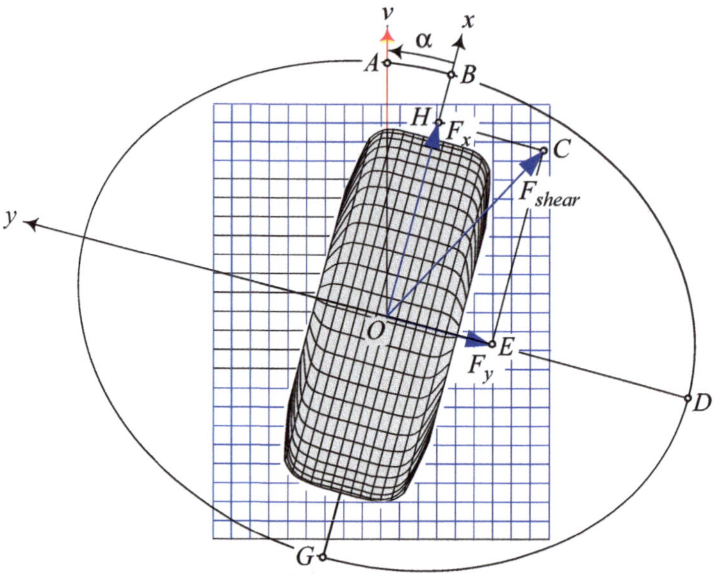

Fig. 1.47 A tire with unsaturated shear tire force on the ground. When the point (F_x, F_y) is inside the friction ellipse of the tire, then the calculated forces F_x, F_y are not saturated

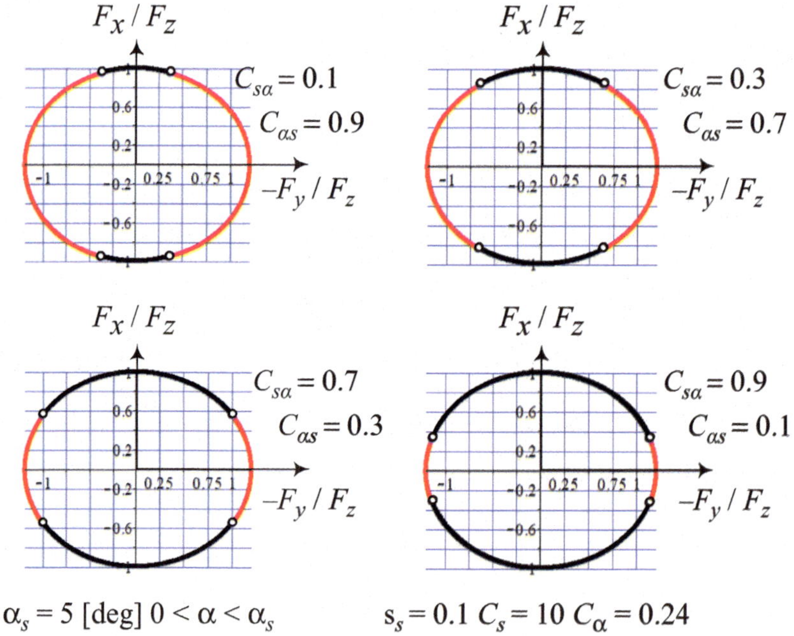

Fig. 1.48 The effects of changing $C_{s\alpha}$ and $C_{\alpha s}$ with the ellipse condition $C_{s\alpha} + C_{\alpha s} = 1$ on the friction limit

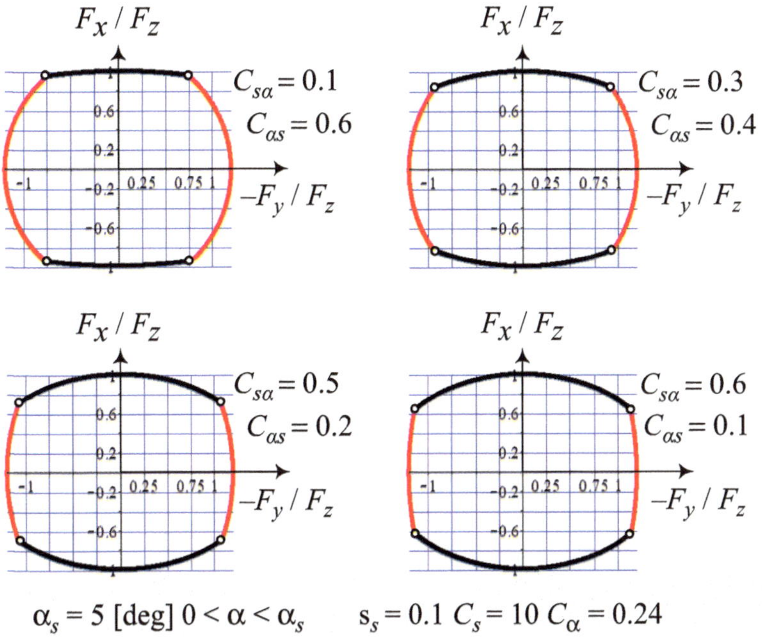

Fig. 1.49 The effects of changing $C_{s\alpha}$ and $C_{\alpha s}$ with condition $C_{s\alpha} + C_{\alpha s} < 1$ on the friction limit

ellipse remains the same as long as the limit condition $C_{s\alpha} + C_{\alpha s} = 1$ is fulfilled. Changing the values of $C_{s\alpha}$, $C_{\alpha s}$ will move the point at which both F_x/F_z and F_y/F_z are saturated. The coordinate of the full saturation point C may be calculated from the Eqs. (1.200) and (1.201) are:

$$\left(\frac{F_{xM}}{F_z}, \frac{F_{yM}}{F_z}\right) = \left(\pm C_s s_s \sqrt{1 - C_{s\alpha}}, \mp C_\alpha \alpha_s \sqrt{1 - C_{\alpha s}}\right) \tag{1.233}$$

To determine the loci of the full saturation points (1.233), we need to substitute $C_{s\alpha} = 1 - C_{\alpha s}$ in $(F_x/F_z, F_y/F_z)$ and eliminate $C_{\alpha s}$ between them. The loci of full saturation points will end up to be the same ellipse (1.227) when $0 < C_{s\alpha} < 1 - C_{\alpha s}$ and $1 - C_{s\alpha} < C_{s\alpha} < 0$.

Example 33 Breaking the ellipse limit condition.

Let us consider a tire with $C_{s\alpha} + C_{\alpha s} \neq 1$ and therefore its force limit breaks the ellipse condition (1.230). Let us consider

$$C_{s\alpha} + C_{\alpha s} = 0.7 \tag{1.234}$$

Figure 1.49 illustrates the friction limit for $C_{s\alpha} = 0.1$, $C_{\alpha s} = 0.6$, and $C_{s\alpha} = 0.3$, $C_{\alpha s} = 0.4$, and $C_{s\alpha} = 0.5$, $C_{\alpha s} = 0.2$, and $C_{s\alpha} = 0.6$, $C_{\alpha s} = 0.1$. The friction limit is not an ellipse anymore when $C_{s\alpha} + C_{\alpha s} \neq 1$, however, the friction limit is still a closed shape. Changing the values of $C_{s\alpha}$, $C_{\alpha s}$ will move the point at which both F_x/F_z and F_{xy}/F_z are saturated as shown in the figure.

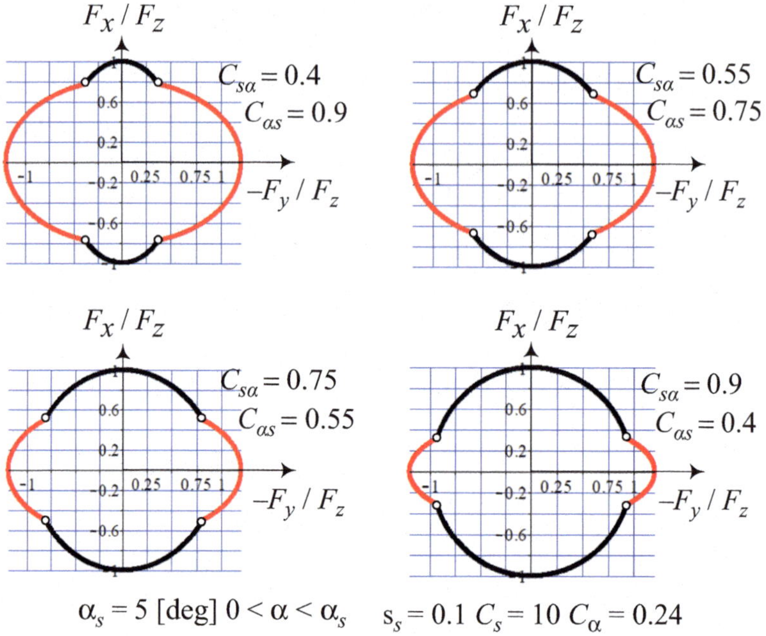

Fig. 1.50 The effects of changing $C_{s\alpha}$ and $C_{\alpha s}$ with condition $C_{s\alpha} + C_{\alpha s} > 1$ on the friction limit

To study the case of $C_{s\alpha} + C_{\alpha s} > 1$, we should consider that there are two conditions to have real numbers for the forces:

$$1 - C_{s\alpha}\left(\frac{\alpha}{\alpha_s}\right)^2 > 0 \tag{1.235}$$

$$1 - C_{\alpha s}\left(\frac{s}{s_s}\right)^2 > 0 \tag{1.236}$$

which puts the following conditions on $C_{s\alpha}$, $C_{\alpha s}$.

$$C_{s\alpha} < \frac{\alpha_s^2}{\alpha^2} \qquad C_{\alpha s} < \frac{s_s^2}{s^2} \tag{1.237}$$

The minimum value of both α_s^2/α^2 and s_s^2/s^2 is one. Therefore, none of $C_{s\alpha}$, $C_{\alpha s}$ can be greater than one while $C_{s\alpha} + C_{\alpha s} > 1$.

Let us try

$$C_{s\alpha} + C_{\alpha s} = 1.3 \tag{1.238}$$

Figure 1.50 illustrates the friction limit for $C_{s\alpha} = 0.4$, $C_{\alpha s} = 0.9$, and $C_{s\alpha} = 0.55$, $C_{\alpha s} = 0.75$, and $C_{s\alpha} = 0.75$, $C_{\alpha s} = 0.55$, and $C_{s\alpha} = 0.9$, $C_{\alpha s} = 0.4$.

Example 34 ★Velocity dependency of combined forces.

Experimental results show that the longitudinal and lateral forces are also velocity dependent as expressed in Eqs. (1.104)–(1.106) for longitudinal force and (1.164)–(1.166) for lateral force when the slips are very small $|\alpha| < \alpha_s$, $|s| < s_s$. In a combined force situation, same model applies and the following equations express how the velocity of a car affects the combined lateral and longitudinal forces.

$$\frac{F_x}{F_z} = \left(C_s\, s - C_{s_1}\, v_x\right)\sqrt{1 - C_{s\alpha}\left(\frac{\alpha}{\alpha_s}\right)^2} \tag{1.239}$$

$$\frac{F_y}{F_z} = \left(-C_\alpha\, \alpha + C_{\alpha_1}\, v_x\right)\sqrt{1 - C_{\alpha s}\left(\frac{s}{s_s}\right)^2} \tag{1.240}$$

Lateral velocity also affects inversely both, longitudinal and lateral forces. The effect of lateral velocity may be included in the force equations as:

$$\frac{F_x}{F_z} = \left(C_s\, s - C_{s_1}\, v_x - C_y\, v_y\right)\sqrt{1 - C_{s\alpha}\left(\frac{\alpha}{\alpha_s}\right)^2} \tag{1.241}$$

$$\frac{F_y}{F_z} = \left(-C_\alpha\, \alpha + C_{\alpha_1}\, v_x + C_y\, v_y\right)\sqrt{1 - C_{\alpha s}\left(\frac{s}{s_s}\right)^2} \tag{1.242}$$

Experiments also show that the saturation values of s_s and α_s reduce with associated velocity.

$$\frac{F_x}{F_z} = \left(C_s\, s - C_{s_1}\, v_x - C_y\, v_y\right)\sqrt{1 - C_{s\alpha}\left(\frac{\alpha}{\alpha_s\left(1 - C_{\alpha y} v_x\right)}\right)^2} \tag{1.243}$$

$$\frac{F_y}{F_z} = \left(-C_\alpha\, \alpha + C_{\alpha_1}\, v_x + C_y\, v_y\right)\sqrt{1 - C_{\alpha s}\left(\frac{s}{s_s\left(1 - C_{sx} v_x\right)}\right)^2} \tag{1.244}$$

Including all of the above velocity dependency, we may use the following models for tire forces.

$$\frac{F_x}{F_z} = \left(C_s\, S\left(s - s_s\left(1 - C_{sx} v_x\right)\right) - C_{s_1}\, v_x - C_y\, v_y\right)$$

$$\times\sqrt{1 - C_{s\alpha}\left(\frac{S\left(\alpha - \alpha_s\right)}{\alpha_s\left(1 - C_{\alpha y} v_x\right)}\right)^2} \tag{1.245}$$

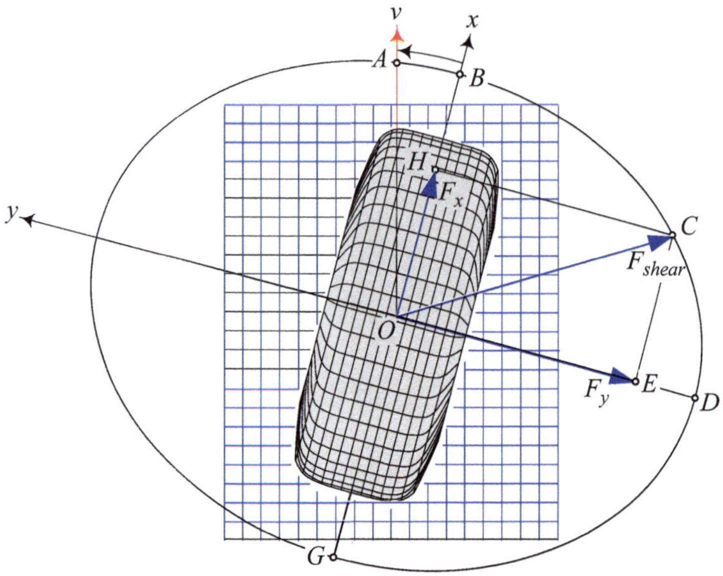

Fig. 1.51 A tire with saturated tire force

$$\frac{F_y}{F_z} = \left(-C_\alpha \, S \left(\alpha - \alpha_s \left(1 - C_{\alpha y} v_x\right)\right) + C_{\alpha_1} \, v_x + C_y \, v_y\right)$$

$$\times \sqrt{1 - C_{\alpha s} \left(\frac{S\,(s - s_s)}{s_s\,(1 - C_{sx} v_x)}\right)^2} \tag{1.246}$$

Example 35 Is a tire sliding longitudinally or laterally?

Figure 1.51 illustrates a tire with saturated tire force. The tire force $\mathbf{F}/F_z = \overrightarrow{OC}$ has the longitudinal component $F_x/F_z = \overrightarrow{OH}$ and the lateral component $F_y/F_z = -\overrightarrow{OE}$. Let us assume that tire parameters C_s, C_α, $C_{s\alpha}$, $C_{\alpha s}$, α_s, s_s are given.

$$s_s = 0.1 \qquad C_s = 10 \qquad C_\alpha = 0.24$$

$$C_{s\alpha} = 0.5 \qquad C_{\alpha s} = 0.5 \qquad \alpha_s = 5\,\text{deg}$$

$$F/F_z = 1.134 \qquad F_x/F_z = 0.5935 \qquad F_y/F_z = 0.966 \tag{1.247}$$

Therefore, the only unknowns in the force equations (1.200) and (1.201) are the longitudinal and side slips s and α.

$$\frac{F_x}{F_z} = C_s \, s \sqrt{1 - C_{s\alpha} \left(\frac{\alpha}{\alpha_s}\right)^2} \tag{1.248}$$

$$\frac{F_y}{F_z} = -C_\alpha \, \alpha \sqrt{1 - C_{\alpha s} \left(\frac{s}{s_s} \right)^2} \tag{1.249}$$

Knowing that for a vector at the limit ellipse, either s or α must be saturated. We may assume that $s = s_s$ and, from F_x / F_z

$$0.5935 = 10 \, (0.1) \sqrt{1 - 0.5 \left(\frac{\alpha}{5} \right)^2} \tag{1.250}$$

we find

$$\alpha = 5.688\,4 \tag{1.251}$$

which is more than α_s and impossible. Therefore we must have $\alpha = \alpha_s$ and calculate s.

$$0.5935 = 10 s \sqrt{1 - 0.5} \tag{1.252}$$

$$s = 0.083934 < s_s \tag{1.253}$$

Therefore, this tire is saturated laterally and unsaturated longitudinally. We may use Eq. (1.201) to examine the results

$$\begin{aligned} \frac{F_y}{F_z} &= -C_\alpha \, \alpha \sqrt{1 - C_{\alpha s} \left(\frac{s}{s_s} \right)^2} \\ &= 0.24 \, (5) \sqrt{1 - 0.5 \left(\frac{0.08934}{0.1} \right)^2} = 0.966 \end{aligned} \tag{1.254}$$

Example 36 Combined force experimental data.

Figure 1.52 illustrates how a sideslip α affects the longitudinal force ratio F_x / F_z as a function of slip ratio s. Figure 1.53 illustrates the effect of sideslip α on the lateral force ratio F_y / F_z as a function of slip ratio s. Figure 1.54 shows how the lateral force F_y / F_z will affect by longitudinal slip s at constant sideslip (Genta 2007; Genta and Morello 2009a,b; Haney 2003).

Example 37 ★TV-shaped model.

To model experimental data of race car tires better, we may use a more rectangular function such as the TV-shaped equation

$$\left(\frac{F_y / F_z}{C_\alpha \alpha_s} \right)^4 + \left(\frac{F_x / F_z}{C_s s_s} \right)^4 = 1 \tag{1.255}$$

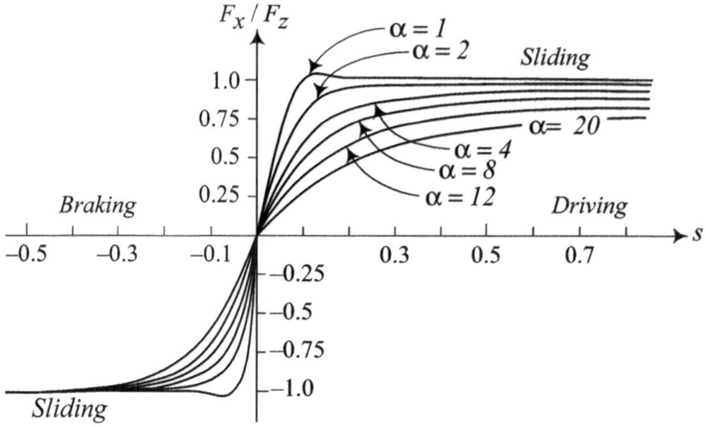

Fig. 1.52 Longitudinal force ratio F_x/F_z as a function of slip ratio s for different sideslip α

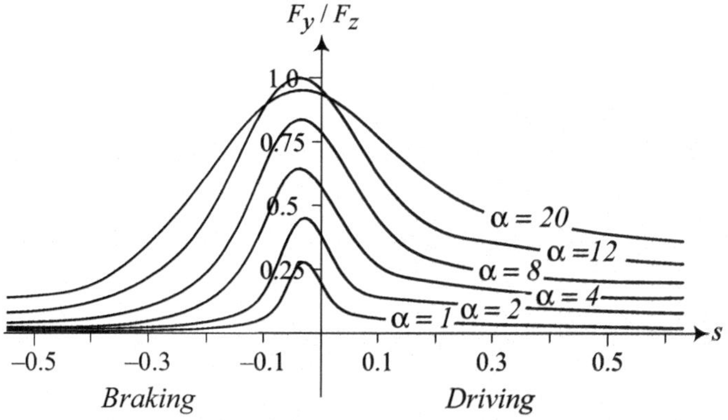

Fig. 1.53 Lateral force ratio F_y/F_z as a function of slip ratio s for different sideslip α

The below suggested force equations for the combined effect of longitudinal slip and sideslip for small values of the longitudinal slip ratio $s < s_s$ and sideslip angle $\alpha < \alpha_s$ although more complicated, matches with experimental data of race car tires better than elliptical model:

$$\frac{F_x}{F_z} = C_s s \sqrt[4]{1 - C_{s\alpha}\left(\frac{\alpha}{\alpha_s}\right)^4} \qquad \alpha < \alpha_s \qquad s < s_s \tag{1.256}$$

$$\frac{F_y}{F_z} = -C_\alpha \alpha \sqrt[4]{1 - C_{\alpha s}\left(\frac{s}{s_s}\right)^4} \qquad \alpha < \alpha_s \qquad s < s_s \tag{1.257}$$

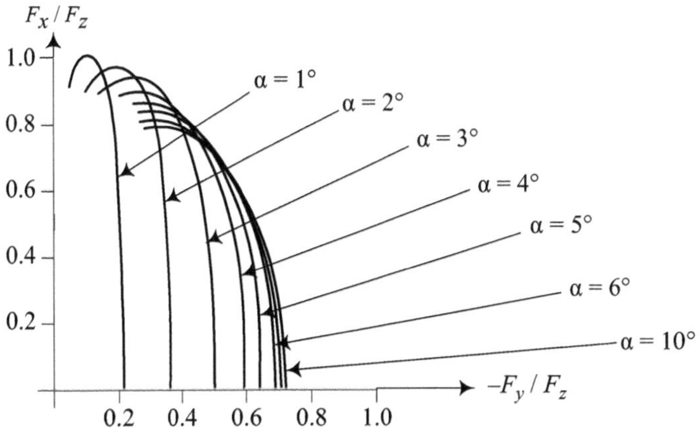

Fig. 1.54 Experimental results on a tire showing how the lateral force F_y/F_z will be affected by longitudinal slip s at constant sideslip angles

where α_s and s_s are respectively the *saturation values* of α and s for the tire, and $C_{s\alpha}$ and $C_{\alpha s}$ are the *tire longitudinal drop TV-factor* for reduction of F_x and F_y in the presence of sideslip α and s. The tire *drop TV-factor* $C_{s\alpha}$ and $C_{\alpha s}$ are determined by experiment.

Substituting the suggested model of F_x and F_y in the TV-shape equation provide us with

$$\left(\frac{\alpha\sqrt[4]{1-C_{\alpha s}\left(\frac{s}{s_s}\right)^4}}{\alpha_s}\right)^4+\left(\frac{s\sqrt[4]{1-C_{s\alpha}\left(\frac{\alpha}{\alpha_s}\right)^4}}{s_s}\right)^4=1 \tag{1.258}$$

and then simplifies to:

$$\frac{\alpha^4}{\alpha_s^4}+\frac{s^4}{s_s^4}-\frac{\alpha^4 s^4}{\alpha_s^4 s_s^4}C_{s\alpha}-\frac{\alpha^4 s^4}{\alpha_s^4 s_s^4}C_{\alpha s}=1 \tag{1.259}$$

On the limit condition, either $s = s_s$ or $\alpha = \alpha_s$. Substituting $s = s_s$ simplifies the above equation to

$$C_{s\alpha}+C_{\alpha s}=1 \tag{1.260}$$

Therefore, if we have $C_{s\alpha}+C_{\alpha s}=1$, then the shape of tire force limit will be the *TV-shaped limit condition* (1.255).

Example 38 The camber effects.

In case camber γ is also included in longitudinal and lateral forces then, the following equations may be used.

$$\frac{F_x}{F_z} = C_s s \sqrt{1 - C_{s\alpha}\left(\frac{\alpha}{\alpha_s}\right)^2 - C_{s\gamma}\left(\frac{\gamma}{\gamma_s}\right)^2} \tag{1.261}$$

$$|\alpha| < \alpha_s \qquad |s| < s_s \qquad |\gamma| < \gamma_s$$

$$\frac{F_y}{F_z} = -C_\alpha \alpha \sqrt{1 - C_{\alpha s}\left(\frac{s}{s_s}\right)^2 - C_{\alpha\gamma}\left(\frac{\gamma}{\gamma_s}\right)^2} - C_\gamma \gamma \tag{1.262}$$

$$|\alpha| < \alpha_s \qquad |s| < s_s \qquad |\gamma| < \gamma_s$$

Introducing a camber γ in a tire will increase the lateral force F_y/F_z by $-C_\gamma \gamma$ and decrease it by square root of $C_{\alpha\gamma}\left(\gamma/\gamma_s\right)^2$. The longitudinal force F_x/F_z will decrease by square root of $C_{s\gamma}\left(\gamma/\gamma_s\right)^2$. The $C_{s\gamma}$ and $C_{\alpha\gamma}$ are the *longitudinal camber drop factor* and *lateral camber drop factor* respectively. The $C_{s\gamma}$ and $C_{\alpha\gamma}$ factors are determined experimentally.

Example 39 Approximate elliptic tire model.

We may expand the elliptical equation in series and approximate it with a second order polynomial.

$$\sqrt{1 - Cx^2} = 1 - \frac{1}{2}Cx^2 + O\left(x^3\right) \tag{1.263}$$

Following this approximation, we may use simpler equations for tire force components.

$$\begin{aligned} \frac{F_x}{F_z} &= C_s\, S\left(s - s_s\right)\sqrt{1 - C_{s\alpha}\left(\frac{S\left(\alpha - \alpha_s\right)}{\alpha_s}\right)^2} \\ &\simeq C_s\, S\left(s - s_s\right)\left(1 - \frac{1}{2}C_{s\alpha}\left(\frac{S\left(\alpha - \alpha_s\right)}{\alpha_s}\right)^2\right) \end{aligned} \tag{1.264}$$

$$\begin{aligned} \frac{F_y}{F_z} &= -C_\alpha S\left(\alpha - \alpha_s\right)\sqrt{1 - C_{\alpha s}\left(\frac{S\left(s - s_s\right)}{s_s}\right)^2} \\ &\simeq -C_\alpha S\left(\alpha - \alpha_s\right)\left(1 - \frac{1}{2}C_{\alpha s}\left(\frac{S\left(s - s_s\right)}{s_s}\right)^2\right) \end{aligned} \tag{1.265}$$

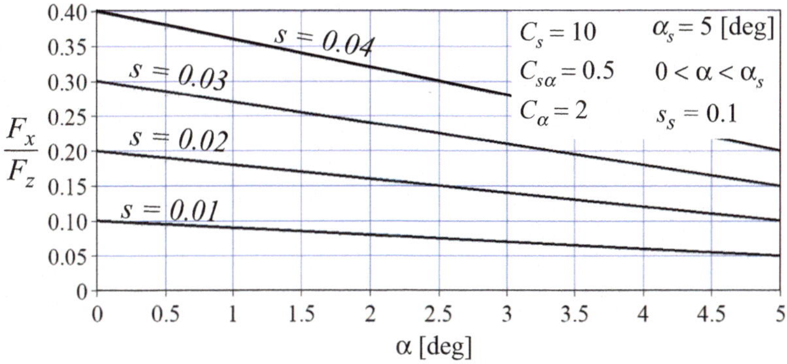

Fig. 1.55 The longitudinal force F_x/F_z drops in the $(\alpha, F_x/F_z)$-plane when sideslip α introduces and increases to a tire with constant longitudinal slip s

In case the tire slips are not saturated, then the equations will be simplifies to:

$$\frac{F_x}{F_z} \simeq C_s\, s \left(1 - \frac{1}{2} C_{s\alpha} \left(\frac{\alpha}{\alpha_s}\right)^2\right) \tag{1.266}$$

$$\frac{F_y}{F_z} \simeq -C_\alpha \alpha \left(1 - \frac{1}{2} C_{\alpha s} \left(\frac{s}{s_s}\right)^2\right) \tag{1.267}$$

1.5.2 Diamond Model

A linear model for interaction of tire forces is the simplest model and acts as the principal model for comparison purposes. We assume the proportional-saturation model for longitudinal and sideslip of tire. When $\alpha = 0$, a small longitudinal slip s generates the longitudinal force $F_x/F_z = C_s s$, and when $s = 0$, a small sideslip angle α generates a lateral force of $F_y/F_z = -C_\alpha \alpha$. When there exists a longitudinal slip s and we also introduce a sideslip α, the longitudinal force will decreases as

$$\frac{F_x}{F_z} = C_s s \left(1 - C_{s\alpha} \frac{|\alpha|}{\alpha_s}\right) \qquad |\alpha| < \alpha_s \qquad |s| < s_s \tag{1.268}$$

where α_s and s_s are respectively the *saturation values* of α and s for the tire, and $C_{s\alpha}$ is the *tire longitudinal drop factor* for reduction of F_x in the presence of sideslip α. The tire *longitudinal drop factor* $C_{s\alpha}$ is determined by experiment for tires and is usually $0.2 < C_{s\alpha} < 0.9$. Figure 1.55 illustrates the longitudinal force reduction in the $(\alpha, F_x/F_z)$-plane when sideslip is introduced to a tire with constant longitudinal slip and then increased.

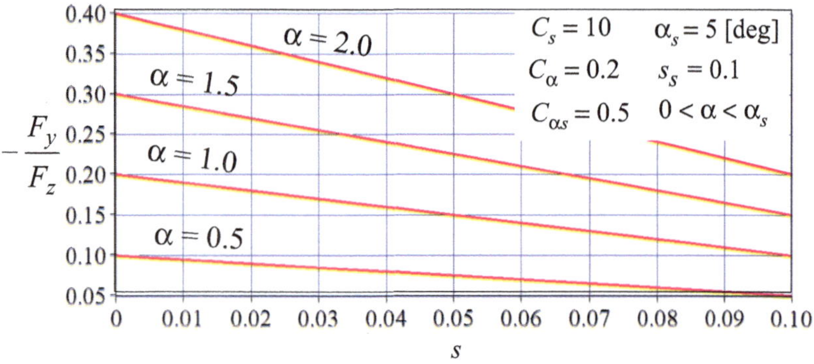

Fig. 1.56 The lateral force $-F_y/F_z$ drops in the $(s, -F_y/F_z)$-plane when longitudinal slip s introduces and increases to a tire with constant sideslip α

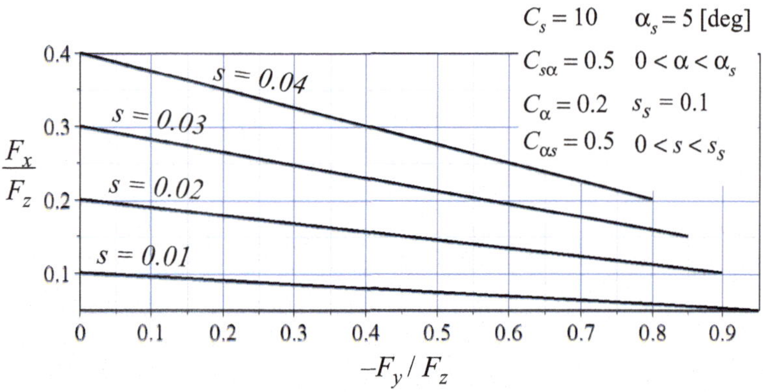

Fig. 1.57 The longitudinal force drops in the $(-F_y, F_x)$ plane, when sideslip introduces and increases to a tire

In the presence of longitudinal slip s, the lateral force will also drop from its level of $F_y = -C_\alpha \alpha$ to

$$\frac{F_y}{F_z} = -C_\alpha \alpha \left(1 - C_{\alpha s} \frac{|s|}{s_s}\right) \qquad |\alpha| < \alpha_s \qquad |s| < s_s \tag{1.269}$$

where $C_{\alpha s}$ is the *tire lateral drop factor* for reduction of F_y in the presence of longitudinal slip s. The tire lateral drop factor $C_{\alpha s}$ is determined by experiment for the tire and is usually $0.2 < C_{\alpha s} < 0.9$. Figure 1.56 illustrates how the lateral force reduces in the $(\alpha, F_y/F_z)$-plane when longitudinal slip is introduced to a tire with constant sideslip and then increased.

Combining Figs. 1.55 and 1.56, we may draw Fig. 1.57 to illustrate the longitudinal force reduction in the $(F_y/F_z, F_x/F_z)$-plane, when sideslip is introduced

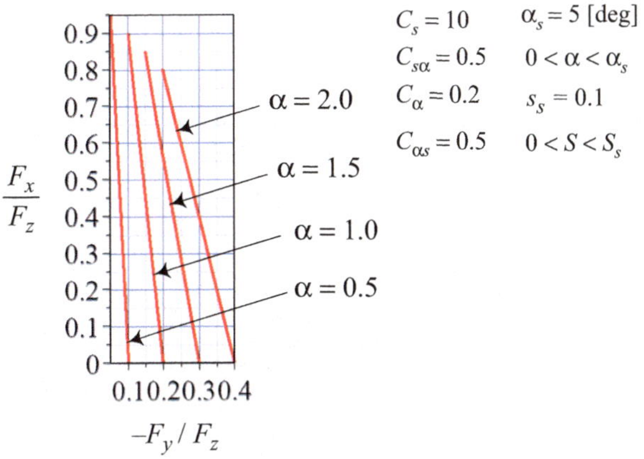

Fig. 1.58 The lateral force drops in the $(-F_y, F_x)$ plane, when longitudinal slip introduces and increases to a tire

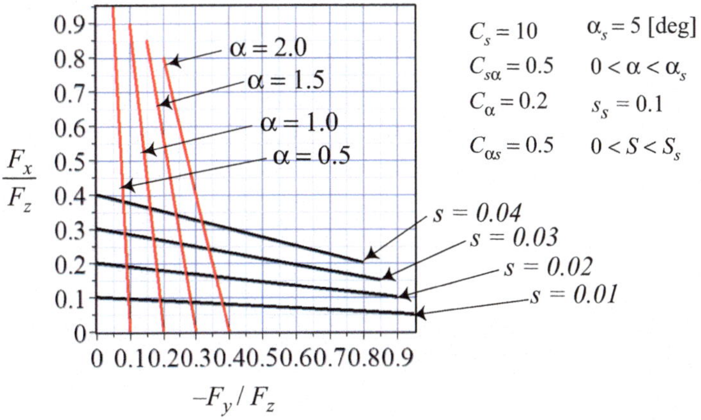

Fig. 1.59 The longitudinal and lateral forces drop in the $(-F_y, F_x)$ plane, when respectively sideslip and longitudinal slip are introduced to a tire

to a tire with constant longitudinal slip and then increased. Similarly, Fig. 1.58 illustrates how the lateral force reduces in the (F_y, F_x)-plane, when longitudinal slip is introduced to a tire with constant sideslip. Figure 1.59 depicts both lateral and longitudinal forces when longitudinal slip and sideslip are introduced in the tire dynamics.

Including positive and negative ranges of α and s and plotting the curves for $-s_s < s < s_s$ and $-\alpha < \alpha < \alpha_s$ for different values of tire drop factors, $C_{\alpha s}$, $C_{s\alpha}$, make the relationship of F_x and F_y as shown in Fig. 1.60 justifying diamond model.

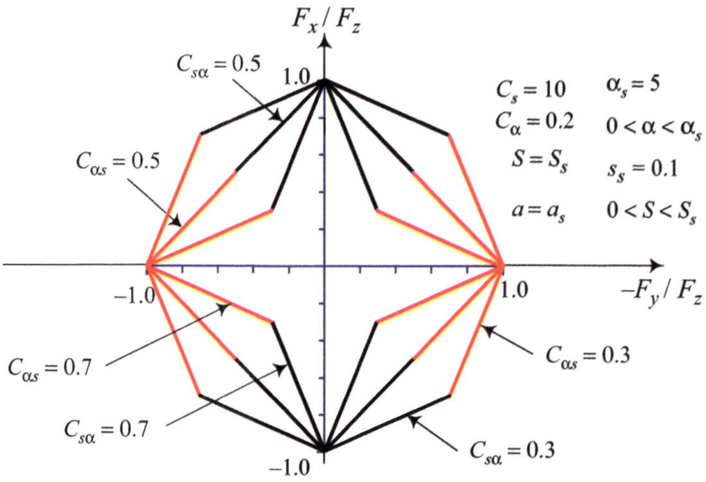

Fig. 1.60 The longitudinal and lateral forces at the limits in the $(-F_y, F_x)$ plane, when respectively sideslip and longitudinal slip are introduced to a tire and increased up to their saturation values

Including the saturation in lateral and longitudinal forces, we define them mathematically by the following equations.

$$\frac{F_x}{F_z} = C_s S (s - s_s) \left(1 - C_{s\alpha} \frac{|\alpha|}{\alpha_s}\right) \tag{1.270}$$

$$\frac{F_y}{F_z} = -C_\alpha S (\alpha - \alpha_s) \left(1 - C_{\alpha s} \frac{|s|}{s_s}\right) \tag{1.271}$$

Figures 1.61 and 1.62 illustrate the diamond proportional-saturation model for longitudinal and lateral forces indicating the effect of sideslip α on longitudinal force F_x/F_z and the effect of longitudinal slip s on lateral force F_y/F_z.

Proof Assume that $\alpha = 0$ and a small longitudinal slip $s < s_s$ generates the longitudinal force $F_x/F_z = C_s s$. When we have longitudinal slip s and then introduce a sideslip α, the longitudinal force will drop. The following linear function is the simplest representation of the phenomenon

$$\frac{F_x}{F_z} = C_s s \left(1 - C_{s\alpha} \frac{|\alpha|}{\alpha_s}\right) \tag{1.272}$$

where α_s is the saturation point of α for the tire. The coefficient $C_{s\alpha}$ is called the *longitudinal drop factor* that indicates the percentage of drop in F_x/F_z from $\alpha = 0$ to $\alpha = \alpha_s$, for a constant s.

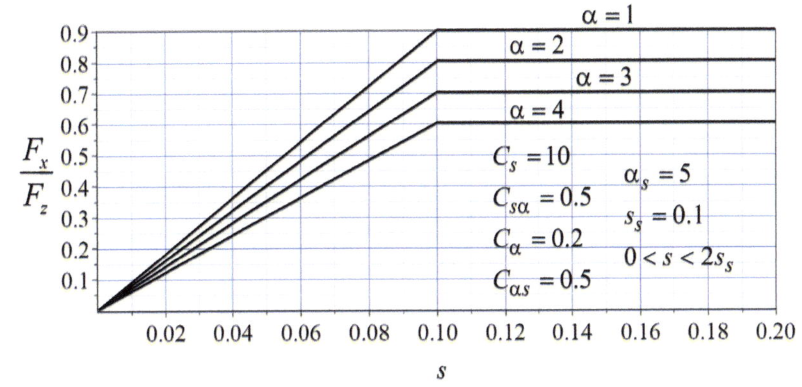

Fig. 1.61 The effect of sideslip angle α on longitudinal force F_x/F_z in saturating model

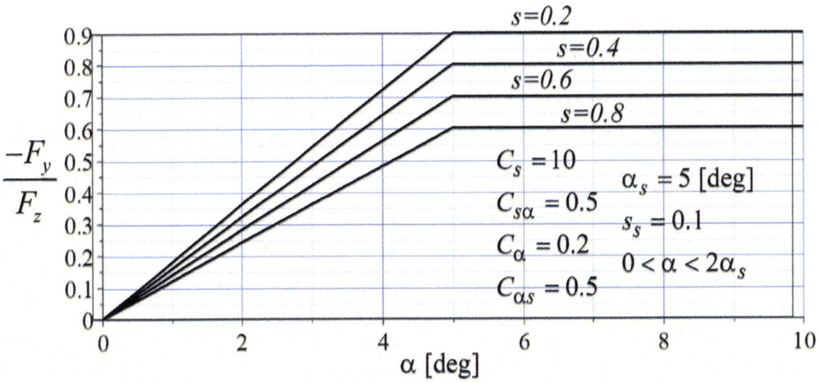

Fig. 1.62 The effect of longitudinal slip ratio s on lateral force $-F_y/F_z$ in saturating model

Similarly, assume that $s = 0$ and a small sideslip angle $\alpha < \alpha_s$ generates a lateral force of $F_y/F_z = -C_\alpha \alpha$. The lateral force will also be dropped when a small longitudinal slip s is added to the dynamics of the tire. The following linear function expresses this phenomenon

$$\frac{F_y}{F_z} = -C_\alpha \alpha \left(1 - C_{\alpha s} \frac{|s|}{s_s}\right) \tag{1.273}$$

where s_s is the saturation point of s for the tire. The coefficient $C_{\alpha s}$ is the *lateral drop factor* that indicates the percentage of drop in F_y/F_z from $s = 0$ to $s = s_s$, for a constant α. Figures 1.57 and 1.58 illustrate how the functions (1.272) and (1.273) work.

Employing linear saturation model for longitudinal and lateral forces, we may use *Saturation* function (1.60) or *Heaviside* function (1.62) to express the forces

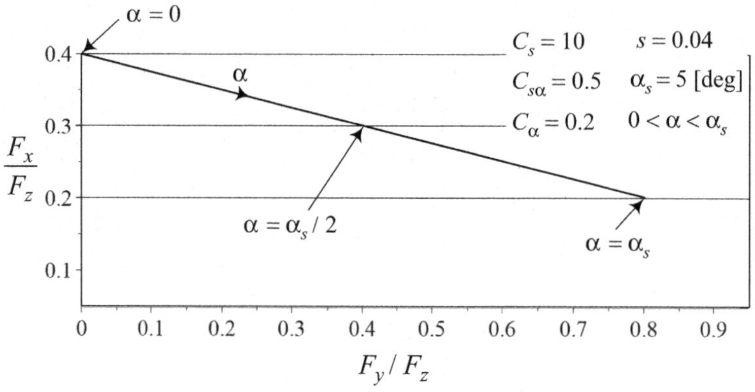

Fig. 1.63 $C_{s\alpha} = 0.5$ means that F_x/F_z drops 50% when α increases from $\alpha = 0$ to $\alpha = \alpha_s$. If $C_{s\alpha} = 0.3$, then F_x/F_z drops 30% when α increases from $\alpha = 0$ to $\alpha = \alpha_s$

in a single equation to cover proportional and saturation parts of F_x/F_z and F_y/F_z. Equations (1.270) and (1.271) express the longitudinal and lateral forces of tires using proportional-saturation models of figures (1.20).

In order to have a unified function for computer simulation, we may rewrite the force equations as:

$$\frac{F_x}{F_z} = C_s S\,(s - s_s)\left(1 - C_{s\alpha}\frac{|\alpha|}{\alpha_s}\right) \tag{1.274}$$

$$\frac{F_y}{F_z} = -C_\alpha S\,(\alpha - \alpha_s)\left(1 - C_{\alpha s}\frac{|s|}{s_s}\right) \tag{1.275}$$

■

Example 40 Meaning of $C_{s\alpha}$.

Figure 1.63 illustrates a sample of the effect of increasing α on reduction of F_x/F_z in Eq. (1.272) from $\alpha = 0$ to $\alpha = \alpha_s$, for a constant $s = 0.04$. When $\alpha = 0$, then $F_x/F_z = 0.4$ for the given values of $C_s = 10$, $\alpha_s = 5$ deg, $C_\alpha = 0.2$. By increasing α the value of F_x/F_z decreases linearly and proportionally. Assuming when $\alpha = \alpha_s$ we have $F_x/F_z = 0.2$, and therefore,

$$\frac{F_{xM}}{F_z} = C_s s_s\left(1 - C_{s\alpha}\frac{|\alpha_s|}{\alpha_s}\right) \tag{1.276}$$

$$0.2 = 0.4\,(1 - C_{s\alpha}) \tag{1.277}$$

$$C_{s\alpha} = 0.5 \tag{1.278}$$

which indicates $C_{s\alpha} = 0.5$ and F_x/F_z drops to $(1 - 0.5) \times 100 = 50\%$.

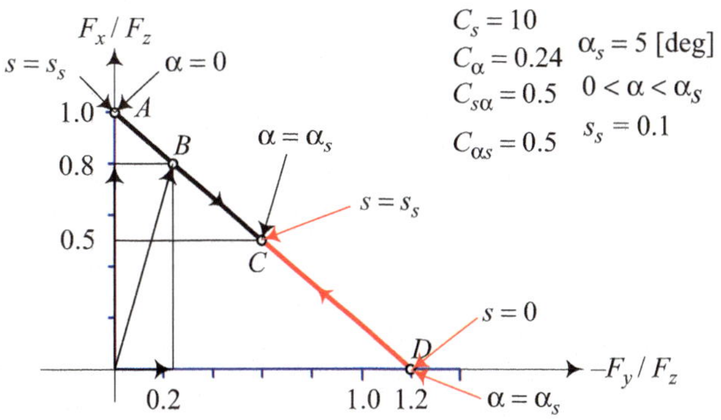

Fig. 1.64 The saturation line. At any point on the saturation line, at least one of s or α is at its saturation level. Only at point C, both s and α are at their saturation levels

In case of $C_{s\alpha} = 0.3$, then F_x/F_z drops to $(1-0.3)\times 100 = 70\%$ when α increases from $\alpha = 0$ to $\alpha = \alpha_s$.

Example 41 The limit slip line.

Let us assume $\alpha = 0$, $s = s_s$, and

$$s_s = 0.1 \qquad C_s = 10 \qquad C_{s\alpha} = 0.5$$
$$\alpha_s = 5\,\text{deg} \qquad C_\alpha = 0.24 \qquad C_{\alpha s} = 0.5 \tag{1.279}$$

therefore, the longitudinal force is saturated and is at its maximum $F_x/F_z = F_{xM}/F_z = C_s\, s_s = 1$. This situation is indicated by point A in Fig. 1.64 at which the tire is generating maximum possible longitudinal force while rolling straight. Increasing sideslip angle α from $\alpha = 0$ to $\alpha = \alpha_s$ will generate lateral force in addition to the longitudinal force while the magnitude of longitudinal force decreases according to Eq. (1.272). The line from point A to point C indicates how the forces F_x/F_z and F_y/F_z will change when α increases from $\alpha = 0$ to $\alpha = \alpha_s$. As long as $s = s_s$ and $0 < \alpha < \alpha_s$, the longitudinal force is saturated and the lateral force is undersaturated

$$\frac{F_x}{F_z} = C_s s_s \left(1 - C_{s\alpha}\frac{\alpha}{\alpha_s}\right) = \frac{F_{xM}}{F_z} \tag{1.280}$$

$$\frac{F_y}{F_z} = -C_\alpha \alpha \left(1 - C_{\alpha s}\right) < \frac{F_{yM}}{F_z} \tag{1.281}$$

At point C where $\alpha = \alpha_s$, the lateral force will also be saturated and the force conditions become

$$\frac{F_x}{F_z} = C_s s_s \left(1 - C_{s\alpha}\right) = \frac{F_{xM}}{F_z} \tag{1.282}$$

$$\frac{F_y}{F_z} = -C_\alpha \alpha_s (1 - C_{\alpha s}) = \frac{F_{yM}}{F_z} \tag{1.283}$$

Because $C_{s\alpha} = 0.5$, the magnitude of F_x/F_z will drop from $F_x/F_z = 1$ at point A to

$$\begin{aligned} \frac{F_x}{F_z} = \frac{F_{xM}}{F_z} &= C_s s_s \left(1 - C_{s\alpha}\frac{\alpha}{\alpha_s}\right) \\ &= 10 \times 0.1 \left(1 - 0.5 \times \frac{5}{5}\right) = 0.5 \end{aligned} \tag{1.284}$$

at point C.

At a nominal point B between A and C, at which $F_x/F_z = 0.8$, we have

$$\frac{F_{xM}}{F_z} = \left(1 - 0.5\frac{\alpha}{5}\right) = 0.8 \tag{1.285}$$

$$\alpha = 2\deg \tag{1.286}$$

and therefore,

$$\begin{aligned} \frac{F_y}{F_z} &= -C_\alpha \alpha \left(1 - C_{\alpha s}\frac{|s|}{s_s}\right) \\ &= -0.24 \times 2 \left(1 - 0.5\left(\frac{s_s}{s_s}\right)\right) = -0.24 \end{aligned} \tag{1.287}$$

The line AC is the limit line of generating longitudinal force by the tire and point C is the absolute terminating point at which both longitudinal and lateral forces are saturated. No matter how much s and α are more than their saturation values of s_s and α_s, the resultant tangential force of the tire at the tireprint will not move from point C.

Now, let us assume $\alpha = \alpha_s$, $s = 0$, for the same tire as (1.279). The lateral force of the tire is saturated and is at its maximum possible value $F_y/F_z = F_{yM}/F_z = -C_\alpha s_\alpha = -1.2$. This situation is indicated by point D in Fig. 1.64 at which the tire is generating the maximum possible lateral force. Increasing slip s from $s = 0$ to $s = s_s$ will generate longitudinal force in addition to the lateral force while the magnitude of lateral force decreases according to Eq. (1.21). The line from point D to point C indicates how the forces F_y/F_z and F_x/F_z will change when s increases from $s = 0$ to $s = s_s$. Therefore, The line AC indicates the situation at which the tire longitudinally slides and laterally sticks to the road, and line DC indicates the situation at which the tire slides laterally and sticks to the ground longitudinally.

Considering both negative and positive maximum forces $\pm F_{xM}/F_z$ and $\pm F_{yM}/F_z$, the limit curves make a closed shape that divides the plane of $(F_x/F_z, F_y/F_z)$ into possible and impossible zones of tangential tire force vectors. Figure 1.65 illustrates a sample of the possible force zone.

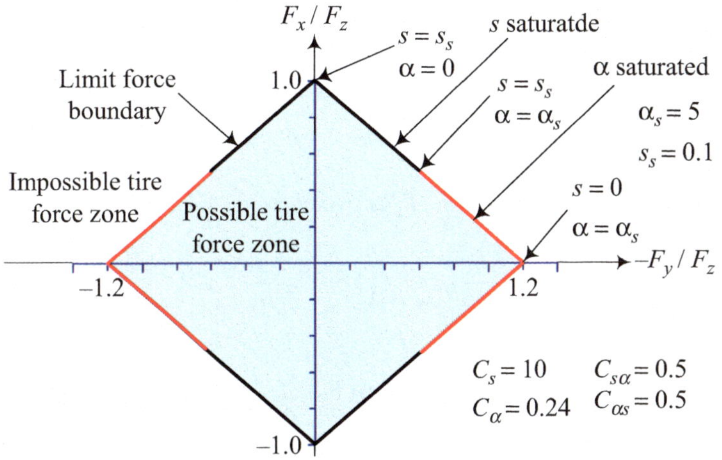

Fig. 1.65 The maximum tangential tire forces is limited into a closed zone surrounded by the limit curves made by F_x/F_z, and $-F_y/F_z$ relationship when at least one of the slips, s, or α to be saturated

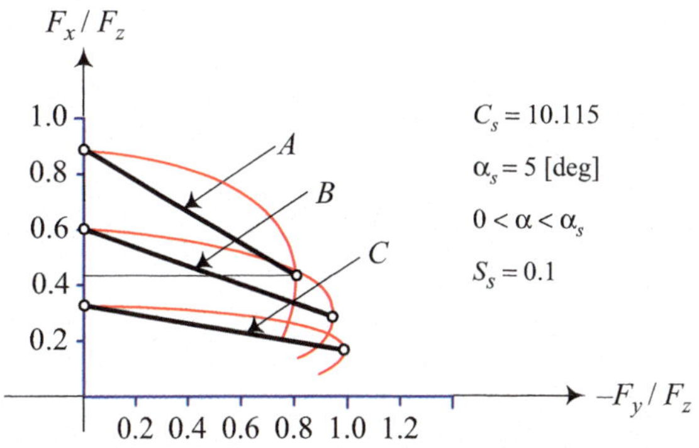

Fig. 1.66 Experimental results of F_x/F_z versus $-F_y/F_z$ for a sample tire and their diamond model approximate

Example 42 Experimental verification.

Experimental data on tires are showing vast and diverse curves for longitudinal and lateral forces for different tires at different conditions. As a result, there is no unique model that covers all tires, all applications, and all conditions. Figure 1.66 illustrates a few experimental data on a sample tire to illustrate how we approximate them with their diamond models. All of these curves belong to unsaturated cases indicating how the combined forces affect each other.

There are two points on every diamond model approximation line. The first point on (F_x/F_z)-axis indicates the initial value of $\alpha = 0$ where there is no sideslip. The second point on the other end of the line indicates the saturation value of α_s. Let us consider the first line indicated by A. The line starts from

$$F_x/F_z = 0.88 \qquad s = 0.087 \qquad C_s = 10.115 \tag{1.288}$$

and ends at

$$F_x/F_z = 0.43 \qquad s = 0.043 \qquad C_s = 10.115 \tag{1.289}$$

assuming $\alpha_s = 5$ deg and using F_x/F_z at two ends, we calculate $C_{s\alpha}$

$$\frac{F_x}{F_z} = C_s s \left(1 - C_{s\alpha}\frac{\alpha}{\alpha_s}\right) \tag{1.290}$$

$$0.29 = 0.6\,(1 - C_{s\alpha}) \tag{1.291}$$

$$C_{s\alpha} = 0.511 \tag{1.292}$$

and therefore, the equation for F_x/F_z when both s and α exist will be

$$\frac{F_x}{F_z} = 10.115s\left(1 - 0.511\frac{\alpha}{5}\right) \tag{1.293}$$

Applying the same method for the second and third experimental curves, we find the following diamond approximations for line B and C, respectively,

$$\frac{F_x}{F_z} = 10.115s\left(1 - 0.516\frac{\alpha}{5}\right) \tag{1.294}$$

$$\frac{F_x}{F_z} = 10.115s\left(1 - 0.512\frac{\alpha}{5}\right) \tag{1.295}$$

indicating that the diamond model is consistent with experiments on this tire.

1.6 Vehicle Kinematics

A body coordinate frame $B(Cxyz)$ is attached to a vehicle at its mass center C. The x-axis is the longitudinal axis passing through C and directed forward. The y-axis goes laterally to the left from the driver's viewpoint. The z-axis makes the coordinate system a right-hand triad as shown in Fig. 1.67. Vehicles are assumed to be symmetric about the (x, z)-plane. When the car is parked on a flat horizontal road, the z-axis is perpendicular to the ground, opposite to the gravitational acceleration $\mathbf{g}$. The equations of motion of vehicles are always expressed in the vehicle body coordinate frame, $B(Cxyz)$, (Jazar 2017).

The orientation of the vehicle is determined by the three angles of *roll angle* φ about the x-axis, *pitch angle* θ about the y-axis, and *yaw angle* ψ about the z-axis.

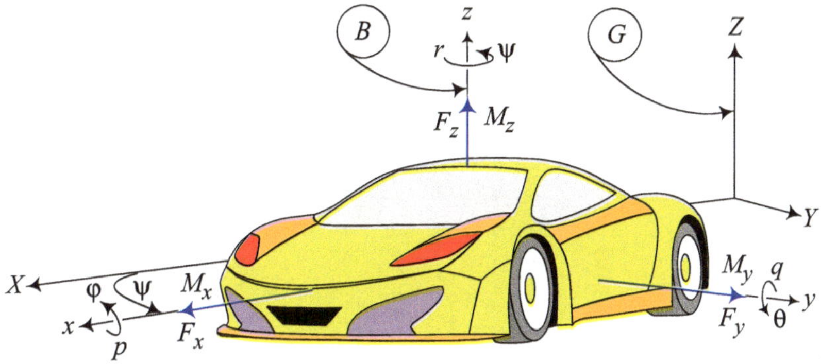

Fig. 1.67 A moving vehicle, indicated by its body coordinate frame B in a global coordinate frame G

The rate of the orientation angles are called *roll rate* p, *pitch rate* q, and *yaw rate* r, respectively.

$$\dot{\varphi} = p \tag{1.296}$$

$$\dot{\theta} = q \tag{1.297}$$

$$\dot{\psi} = r \tag{1.298}$$

The *vehicle force system* $(\mathbf{F}, \mathbf{M})$, expressed in body coordinate is the resultant of external forces and moments that vehicle receives from the ground and environment (Ellis 1994; Dixon 1996).

$${}^{B}\mathbf{F} = F_x\hat{\imath} + F_y\hat{\jmath} + F_z\hat{k} \tag{1.299}$$

$${}^{B}\mathbf{M} = M_x\hat{\imath} + M_y\hat{\jmath} + M_z\hat{k} \tag{1.300}$$

The components of the vehicle force system that are shown in Fig. 1.67 have special names and importance.

1. The F_x is the *longitudinal force* acting on the x-axis. The longitudinal force may also be called *forward force*, *traction force*, or *brake force*. Traction or braking force of tires and the aerodynamic forces on the vehicle make F_x.
2. The F_y is the *lateral force* leftward on the y-axis. Imbalance lateral force on front and rear tires is needed to generate a yaw moment to turn a vehicle. Steering is the main way to generate F_y.
3. The F_z is vertical to the ground plane and is called *vehicle load*. Vehicle load affects the limit and force capacity of tires.
4. The M_x is the *roll moment* about the x-axis. The roll moment is also called the *longitudinal moment*, *bank moment*, *tilting torque*, or *overturning moment*.
5. The M_y is the *pitch moment* or lateral moment about the y-axis.

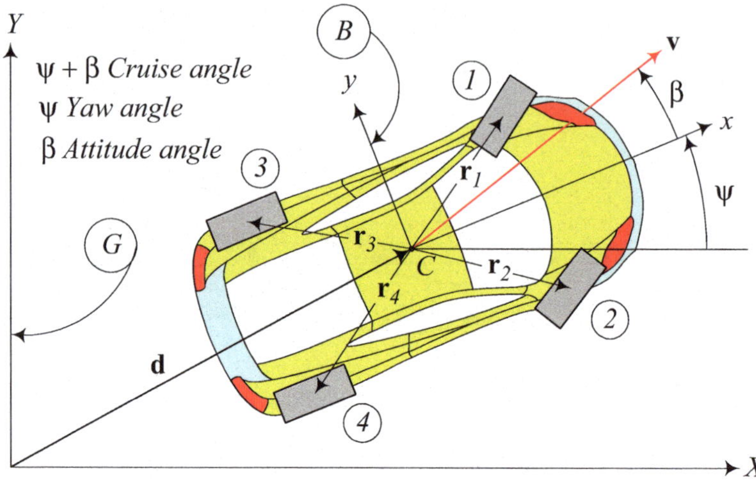

Fig. 1.68 Top view of a moving vehicle to show the yaw angle ψ between the X and x axes, the attitude angle β between the x-axis and the velocity vector $\mathbf{v}$, at the mass center C, and the cruise angle $\beta + \psi$ between the X-axis and the velocity vector $\mathbf{v}$

6. The M_z is the *yaw moment* or upward moment about the z-axis. The yaw moment may also be called the *aligning moment*. Yaw moment along with lateral forces would make a vehicle turn.

We determine the position and orientation of a vehicle by determining the position and orientation of the *body coordinate frame* $B(Cxyz)$ with respect to a fixed global coordinate frame $G(OXYZ)$. Figure 1.67 shows how a moving vehicle is indicated by a body frame B in a global frame G (Dieter et al. 2018; Dukkipati et al. 2008).

The angle ψ between the X and x axes measured from X to x about Z is the *yaw angle* or *heading angle* ψ. A velocity vector $\mathbf{v}$ of the vehicle at its mass center C makes an angle β with the body x-axis, measured from x to $\mathbf{v}$ about z, which is called *vehicle sideslip angle* or *attitude angle*. Therefore, the vehicle's velocity vector $\mathbf{v}$ makes an angle $\beta + \psi$ with the global X-axis, measured from X to $\mathbf{v}$ about Z that is called the *cruise angle*. A positive configuration of these angles is shown in Fig. 1.68.

$$^{B}\mathbf{v} = v \cos\beta \; {}^{B}\hat{\imath} + v \sin\beta \; {}^{B}\hat{\jmath} = v_x \; {}^{B}\hat{\imath} + v_y \; {}^{B}\hat{\jmath} \tag{1.301}$$

The definition of β for a vehicle in (1.301) is similar to the definition of β for tires in (1.123) and (1.128). To make it distinguishable between the wheel-body sideslip angle (1.129) in C-frame and the vehicle sideslip angle (1.301) in B-frame, we usually refer to β in (1.301) as vehicle sideslip or vehicle attitude angle.

The numbering of tires of a vehicle starts from the first axle at the front. The front left wheel is number 1, and then the front right wheel would be number 2. Then the

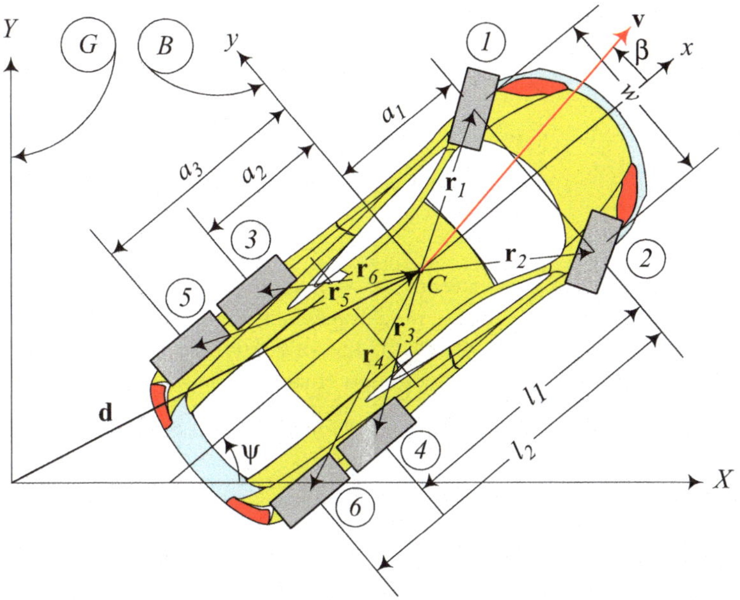

Fig. 1.69 A six-wheel passenger car wheel numbering, relative position, and orientation of $B(x, y, z)$ and coordinate $G(x, y, z)$ frames

left wheel on second axle would be number 3 and the tire on second axle on the right will be number 4. Numbering increases sequentially on the left and right of the next axles. Therefore, the tires with odd numbers are on the left and even numbers on the right. Assuming each axle has only two wheels, the number of a wheel indicates the axle as well as the left or right one.

Besides a single number indicator for wheels, there are literatures in which they use axle number and add left-right wheels, such that the position vector of the wheels 1 to 6 of a three axle vehicle is numbered as: $\mathbf{r}_{1l}$, $\mathbf{r}_{1r}$, $\mathbf{r}_{2l}$, $\mathbf{r}_{2r}$, $\mathbf{r}_{3l}$, $\mathbf{r}_{3r}$, or $\mathbf{r}_{11}$, $\mathbf{r}_{12}$, $\mathbf{r}_{21}$, $\mathbf{r}_{22}$, $\mathbf{r}_{31}$, $\mathbf{r}_{32}$. Numbering of a four-wheel and a six-wheel vehicle are shown in Figs. 1.68 and 1.69, respectively.

Each wheel is indicated by a position vector ${}^B\mathbf{r}_i$, expressed in the body coordinate frame B

$$ {}^B\mathbf{r}_i = x_i i + y_i j + z_i k \tag{1.302} $$

and its velocity will be indicated by ${}^B\mathbf{v}_i$, expressed in the body coordinate frame B

$$ {}^B\mathbf{v}_i = {}^B\mathbf{v} + {}_G\boldsymbol{\omega}_B \times {}^B\mathbf{r}_i = \begin{bmatrix} v\cos\beta - r y_i \\ v\sin\beta + r x_i \\ 0 \end{bmatrix} \tag{1.303} $$

and the G-expression velocity ${}^G\mathbf{v}_i$ of tire number i is:

$$\begin{aligned} {}^G\mathbf{v}_i &= {}_G\boldsymbol{\omega}_B \times \left({}^G\mathbf{r}_i - {}^G\mathbf{d}_B\right) + {}^G\dot{\mathbf{d}}_B \\ &= {}^G R_B \, {}^B\mathbf{v}_i = \begin{bmatrix} -r\left(y_i \cos\psi + x_i \sin\psi\right) + v\cos\left(\beta+\psi\right) \\ r\left(x_i \cos\psi - y_i \sin\psi\right) + v\sin\left(\beta+\psi\right) \\ 0 \end{bmatrix} \end{aligned} \tag{1.304}$$

where ${}_G\boldsymbol{\omega}_B$ is the angular velocity of the vehicle in G-frame, ${}^G\mathbf{d}_B$ is the position vector of the origin of the B-frame in G-frame as shown in Fig. 1.68. For a planar vehicle model, ${}_G\boldsymbol{\omega}_B = r\,{}^G\hat{k}$ where r is the yaw rate $r = \omega_z$ and ${}^G\hat{k}$ is the unit vector of the Z-axis in the global coordinate frame.

Proof Let us show the coordinates of the origin of the B-frame in G-frame by ${}^G\mathbf{d}$

$${}^G\mathbf{d} = \begin{bmatrix} X_C \\ Y_C \end{bmatrix} \tag{1.305}$$

The transformation matrix between B-frame and G-frame ${}^G R_B$ is a rotation matrix by the yaw angle ψ.

$${}^G R_B = \begin{bmatrix} \cos\psi & -\sin\psi \\ \sin\psi & \cos\psi \end{bmatrix} \tag{1.306}$$

Therefore the position vector of tires in the G-frame will be

$$\begin{aligned} {}^G\mathbf{r}_i &= {}^G\mathbf{d} + {}^G R_B \, {}^B\mathbf{r}_i \\ &= \begin{bmatrix} X_C \\ Y_C \end{bmatrix} + \begin{bmatrix} \cos\psi & -\sin\psi \\ \sin\psi & \cos\psi \end{bmatrix} \begin{bmatrix} x_i \\ y_i \end{bmatrix} \\ &= \begin{bmatrix} X_C + x_i \cos\psi - y_i \sin\psi \\ Y_C + y_i \cos\psi + x_i \sin\psi \end{bmatrix} \end{aligned} \tag{1.307}$$

Assuming different front track w_f and rear track w_r, the position vectors of the wheels of the four-wheel vehicle in Fig. 1.68 are

$$\begin{aligned} {}^B\mathbf{r}_1 &= \begin{bmatrix} a_1 \\ \frac{1}{2}w_f \end{bmatrix} & {}^B\mathbf{r}_2 &= \begin{bmatrix} a_1 \\ -\frac{1}{2}w_f \end{bmatrix} \\ {}^B\mathbf{r}_3 &= \begin{bmatrix} -a_2 \\ \frac{1}{2}w_r \end{bmatrix} & {}^B\mathbf{r}_4 &= \begin{bmatrix} -a_2 \\ -\frac{1}{2}w_r \end{bmatrix} \end{aligned} \tag{1.308}$$

The velocity of the vehicle $\mathbf{v}$ in B-frame is

$$ {}^{B}\mathbf{v} = \begin{bmatrix} v\cos\beta \\ v\sin\beta \end{bmatrix} \tag{1.309} $$

Having the velocity ${}^{B}\mathbf{v}$ and the angular velocity ${}^{B}_{G}\boldsymbol{\omega}_{B} = r\ {}^{B}\hat{k}$ of the vehicle at the mass center C, we are able to calculate the velocity of any other point of the vehicle. The velocity of tires will be needed in calculating traction force and torques.

$$ \begin{aligned} {}^{B}_{G}\boldsymbol{\omega}_{B} &= {}^{G}R_{B}\ {}_{G}\boldsymbol{\omega}_{B} \\ &= \begin{bmatrix} \cos\psi & -\sin\psi & 0 \\ \sin\psi & \cos\psi & 0 \\ 0 & 0 & 1 \end{bmatrix} \times \begin{bmatrix} 0 \\ 0 \\ r \end{bmatrix} = \begin{bmatrix} 0 \\ 0 \\ r \end{bmatrix} \end{aligned} \tag{1.310} $$

$$ \begin{aligned} {}^{B}\mathbf{v}_{i} &= {}^{B}\mathbf{v} + {}^{B}_{G}\boldsymbol{\omega}_{B} \times {}^{B}\mathbf{r}_{i} \\ &= \begin{bmatrix} v\cos\beta \\ v\sin\beta \\ 0 \end{bmatrix} + \begin{bmatrix} 0 \\ 0 \\ r \end{bmatrix} \times \begin{bmatrix} x_i \\ y_i \\ 0 \end{bmatrix} = \begin{bmatrix} v\cos\beta - r y_i \\ v\sin\beta + r x_i \\ 0 \end{bmatrix} \end{aligned} \tag{1.311} $$

The vehicle velocity ${}^{B}\mathbf{v}$ after transformation to G-frame will be

$$ \begin{aligned} {}^{G}\mathbf{v} &= {}^{G}R_{B}\ {}^{B}\mathbf{v} = \begin{bmatrix} \cos\psi & -\sin\psi \\ \sin\psi & \cos\psi \end{bmatrix} \begin{bmatrix} v\cos\beta \\ v\sin\beta \end{bmatrix} \\ &= \begin{bmatrix} v\cos(\beta+\psi) \\ v\sin(\beta+\psi) \end{bmatrix} = {}^{G}\dot{\mathbf{d}} \end{aligned} \tag{1.312} $$

Therefore, the global velocity of the center of the wheel number i is:

$$ \begin{aligned} {}^{G}\mathbf{v}_{i} &= {}_{G}\boldsymbol{\omega}_{B} \times \left({}^{G}\mathbf{r}_{i} - {}^{G}\mathbf{d}_{B} \right) + {}^{G}\dot{\mathbf{d}}_{B} \\ &= \begin{bmatrix} 0 \\ 0 \\ r \end{bmatrix} \times \left(\begin{bmatrix} X_C - y_i\sin\psi + x_i\cos\psi \\ Y_C + y_i\cos\psi + x_i\sin\psi \\ 0 \end{bmatrix} - \begin{bmatrix} X_C \\ Y_C \\ 0 \end{bmatrix} \right) \\ &\quad + \begin{bmatrix} v\cos(\beta+\psi) \\ v\sin(\beta+\psi) \\ 0 \end{bmatrix} \\ &= \begin{bmatrix} -r\left(y_i\cos\psi + x_i\sin\psi\right) + v\cos(\beta+\psi) \\ r\left(x_i\cos\psi - y_i\sin\psi\right) + v\sin(\beta+\psi) \\ 0 \end{bmatrix} \end{aligned} \tag{1.313} $$

■

Example 43 Wheel numbers, their positions and velocity.

Figure 1.69 depicts a six-wheel passenger car. The velocity of the car at its mass center is indicated by **v**. The wheel numbers are indicated next to each wheel. The front left wheel is wheel number 1, and the front right wheel is number 2. Moving to the second axle, we have the wheels numbered 3 and 4. The left wheel of the third axle gets number 5, and the right wheel gets number 6.

Let us assume the global location vector of the car's mass center is given by

$$ {}^{G}\mathbf{d}_B = \begin{bmatrix} X_C \\ Y_C \end{bmatrix} \tag{1.314} $$

and the body velocity ${}^{B}\mathbf{v}$ at the mass center C and angular velocity of the car ${}_{G}\omega_B$ by

$$ {}^{B}\mathbf{v} = \begin{bmatrix} v_x \\ v_y \end{bmatrix} = \begin{bmatrix} v\cos\beta \\ v\sin\beta \end{bmatrix} \qquad {}_{G}\boldsymbol{\omega}_B = \begin{bmatrix} 0 \\ 0 \\ \dot{\psi} \end{bmatrix} = \begin{bmatrix} 0 \\ 0 \\ r \end{bmatrix} \tag{1.315} $$

The body position vectors of the wheels are

$$ {}^{B}\mathbf{r}_1 = \begin{bmatrix} a_1 \\ w/2 \end{bmatrix} \qquad {}^{B}\mathbf{r}_2 = \begin{bmatrix} a_1 \\ -w/2 \end{bmatrix} \tag{1.316} $$

$$ {}^{B}\mathbf{r}_3 = \begin{bmatrix} -a_2 \\ w/2 \end{bmatrix} \qquad {}^{B}\mathbf{r}_4 = \begin{bmatrix} -a_2 \\ -w/2 \end{bmatrix} \tag{1.317} $$

$$ {}^{B}\mathbf{r}_5 = \begin{bmatrix} -a_3 \\ w/2 \end{bmatrix} \qquad {}^{B}\mathbf{r}_6 = \begin{bmatrix} -a_3 \\ -w/2 \end{bmatrix} \tag{1.318} $$

Employing the transformation matrix ${}^{G}R_B$ to go from B to G coordinate frame,

$$ {}^{G}R_B = \begin{bmatrix} \cos\psi & -\sin\psi \\ \sin\psi & \cos\psi \end{bmatrix} \tag{1.319} $$

we may determine the global position of the wheels:

$$ {}^{G}\mathbf{r}_1 = {}^{G}\mathbf{d} + {}^{G}R_B\,{}^{B}\mathbf{r}_1 = \begin{bmatrix} X_C - \frac{1}{2}w\sin\psi + a_1\cos\psi \\ Y_C + \frac{1}{2}w\cos\psi + a_1\sin\psi \end{bmatrix} \tag{1.320} $$

$$ {}^{G}\mathbf{r}_2 = {}^{G}\mathbf{d} + {}^{G}R_B\,{}^{B}\mathbf{r}_2 = \begin{bmatrix} X_C + \frac{1}{2}w\sin\psi + a_1\cos\psi \\ Y_C - \frac{1}{2}w\cos\psi + a_1\sin\psi \end{bmatrix} \tag{1.321} $$

$$
{}^{G}\mathbf{r}_3 = {}^{G}\mathbf{d} + {}^{G}R_B \, {}^{B}\mathbf{r}_3 = \begin{bmatrix} X_C - \frac{1}{2}w\sin\psi - a_2\cos\psi \\ Y_C + \frac{1}{2}w\cos\psi - a_2\sin\psi \end{bmatrix} \tag{1.322}
$$

$$
{}^{G}\mathbf{r}_4 = {}^{G}\mathbf{d} + {}^{G}R_B \, {}^{B}\mathbf{r}_4 = \begin{bmatrix} X_C + \frac{1}{2}w\sin\psi - a_2\cos\psi \\ Y_C - \frac{1}{2}w\cos\psi - a_2\sin\psi \end{bmatrix} \tag{1.323}
$$

$$
{}^{G}\mathbf{r}_5 = {}^{G}\mathbf{d} + {}^{G}R_B \, {}^{B}\mathbf{r}_5 = \begin{bmatrix} X_C - \frac{1}{2}w\sin\psi - a_3\cos\psi \\ Y_C + \frac{1}{2}w\cos\psi - a_3\sin\psi \end{bmatrix} \tag{1.324}
$$

$$
{}^{G}\mathbf{r}_6 = {}^{G}\mathbf{d} + {}^{G}R_B \, {}^{B}\mathbf{r}_6 = \begin{bmatrix} X_C + \frac{1}{2}w\sin\psi - a_3\cos\psi \\ Y_C - \frac{1}{2}w\cos\psi - a_3\sin\psi \end{bmatrix} \tag{1.325}
$$

The velocity of C in G-frame is:

$$
\begin{aligned}
{}^{G}\mathbf{v} = {}^{G}R_B \, {}^{B}\mathbf{v} &= \begin{bmatrix} \cos\psi & -\sin\psi \\ \sin\psi & \cos\psi \end{bmatrix} \begin{bmatrix} v\cos\beta \\ v\sin\beta \end{bmatrix} \\
&= \begin{bmatrix} v\cos(\beta+\psi) \\ v\sin(\beta+\psi) \end{bmatrix} = {}^{G}\dot{\mathbf{d}}
\end{aligned} \tag{1.326}
$$

The velocity of a point P at ${}^{G}\mathbf{r}_P$ will be calculated by

$$
{}^{G}\mathbf{v}_P = {}_{G}\boldsymbol{\omega}_B \times \left({}^{G}\mathbf{r}_P - {}^{G}\mathbf{d}_B\right) + {}^{G}\dot{\mathbf{d}}_B \tag{1.327}
$$

As an example, the global velocity of the center of the wheel number 1 is:

$$
\begin{aligned}
{}^{G}\mathbf{v}_1 &= {}_{G}\boldsymbol{\omega}_B \times \left({}^{G}\mathbf{r}_1 - {}^{G}\mathbf{d}_B\right) + {}^{G}\dot{\mathbf{d}}_B \\
&= \begin{bmatrix} 0 \\ 0 \\ r \end{bmatrix} \times \left(\begin{bmatrix} X_C - \frac{1}{2}w\sin\psi + a_1\cos\psi \\ Y_C + \frac{1}{2}w\cos\psi + a_1\sin\psi \\ 0 \end{bmatrix} - \begin{bmatrix} X_C \\ Y_C \\ 0 \end{bmatrix} \right) \\
&\quad + \begin{bmatrix} v\cos(\beta+\psi) \\ v\sin(\beta+\psi) \\ 0 \end{bmatrix} \\
&= \begin{bmatrix} v\cos(\beta+\psi) - r\left(\frac{1}{2}w\cos\psi + a_1\sin\psi\right) \\ v\sin(\beta+\psi) + r\left(a_1\cos\psi - \dfrac{1}{2}w\sin\psi\right) \\ 0 \end{bmatrix}
\end{aligned} \tag{1.328}
$$

Having ${}^G\mathbf{v}$,we may calculate the velocity of tires in B-frame by transformation.

$$
{}^B\mathbf{v}_1 = {}^G R_B^T \, {}^G\mathbf{v}_1 = \begin{bmatrix} \cos\psi & -\sin\psi & 0 \\ \sin\psi & \cos\psi & 0 \\ 0 & 0 & 1 \end{bmatrix}^T \times
$$
$$
\begin{bmatrix} v\cos(\beta+\psi) - r\left(a_1 \sin\psi + \frac{w}{2}\cos\psi\right) \\ v\sin(\beta+\psi) + r\left(a_1\cos\psi - \frac{w}{2}\sin\psi\right) \\ 0 \end{bmatrix} = \begin{bmatrix} v\cos\beta - \frac{w}{2}r \\ v\sin\beta + a_1 r \\ 0 \end{bmatrix} \tag{1.329}
$$

As an example, consider a car as shown in Fig. 1.69, having the following data:

$$
{}^G\mathbf{d}_B = \begin{bmatrix} 25 \\ 5 \end{bmatrix} \text{m} \tag{1.330}
$$

$$
\dot{\psi} = r = 0.1 \,\text{rad/s} \qquad {}^B v_C = 20 \,\text{m/s} \tag{1.331}
$$

$$
\beta = 0.2 \,\text{rad} \qquad \psi = 0.3 \,\text{rad} \tag{1.332}
$$

$$
a_1 = 1.2 \,\text{m} \qquad a_2 = 1.5 \,\text{m} \qquad a_3 = 2.1 \,\text{m} \qquad w = 1.4 \,\text{m} \tag{1.333}
$$

The velocity vector of C in body coordinate frame is:

$$
{}^B\mathbf{v} = \begin{bmatrix} v\cos\beta \\ v\sin\beta \end{bmatrix} = \begin{bmatrix} 20\cos 0.2 \\ 20\sin 0.2 \end{bmatrix} = \begin{bmatrix} 19.601 \\ 3.9734 \end{bmatrix} \text{m/s} \tag{1.334}
$$

The transformation matrix from B to G coordinate frame is:

$$
{}^G R_B = \begin{bmatrix} \cos\psi & -\sin\psi \\ \sin\psi & \cos\psi \end{bmatrix} = \begin{bmatrix} 0.95534 & -0.29552 \\ 0.29552 & 0.95534 \end{bmatrix} \tag{1.335}
$$

The global position of the wheels is:

$$
{}^G\mathbf{r}_i = {}^G\mathbf{d} + {}^G R_B \, {}^B\mathbf{r}_i \tag{1.336}
$$

The global velocity of C is:

$$
{}^G\mathbf{v} = {}^G R_B \, {}^B\mathbf{v} = \begin{bmatrix} 17.551 \\ 9.5884 \end{bmatrix} = {}^G\dot{\mathbf{d}} \tag{1.337}
$$

The global velocity of the center of the wheel number 1 is:

$$
{}^G\mathbf{v}_1 = {}_G\boldsymbol{\omega}_B \times \left({}^G\mathbf{r}_1 - {}^G\mathbf{d}_B\right) + {}^G\dot{\mathbf{d}}_B = \begin{bmatrix} 17.449 \\ 9.6824 \end{bmatrix} \tag{1.338}
$$

Example 44 Velocity distribution of a vehicle.

Consider a general point P at ${}^B\mathbf{r}_P = x\hat{\imath} + y\hat{\jmath}$ of the vehicle in Fig. 1.69 along with

$$ {}^B\mathbf{r}_P = \begin{bmatrix} x \\ y \end{bmatrix} \text{m} \qquad {}^G\mathbf{d}_B = \begin{bmatrix} 25 \\ 5 \end{bmatrix} \text{m} \tag{1.339} $$

$$ \dot{\psi} = r = 0.1\,\text{rad/s} \qquad {}^B v_C = 20\,\text{m/s} \tag{1.340} $$

$$ \beta = 0.2\,\text{rad} \qquad \psi = 0.3\,\text{rad} \tag{1.341} $$

$$ a_1 = 1.2\,\text{m} \qquad a_2 = 1.5\,\text{m} \qquad a_3 = 2.1\,\text{m} \qquad w = 1.4\,\text{m} \tag{1.342} $$

$$ {}^G R_B = \begin{bmatrix} \cos\psi & -\sin\psi \\ \sin\psi & \cos\psi \end{bmatrix} = \begin{bmatrix} 0.95534 & -0.29552 \\ 0.29552 & 0.95534 \end{bmatrix} \tag{1.343} $$

$$ {}^B\mathbf{v} = \begin{bmatrix} v\cos\beta \\ v\sin\beta \end{bmatrix} = \begin{bmatrix} 19.601 \\ 3.9734 \end{bmatrix} \text{m/s} \tag{1.344} $$

$$ {}^G\mathbf{v} = {}^G R_B\, {}^B\mathbf{v} = \begin{bmatrix} 17.551 \\ 9.5884 \end{bmatrix} = {}^G\dot{\mathbf{d}} \tag{1.345} $$

The global coordinates of P are

$$ {}^G\mathbf{r}_P = {}^G\mathbf{d} + {}^G R_B\, {}^B\mathbf{r}_P = \begin{bmatrix} 0.95534x - 0.29552y + 25 \\ 0.29552x + 0.95534y + 5 \end{bmatrix} \tag{1.346} $$

and the global velocity components at P are:

$$ \begin{aligned} {}^G\mathbf{v}_P &= {}_G\boldsymbol{\omega}_B \times \left({}^G\mathbf{r}_P - {}^G\mathbf{d}_B\right) + {}^G\dot{\mathbf{d}}_B \\ &= \begin{bmatrix} -2.9552 \times 10^{-2}x - 9.5534 \times 10^{-2}y + 17.551 \\ 9.5534 \times 10^{-2}x - 2.9552 \times 10^{-2}y + 9.5884 \\ 0 \end{bmatrix} \end{aligned} \tag{1.347} $$

$$ \begin{aligned} {}^B\mathbf{v} &= {}^G R_B^T\, {}^G\mathbf{v} = \begin{bmatrix} 0.95534 & -0.29552 & 0 \\ 0.29552 & 0.95534 & 0 \\ 0 & 0 & 1 \end{bmatrix}^T {}^G\mathbf{v} \\ &= \begin{bmatrix} 19.601 - 0.1y \\ 3.9735 + 0.1 \\ 0 \end{bmatrix} \end{aligned} \tag{1.348} $$

The velocity distribution is a linear function of the local coordinates in B-frame. Therefore, having the velocity of any two points of the vehicle is enough to calculate the velocity of any other point of the vehicle.

Example 45 Global vehicle path.

When we find the translational and rotational velocities of a vehicle, v_X, v_Y, r, we may find the global path of motion of the vehicle by integration.

$$\psi = \int \dot{\psi}\, dt = \psi_0 + \int r\, dt$$

$$^G\mathbf{v} = \begin{bmatrix} \dot{Y} \\ \dot{X} \end{bmatrix} = {}^G R_B\, {}^B\mathbf{v} \tag{1.349}$$

$$X = \int \dot{X}\, dt = \int \left(v_x \cos\psi - v_y \sin\psi\right) dt \tag{1.350}$$

$$Y = \int \dot{Y}\, dt = \int \left(v_x \sin\psi + v_y \cos\psi\right) dt \tag{1.351}$$

1.7 Weight Transfer

Tire longitudinal and lateral forces are proportional to the load on the tire. The load on tires of a vehicle is a combination of weight of the vehicle and vehicle acceleration, as well as orientation of the vehicle on the road. In this section we review how the normal force F_z on a tire of a moving vehicle is calculated.

1.7.1 Longitudinally Accelerating Vehicle

We assume vehicles are longitudinally symmetric. Therefore, the left and right sides of vehicles are under same vertical force in level conditions. When a vehicle is speeding with acceleration a_x on a level road, as shown in Fig. 1.70, the vertical forces under the front and rear tires are:

$$F_{z_1} = F_{z_2} = \frac{1}{2}mg\frac{a_2}{l} - \frac{1}{2}ma_x\frac{h}{l} \tag{1.352}$$

$$F_{z_3} = F_{z_4} = \frac{1}{2}mg\frac{a_1}{l} + \frac{1}{2}ma_x\frac{h}{l} \tag{1.353}$$

$$l = a_1 + a_2 \tag{1.354}$$

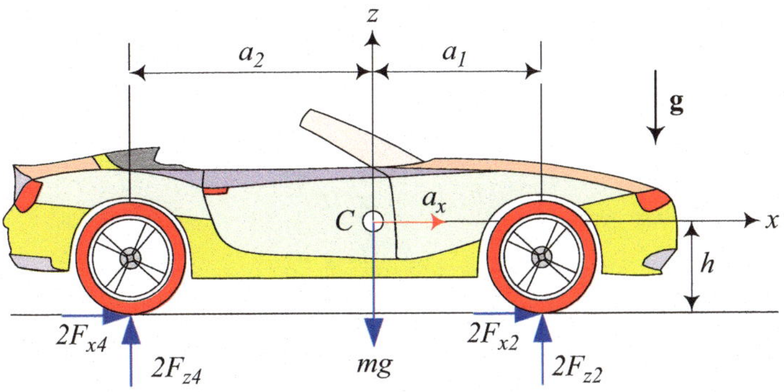

Fig. 1.70 An accelerating car on a level road

The first terms, $\frac{1}{2}mg\frac{a_2}{l}$ and $\frac{1}{2}mg\frac{a_1}{l}$, are the *static parts*, and the second terms, $\pm\frac{1}{2}ma_X\frac{h}{l}$, are *dynamic parts* of the vertical forces. The parameters a_1 and a_2 are longitudinal distances between front and rear axles respectively from the mass center C. The height of the mass center C from the ground is indicated by h (Dieter et al. 2018).

Proof The vehicle is considered as a rigid body that moves along a horizontal road. The force at the tireprint of each tire may be decomposed into a vertical F_z and a longitudinal force F_x. The Newton's equation of motion for the accelerating vehicle in the x-direction and two static equilibrium equations in y- and z-directions is:

$$\sum F_X = ma_X \tag{1.355}$$

$$\sum F_Z = 0 \tag{1.356}$$

$$\sum M_Y = 0 \tag{1.357}$$

Expanding the equations produces three equations for four unknowns F_{x_2}, F_{x_4}, F_{z_2}, F_{z_4}.

$$2F_{x_2} + 2F_{x_4} = ma_X \tag{1.358}$$

$$2F_{z_2} + 2F_{z_4} - mg = 0 \tag{1.359}$$

$$-2F_{z_2}a_1 + 2F_{z_4}a_2 - 2\left(F_{x_2} + F_{x_4}\right)h = 0 \tag{1.360}$$

We may substitute $\left(F_{x_2} + F_{x_4}\right)$ from the first equation in the third equations to solve for the normal forces F_{z_2}, F_{z_4} from second and third equations

$$F_{z_1} = F_{z_2} = \left(F_{z_2}\right)st + \left(F_{z2}\right)_{dyn} = \frac{1}{2}mg\frac{a_2}{l} - \frac{1}{2}ma_X\frac{h}{l} \tag{1.361}$$

$$F_{z_3} = F_{z_4} = \left(F_{z_4}\right)_{st} + \left(F_{z_4}\right)_{dyn} = \frac{1}{2}mg\frac{a_1}{l} + \frac{1}{2}ma_x\frac{h}{l} \tag{1.362}$$

The static parts

$$\left(F_{z_2}\right)_{st} = \frac{1}{2}mg\frac{a_2}{l} \tag{1.363}$$

$$\left(F_{z_4}\right)_{st} = \frac{1}{2}mg\frac{a_1}{l} \tag{1.364}$$

are weight distribution for a stationary car and depend on the horizontal position of the mass center. However, the dynamic parts

$$\left(F_{z_2}\right)_{dyn} = -\frac{1}{2}ma_x\frac{h}{l} \tag{1.365}$$

$$\left(F_{z_4}\right)_{dyn} = \frac{1}{2}ma_x\frac{h}{l} \tag{1.366}$$

indicate the weight distribution because of horizontal acceleration, and depend on the vertical position of the mass center.

When accelerating $a_x > 0$, the normal forces under the front tires are less than the static load, and under the rear tires are more than the static load. ■

Example 46 Front and rear wheel drive accelerating on a level road.

When the car is front-wheel-drive, then $F_{x_3} = F_{x_4} = 0$. Equations (1.355)–(1.357) will provide us with the same vertical tireprint forces as (1.352) and (1.353). However, the required horizontal force, $2F_{x_2}$, to achieve the same acceleration, a_x, must be provided solely by the front wheels.

$$2F_{x_2} = ma_x \tag{1.367}$$

If a car is rear-wheel drive, then $F_{x_1} = F_{x_2} = 0$ and the required force to achieve the acceleration, a_x, must be provided only by the rear wheels.

$$2F_{x_4} = ma_x \tag{1.368}$$

The vertical force under the wheels will still be the same as (1.352) and (1.353).

Example 47 Maximum acceleration on a level road.

The maximum acceleration of a car is proportional to the friction under its tires. We assume the friction coefficients at the front and rear tires are equal and all tires reach their maximum traction at the same time.

$$F_{x_i} = \pm\mu_x F_{z_i} \tag{1.369}$$

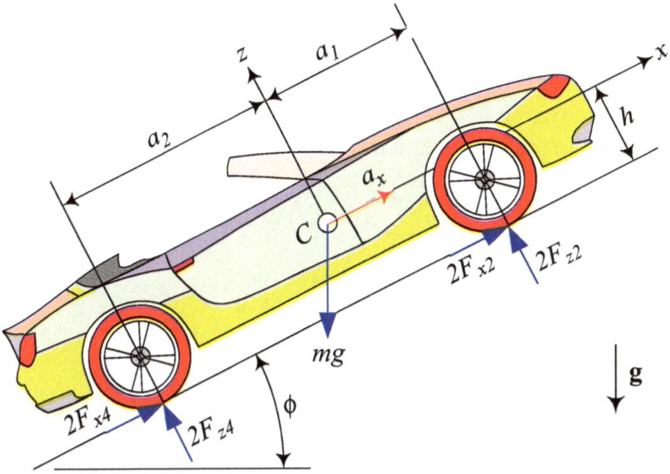

Fig. 1.71 An accelerating car on inclined pavement

Also, we assume the engine has infinite power to produce as much power as needed. Newton's equation (1.352) can now be written as

$$ma_X = \pm\mu_x \sum_{i=1}^{4} F_{z_i} \tag{1.370}$$

Substituting F_{z_i} from (1.361) and (1.362) results in

$$a_X = \pm\mu_x g \tag{1.371}$$

Therefore, the maximum acceleration and deceleration depend directly on the friction coefficient.

$$\mu_x = C_s\, s_s \tag{1.372}$$

Example 48 Accelerating vehicles on an inclined road.

When a symmetric vehicle is accelerating on an inclined pavement with angle ϕ as shown in Fig. 1.71, the vertical force under each of the front and rear axles, F_{z_1}, F_{z_2}, F_{z_3}, F_{z_4}, would be:

$$F_{z_1} = F_{z_2} = \frac{1}{2}mg\left(\frac{a_2}{l}\cos\phi - \frac{h}{l}\sin\phi\right) - \frac{1}{2}ma_X\frac{h}{l} \tag{1.373}$$

$$F_{z_3} = F_{z_4} = \frac{1}{2}mg\left(\frac{a_1}{l}\cos\phi + \frac{h}{l}\sin\phi\right) + \frac{1}{2}ma_X\frac{h}{l} \tag{1.374}$$

$$l = a_1 + a_2$$

The dynamic parts, $\pm\frac{1}{2}ma_X x\frac{h}{l}$, depend on acceleration a_x and height h of mass center C, while the static parts depend on the slope angle ϕ as well as the longitudinal and vertical positions of the mass center.

The Newton's equation in x-direction and two static equilibrium equations must be solved together to find the equation of motion and ground reaction forces.

$$\sum F_X = ma_X \tag{1.375}$$

$$\sum F_Z = 0 \tag{1.376}$$

$$\sum M_Y = 0 \tag{1.377}$$

Expanding these equations produces three equations for four unknowns F_{x_1}, F_{x_2}, F_{z_1}, F_{z_2}.

$$2F_{x_2} + 2F_{x_4} - mg\sin\phi = ma_X \tag{1.378}$$

$$2F_{z_2} + 2F_{z_4} - mg\cos\phi = 0 \tag{1.379}$$

$$-2F_{z_2}a_1 + 2F_{z_4}a_2 - 2\left(F_{x_2} + F_{x_4}\right)h = 0 \tag{1.380}$$

It is possible to eliminate $\left(F_{x_2} + F_{x_4}\right)$ between the first and third equations, and solve for the vertical forces F_{z_1}, F_{z_2}, F_{z_3}, F_{z_4}.

$$\begin{aligned} F_{z_2} &= F_{z_1} = \left(F_{z_2}\right)_{st} + \left(F_{z_2}\right)_{dyn} \\ &= \frac{1}{2}mg\left(\frac{a_2}{l}\cos\phi - \frac{h}{l}\sin\phi\right) - \frac{1}{2}ma_X\frac{h}{l} \end{aligned} \tag{1.381}$$

$$\begin{aligned} F_{z_4} &= F_{z_3} = \left(F_{z_4}\right)_{st} + \left(F_{z_4}\right)_{dyn} \\ &= \frac{1}{2}mg\left(\frac{a_1}{l}\cos\phi + \frac{h}{l}\sin\phi\right) + \frac{1}{2}ma_X\frac{h}{l} \end{aligned} \tag{1.382}$$

Example 49 Maximum acceleration on an inclined road.

The maximum acceleration depends on the friction under the tires. Let us assume the friction coefficients at the front and rear tires are equal. Then, the front and rear traction forces are

$$F_{x_i} \le \mu_x F_{z_i} \tag{1.383}$$

If we assume the front and rear wheels reach their traction limits at the same time, then

$$F_{x_i} = \pm\mu_x F_{z_i} \tag{1.384}$$

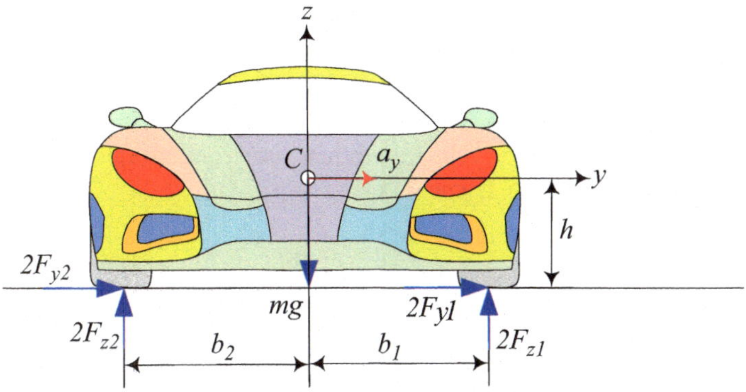

Fig. 1.72 A vehicle on lateral acceleration

and we may rewrite Newton's equation (1.375) as

$$ma_{x_M} = \pm\mu_x \sum_{i=1}^{4} F_{z_i} - mg\sin\phi \tag{1.385}$$

where a_{x_M} is the maximum achievable acceleration.

Substituting $\sum_{i=1}^{4} F_{z_i}$ from (1.381) and (1.382) results in

$$\frac{a_{x_M}}{g} = \pm\mu_x \cos\phi - \sin\phi \tag{1.386}$$

Accelerating on an uphill road and braking on a downhill road are the extreme cases in which the car can stall, $a_x = 0$. In these cases, the car can move only if

$$\mu_x \geq |\tan\phi| \tag{1.387}$$

Following the directions of the body coordinate frame of the vehicle, uphill road should be assigned by $\phi < 0$ and downhill by $\phi > 0$, as the slope angle ϕ should be measured about the y-axis. However, in this section we have used the absolute value of the slope angle for calculations.

1.7.2 Laterally Accelerating Vehicle

Figure 1.72 illustrates a vehicle on lateral acceleration. The vertical force under the left and right tires, F_{z_1}, F_{z_2}, F_{z_3}, F_{z_4}, are:

$$F_{z_1} = F_{z_3} = \frac{1}{2}mg\frac{b_2}{w} - \frac{1}{2}ma_Y\frac{h}{w} \tag{1.388}$$

$$F_{z_2} = F_{z_4} = \frac{1}{2} mg \frac{b_1}{w} + \frac{1}{2} ma_Y \frac{h}{w} \tag{1.389}$$

$$w = b_1 + b_2 \tag{1.390}$$

Proof Assuming a symmetric vehicle, we may use equilibrium and equation of motion in the body coordinate frame $(Cxyz)$

$$\sum F_Y = ma_Y \tag{1.391}$$

$$\sum F_Z = 0 \tag{1.392}$$

$$\sum M_X = 0 \tag{1.393}$$

we can write

$$2F_{y_1} + 2F_{y_2} = ma_Y \tag{1.394}$$

$$2F_{z_1} + 2F_{z_2} - mg = 0 \tag{1.395}$$

$$2F_{z_1} b_1 - 2F_{z_2} b_2 + 2\left(F_{y_1} + F_{y_2}\right) h = 0 \tag{1.396}$$

Substituting $\left(F_{y_1} + F_{y_2}\right)$ from (1.394) into (1.396) enables us to solve for F_{z_i}.

$$F_{z_1} = F_{z_3} = \frac{1}{2} mg \frac{b_2}{w} - \frac{1}{2} ma_Y \frac{h}{w} \tag{1.397}$$

$$F_{z_2} = F_{z_4} = \frac{1}{2} mg \frac{b_1}{w} + \frac{1}{2} ma_Y \frac{h}{w} \tag{1.398}$$

■

Example 50 Vehicle on a banked road.

Figure 1.73 illustrates the effect of a bank angle ϕ on the load distribution of a vehicle. A bank road causes the load on the lower tires to increase and the load on the upper tires to decrease. The tire reaction forces are:

$$F_{z_1} = \frac{1}{2} \frac{mg}{w} \left(b_2 \cos\phi - h \sin\phi\right) \tag{1.399}$$

$$F_{z_2} = \frac{1}{2} \frac{mg}{w} \left(b_1 \cos\phi + h \sin\phi\right) \tag{1.400}$$

$$w = b_1 + b_2 \tag{1.401}$$

The maximum bank angle is

$$\tan\phi_M = \mu_y \tag{1.402}$$

at which the car will slide down laterally.

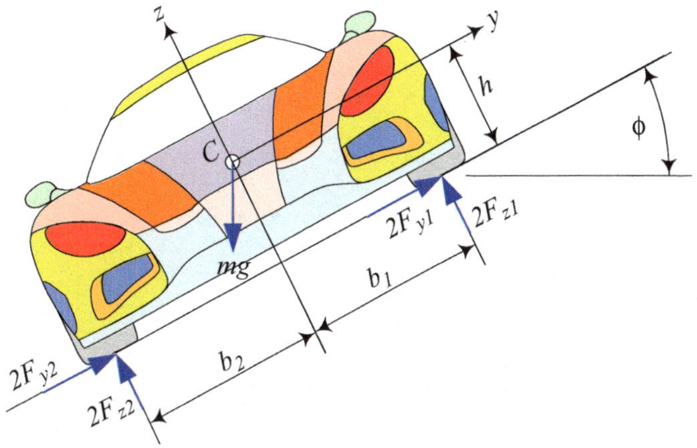

Fig. 1.73 Vertical forces to the ground under the uphill and downhill tires of a vehicle, on banked road

Starting with equilibrium equations in the body coordinate frame $(Cxyz)$

$$\sum F_Y = 0 \tag{1.403}$$

$$\sum F_Z = 0 \tag{1.404}$$

$$\sum M_X = 0 \tag{1.405}$$

we write

$$2F_{y_1} + 2F_{y_2} - mg\sin\phi = 0 \tag{1.406}$$

$$2F_{z_1} + 2F_{z_2} - mg\cos\phi = 0 \tag{1.407}$$

$$2F_{z_1}b_1 - 2F_{z_2}b_2 + 2\left(F_{y_1} + F_{y_2}\right)h = 0 \tag{1.408}$$

We assumed the force under the lower tires, front and rear, are equal, and also the forces under the upper tires, front and rear, are equal. To calculate the reaction forces under each tire, we may assume the overall lateral force $F_{y_1} + F_{y_2}$ as an unknown. The solution of these equations provide the lateral and reaction forces under the upper and lower tires.

$$F_{z_1} = F_{z_3} = \frac{1}{2}mg\frac{b_2}{w}\cos\phi - \frac{1}{2}mg\frac{h}{w}\sin\phi \tag{1.409}$$

$$F_{z_2} = F_{z_4} = \frac{1}{2}mg\frac{b_1}{w}\cos\phi + \frac{1}{2}mg\frac{h}{w}\sin\phi \tag{1.410}$$

$$F_{y_i} = \frac{1}{4}mg\sin\phi \tag{1.411}$$

At the ultimate angle $\phi = \phi_M$, all wheels will begin to slide simultaneously and therefore,

$$F_{y_i} = \mu_{y_i} F_{z_i} \tag{1.412}$$

The equilibrium equations show that

$$2\mu_{y_1} F_{z_1} + 2\mu_{y_2} F_{z_2} - mg \sin\phi = 0 \tag{1.413}$$

$$2F_{z_1} + 2F_{z_2} - mg \cos\phi = 0 \tag{1.414}$$

$$2F_{z_1} b_1 - 2F_{z_2} b_2 + 2\left(\mu_{y_1} F_{z_1} + \mu_{y_2} F_{z_2}\right) h = 0 \tag{1.415}$$

Assuming all friction coefficients are equal

$$\mu_{y_i} = \mu_y \tag{1.416}$$

provides us with

$$F_{z_1} = F_{z_3} = \frac{1}{2} mg \frac{b_2}{w} \cos\phi_M - \frac{1}{2} mg \frac{h}{w} \sin\phi_M \tag{1.417}$$

$$F_{z_2} = F_{z_4} = \frac{1}{2} mg \frac{b_1}{w} \cos\phi_M + \frac{1}{2} mg \frac{h}{w} \sin\phi_M \tag{1.418}$$

$$\tan\phi_M = \mu_y \tag{1.419}$$

These calculations are correct as long as

$$\tan\phi_M \le \frac{b_2}{h} \tag{1.420}$$

$$\mu_y \le \frac{b_2}{h} \tag{1.421}$$

If the lateral friction μ_y is higher than b_2/h, then the car will roll downhill. To increase the capability of a car moving on a banked road, the car should be as wide as possible with a mass center as low as possible.

Example 51 Tire forces of a parked car in a banked road.

A car having

$$m = 980\,\text{kg} \qquad h = 0.6\,\text{m} \qquad w = 1.52\,\text{m} \qquad b_1 = b_2 \tag{1.422}$$

is parked on a banked road with $\phi = 4$ deg. The forces under the lower and upper tires of the car are:

$$2F_{z_1} = 2265.2\,\text{N} \qquad 2F_{z_2} = 2529.9\,\text{N} \tag{1.423}$$

$$2F_{y_1} + 2F_{y_2} = 335.3\,\text{N} \tag{1.424}$$

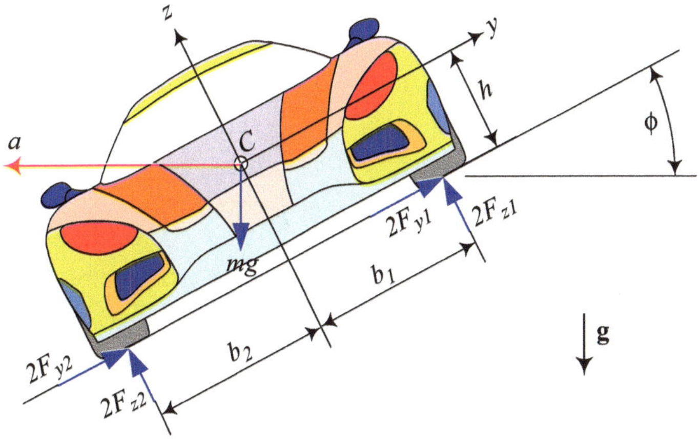

Fig. 1.74 A moving car on a round banked road

The ratio of the uphill force $2F_{z_1}$ to downhill force $2F_{z_2}$ depends on only the mass center location.

$$\frac{F_{z_1}}{F_{z_2}} = \frac{b_2 \cos\phi - h \sin\phi}{b_1 \cos\phi + h \sin\phi} \tag{1.425}$$

Assuming a symmetric car with $b_1 = b_2 = w/2$ simplifies the equation to

$$\frac{F_{z_1}}{F_{z_2}} = \frac{w \cos\phi - 2h \sin\phi}{w \cos\phi + 2h \sin\phi} \tag{1.426}$$

Example 52 Vehicle on a banked round road.

When a vehicle of mass m is moving with speed v on a flat round path of radius R, the direction of the wheels lateral force are inward and provide the required centripetal acceleration.

$$\left(2F_{y_1} + 2F_{y_2}\right) \cos\phi = m\frac{v^2}{R} \tag{1.427}$$

Knowing that the wheels' lateral forces are limited by the maximum friction force between tire and road, we conclude that there is a maximum speed v_M at which the required lateral force will not be produced by tires and vehicle slides out of the road.

To have a safe road, we have to design round roads such that vehicle do not need any wheel lateral force to provide the required centripetal force at the recommended speed. Designing roads with a bank angle is a good approximate solution; so a component of weight force provides the required centripetal force. Figure 1.74 illustrates a moving car on a round banked road. Assuming equal vertical force under each tire, the balance of the applied forces on the vehicle in the body coordinate frame provides us with

$$mg\cos\phi - 4F_{z_1} - m\frac{v^2}{R}\sin\phi = 0 \tag{1.428}$$

$$mg\sin\phi - m\frac{v^2}{R}\cos\phi = 0 \tag{1.429}$$

The second equation indicates the required bank angle as a function of speed

$$\tan\phi = \frac{v^2}{Rg} \tag{1.430}$$

The bank angle is independent of the vehicle mass m and is a function of the road radius of turn R and velocity of the vehicle v. Assuming that the road radius of turn R is not variable, the bank angle must ideally vary with the speed of the vehicle. To design a traditionally fixed road, we have to decide about the proper velocity of vehicles on the road and calculate the bank angle based on (1.430). Because the angle is only a function of the vehicle velocity, the proper banked road works well for all types of vehicles as long as they keep their velocity as recommended. Any lower or higher speed would respectively need some positive or negative lateral force to be generated by tires. The lack or excessive tire lateral force will be provided by steering and sideslip angles of vehicles, or by roll and camber angles of bicycles and motorcycles.

1.7.3 Longitudinally and Laterally Accelerating Vehicle

Figure 1.75 illustrates a vehicle on longitudinal and lateral acceleration a_x and a_y. The tire forces of the vehicle on a flat ground are equal to:

$$F_{z_1} = m\left(\frac{a_2}{l}g - \frac{h}{l}a_X\right)\frac{b_{2f}}{w_f} - m\left(\frac{a_2}{l}g - \frac{h}{l}a_X\right)\frac{h}{w_f}\frac{a_Y}{g} \tag{1.431}$$

$$F_{z_2} = m\left(\frac{a_2}{l}g - \frac{h}{l}a_X\right)\frac{b_{1f}}{w_f} + m\left(\frac{a_2}{l}g - \frac{h}{l}a_X\right)\frac{h}{w_f}\frac{a_Y}{g} \tag{1.432}$$

$$F_{z_3} = m\left(\frac{a_1}{l}g + \frac{h}{l}a_X\right)\frac{b_{2r}}{w_r} - m\left(\frac{a_1}{l}g + \frac{h}{l}a_X\right)\frac{h}{w_r}\frac{a_Y}{g} \tag{1.433}$$

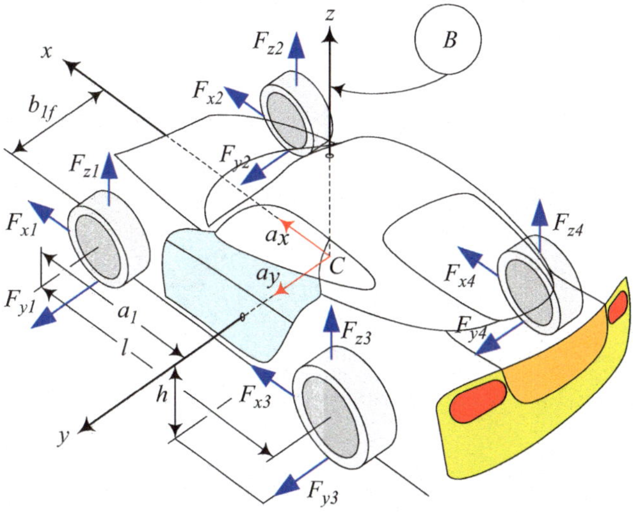

Fig. 1.75 A vehicle on longitudinal and lateral acceleration a_x and a_y

$$F_{z_4} = m\left(\frac{a_1}{l}g + \frac{h}{l}a_X\right)\frac{b_{1r}}{w_r} + m\left(\frac{a_1}{l}g + \frac{h}{l}a_X\right)\frac{h}{w_r}\frac{a_Y}{g} \tag{1.434}$$

$$w_f = b_{1f} + b_{2f} \tag{1.435}$$

$$w_r = b_{1r} + b_{2r} \tag{1.436}$$

Proof Let us use the tire numbers according to the recommendation in previous section. The front left wheel is number 1, and then the front right wheel would be number 2. The left wheel on second axle is number 3 and the second axle on the right will be number 4.

Figure 1.76 illustrates a vehicle on a flat ground with forward acceleration a_x. The forces F_{zf}, F_{zr}, on front and rear axles of the vehicle from (1.352) to (1.353) are:

$$F_{z_f} = mg\frac{a_2}{l} - ma_X\frac{h}{l} \tag{1.437}$$

$$F_{z_r} = mg\frac{a_1}{l} + ma_X\frac{h}{l} \tag{1.438}$$

$$l = a_1 + a_2 \tag{1.439}$$

Figure 1.77 depicts the front view of the same vehicle when the load on the front axle is as (1.437) and the vehicle has a lateral acceleration of a_y. Using (1.437), we introduce a virtual front mass m_{vf}

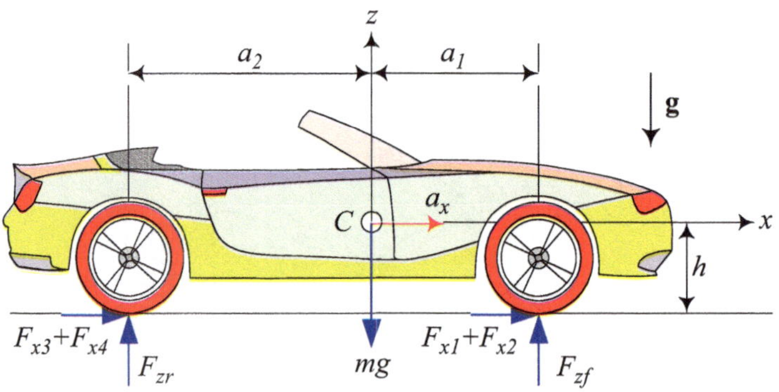

Fig. 1.76 A vehicle on a flat ground with forward acceleration a_x

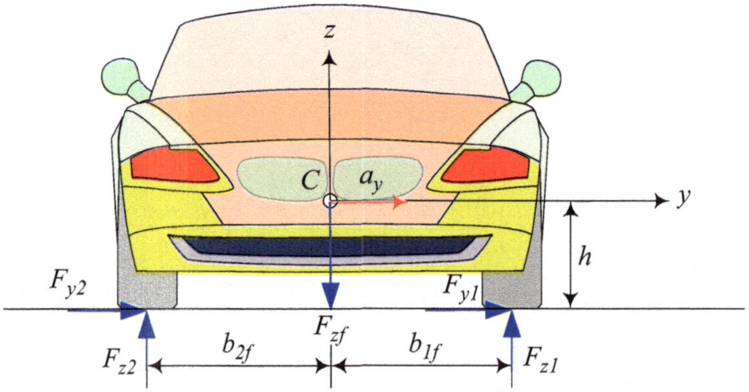

Fig. 1.77 A vehicle with a vertical load of F_{zf} on front axle and on a lateral acceleration of a_y

$$m_{vf} = \frac{F_{z_f}}{g} = \frac{1}{g}\left(mg\frac{a_2}{l} - ma_X\frac{h}{l}\right) \tag{1.440}$$

Employing (1.388)–(1.389) we find the vertical force under front wheels z_1 and z_2 as:

$$\begin{aligned} F_{z_1} &= m_{vf}\, g\frac{b_{2f}}{w_f} - m_{vf}\, a_Y\frac{h}{w_f} \\ &= \left(mg\frac{a_2}{l} - ma_x\frac{h}{l}\right)\left(\frac{b_{2f}}{w_f} - \frac{a_Y}{g}\frac{h}{w_f}\right) \\ &= m\left(\frac{a_2}{l}g - \frac{h}{l}a_X\right)\frac{b_{2f}}{w_f} - m\left(\frac{a_2}{l}g - \frac{h}{l}a_X\right)\frac{h}{w_f}\frac{a_Y}{g} \end{aligned} \tag{1.441}$$

$$F_{z_2} = m_{vf}\, g \frac{b_{1f}}{w_f} + m_{vf}\, a_Y \frac{h}{w_f}$$
$$= \left(mg\frac{a_2}{l} - ma_X\frac{h}{l}\right)\left(\frac{b_{1f}}{w_f} + \frac{a_Y}{g}\frac{h}{w_f}\right)$$
$$= m\left(\frac{a_2}{l}g - \frac{h}{l}a_X\right)\frac{b_{1f}}{w_f} + m\left(\frac{a_2}{l}g - \frac{h}{l}a_X\right)\frac{h}{w_f}\frac{a_Y}{g} \tag{1.442}$$

$$w_f = b_{1f} + b_{2f} \tag{1.443}$$

Similarly, we use (1.438) to introduce a virtual rear mass m_{vr}

$$m_{vr} = \frac{F_{z_r}}{g} = \frac{1}{g}\left(mg\frac{a_1}{l} + ma_X\frac{h}{l}\right) \tag{1.444}$$

and use (1.388)–(1.389) to find the vertical force under rear wheels z_3 and z_4 as:

$$F_{z_3} = m_{vr}\, g \frac{b_{2r}}{w_r} - m_{vr}\, a_Y \frac{h}{w_r}$$
$$= \left(mg\frac{a_1}{l} + ma_X\frac{h}{l}\right)\left(\frac{b_{2r}}{w_r} - \frac{a_Y}{g}\frac{h}{w_r}\right)$$
$$= m\left(\frac{a_1}{l}g + \frac{h}{l}a_X\right)\frac{b_{2r}}{w_r} - m\left(\frac{a_1}{l}g + \frac{h}{l}a_X\right)\frac{h}{w_r}\frac{a_Y}{g} \tag{1.445}$$

$$F_{z_4} = F_{z_f}\frac{b_{1r}}{w_r} + \frac{F_{z_r}}{g}a_Y\frac{h}{w_r}$$
$$= \left(mg\frac{a_1}{l} + ma_X\frac{h}{l}\right)\left(\frac{b_{1r}}{w_r} + \frac{a_Y}{g}\frac{h}{w_r}\right)$$
$$= m\left(\frac{a_1}{l}g + \frac{h}{l}a_X\right)\frac{b_{1r}}{w_r} + m\left(\frac{a_1}{l}g + \frac{h}{l}a_X\right)\frac{h}{w_r}\frac{a_Y}{g} \tag{1.446}$$

$$w_r = b_{1r} + b_{2r} \tag{1.447}$$

There were few assumptions in this analysis that make the results to be approximately correct.

1. We started from Eqs. (1.352) to (1.353) to determine the vertical load on the front and rear axles. These equations assume the left and right wheels are under the same forces.
2. We used (1.437) and (1.438) and defined virtual masses at front and rear parts of the vehicle. This is only an assumption to be able to connect the effects of longitudinal and lateral accelerations a_X and a_Y.
3. The sprung and unsprung masses are assumed rigidly connected. ■

Example 53 Linearized equations for tires' load.

Under normal driving conditions, vehicles usually have very low lateral and longitudinal accelerations. In case the term including $a_x a_y$ may be ignored compared to other terms, the set of Eqs. (1.431)–(1.434) for vertical forces of tires will become linear.

$$F_{z_1} = \frac{1}{l}\frac{m}{w_f}\left(g a_2 b_{2f} - h b_{2f} a_X - h a_2 a_Y\right) \tag{1.448}$$

$$F_{z_2} = \frac{1}{l}\frac{m}{w_f}\left(g a_2 b_{1f} - h b_{1f} a_X + h a_2 a_Y\right) \tag{1.449}$$

$$F_{z_3} = \frac{1}{l}\frac{m}{w_r}\left(g a_1 b_{2r} + h b_{2r} a_X - h a_1 a_Y\right) \tag{1.450}$$

$$F_{z_4} = \frac{1}{l}\frac{m}{w_r}\left(g a_1 b_{1r} + h b_{1r} a_X + h a_1 a_Y\right) \tag{1.451}$$

Example 54 The problem with exact equations.

Consider the rigid vehicle in Fig. 1.75. There will be six equations of motions:

$$\sum F_X = m a_X \qquad \sum F_Y = m a_Y \qquad \sum F_Z = 0$$
$$\sum M_X = 0 \qquad \sum M_X = 0 \qquad \sum M_X = 0 \tag{1.452}$$

which will be expanded to

$$F_{x_1} + F_{x_2} + F_{x_3} + F_{x_4} = m a_X \tag{1.453}$$

$$F_{y_1} + F_{y_2} + F_{y_3} + F_{y_4} = m a_Y \tag{1.454}$$

$$F_{z_1} + F_{z_2} + F_{z_3} + F_{z_4} - mg = 0 \tag{1.455}$$

$$\begin{aligned} &-\left(F_{z_1} + F_{z_2}\right) a_1 + \left(F_{z_3} + F_{z_4}\right) a_2 \\ &\quad -\left(F_{x_1} + F_{x_2} + F_{x_3} + F_{x_4}\right) h = I_y \dot{q} \end{aligned} \tag{1.456}$$

$$\begin{aligned} &F_{z_1} b_{1f} - F_{z_2} b_{2f} + F_{z_3} b_{1r} - F_{z_4} b_{2r} \\ &\quad + \left(F_{y_1} + F_{y_2} + F_{y_3} + F_{y_4}\right) h = I_x \dot{p} \end{aligned} \tag{1.457}$$

$$\begin{aligned} &F_{y_1} a_1 + F_{y_2} a_1 - F_{y_3} a_2 - F_{y_4} a_2 \\ &- F_{x_1} b_{1f} + F_{x_2} b_{2f} - F_{x_3} b_{1r} + F_{x_4} b_{2r} = I_z \dot{r} \end{aligned} \tag{1.458}$$

Assuming every parameter and variables are given except the vertical forces F_{z_1}, F_{z_2}, F_{z_3}, F_{z_4}, we only have three equations (1.455), (1.456), (1.457) to determine four unknowns. This system is undetermined and is not solvable unless a constraint being introduced into the system of equations.

$$\begin{bmatrix} 1 & 1 & 1 & 1 \\ -a_1 & -a_1 & a_2 & a_2 \\ b_{1f} & -b_{2f} & b_{1r} & -b_{2r} \end{bmatrix} \begin{bmatrix} F_{z_1} \\ F_{z_2} \\ F_{z_3} \\ F_{z_4} \end{bmatrix} = \begin{bmatrix} mg \\ ma_X h + I_y \dot{q} \\ -ma_Y h + I_x \dot{p} \end{bmatrix} \tag{1.459}$$

If the calculated forces in Eqs. (1.431)–(1.434) are correct, then they must satisfy Eq. (1.459) when $\dot{q} = 0$, $\dot{p} = 0$. Substituting the vertical forces F_{z_1}, F_{z_2}, F_{z_3}, F_{z_4} from (1.431) to (1.434) into Eq. (1.459) indicates that all three equations will be satisfied.

Example 55 Weight transfer in local coordinate frame.

Equations (1.431)–(1.434) determine the load transfer on vehicle's wheels because of longitudinal and lateral accelerations a_X and a_Y, both expressed in the global coordinate frame. We will see that the equations of motion of vehicles are better to be defined in the body coordinate frame. In the body coordinate frame, we have

$$a_X = \dot{v}_x - r\,v_y \tag{1.460}$$

$$a_Y = \dot{v}_y + r\,v_x \tag{1.461}$$

Therefore, the weight transfer equations (1.431)–(1.434) must be modified as below to be applied.

$$\begin{aligned} F_{z_1} &= m\left(\frac{a_2}{l}g - \frac{h}{l}\left(\dot{v}_x - r\,v_y\right)\right)\frac{b_{2f}}{w_f} \\ &\quad -m\left(\frac{a_2}{l}g - \frac{h}{l}\left(\dot{v}_x - r\,v_y\right)\right)\frac{h}{w_f}\frac{\dot{v}_y + r\,v_x}{g} \end{aligned} \tag{1.462}$$

$$\begin{aligned} F_{z_2} &= m\left(\frac{a_2}{l}g - \frac{h}{l}\left(\dot{v}_x - r\,v_y\right)\right)\frac{b_{1f}}{w_f} \\ &\quad +m\left(\frac{a_2}{l}g - \frac{h}{l}\left(\dot{v}_x - r\,v_y\right)\right)\frac{h}{w_f}\frac{\dot{v}_y + r\,v_x}{g} \end{aligned} \tag{1.463}$$

$$\begin{aligned} F_{z_3} &= m\left(\frac{a_1}{l}g + \frac{h}{l}\left(\dot{v}_x - r\,v_y\right)\right)\frac{b_{2r}}{w_r} \\ &\quad -m\left(\frac{a_1}{l}g + \frac{h}{l}\left(\dot{v}_x - r\,v_y\right)\right)\frac{h}{w_r}\frac{\dot{v}_y + r\,v_x}{g} \end{aligned} \tag{1.464}$$

$$F_{z4} = m\left(\frac{a_1}{l}g + \frac{h}{l}\left(\dot{v}_x - r\,v_y\right)\right)\frac{b_{1r}}{w_r}$$
$$+\,m\left(\frac{a_1}{l}g + \frac{h}{l}\left(\dot{v}_x - r\,v_y\right)\right)\frac{h}{w_r}\frac{\dot{v}_y + r\,v_x}{g} \tag{1.465}$$

$$w_f = b_{1f} + b_{2f} \tag{1.466}$$

$$w_r = b_{1r} + b_{2r} \tag{1.467}$$

1.8 Chapter Summary

The dynamic performance of a vehicle is mainly determined by the interaction of its tires and road. A vehicle can only move and maneuver by the force systems generated under the tires. In this chapter, we introduce the required coordinate frames to determine the location and orientation of tires in the vehicle body coordinate frame; the mathematical equation to calculate longitudinal and lateral forces; and individual equations needed to develop dynamic equations of vehicles in the following chapters.

The resultant force system that a tire receives from the ground is at the center of the tireprint and can be decomposed along x_t, y_t, and z_t axes of the tire coordinate frame T. The interaction of a tire with road generates a three-dimensional ($3D$) force system including three forces and three moments. The force system at the tireprint of a loaded, rolling, steered, cambered tire includes: forward force F_x, lateral force F_y, vertical force F_z, aligning moment M_z, roll moment M_x, and pitch moment M_y. The forward force F_x and lateral force F_y are the most significant forces in vehicle maneuvering. To accelerate or brake a vehicle, a longitudinal force must be developed between the tire and the ground. When a torque T is applied to the spin axis of a tire, longitudinal slip ratio s occurs and a longitudinal force F_x is generated at the tireprint proportional to s. The tire lateral force F_y is a function of two angles of the tire: sideslip angle α and camber angle γ. The F_x and F_y take the tire load F_z, sideslip α, longitudinal slip s, and the camber angle γ as input.

We adopt the proportional-saturation model for longitudinal and lateral slips of tire. When $\alpha = 0$, a small longitudinal slip $s < s_s$ generates the longitudinal force $F_x/F_z = C_s s$, and when $s = 0$, a small sideslip angle $\alpha < \alpha_s$ generates a lateral force of $F_y/F_z = -C_\alpha \alpha$. When there exists a longitudinal slip $s < s_s$ and then we also introduce a sideslip $\alpha < \alpha_s$, the longitudinal force will reduce. Similarly, when there exists a longitudinal slip $s < s_s$, the lateral force will drop. The elliptic mathematical model introduces the analytical expression of the interaction F_x/F_z and F_y/F_z.

Because the longitudinal and lateral forces are affected by the vertical force F_z on the tire, there must be a model to calculate the weight transfer during forward and lateral acceleration. Such equations are calculated in this chapter.

1.9 Key Symbols

$a \equiv \ddot{x}$	Acceleration
a, b	Semiaxes of tire print area A
a_i	Longitudinal distance of axle i from mass center
	Axle number starting from the front first axle
a_x	Forward acceleration
a_y	Lateral acceleration
a_{xM}	Maximum forward acceleration
a_{yM}	Maximum lateral acceleration
a_{x_α}	Aligning arm
a_{x_γ}	Camber trail
a_{y_γ}	Camber arm
A, A_P	Tireprint area
b	Lateral distance of a wheel from longitudinal x-axis
B	Vehicle body coordinate frame
C	Wheel-body coordinate frame, mass center
C_{M_α}	Sideslip aligning moment coefficient
C_{M_γ}	Camber aligning moment coefficient
C_s	Longitudinal slip coefficient
$C_{\dot{s}}$	Longitudinal force drop coefficient with slip rate
$C_{s\alpha}$	Tire longitudinal force drop factor
$C_{s\gamma}$	Tire longitudinal-camber force drop factor
C_{s_1}, C_{s_2}	Velocity drop coefficient in longitudinal force
C_{s_x}, C_{s_y}	Longitudinal and lateral slip coefficients
C_{y_γ}	Camber arm coefficient
C_α	Sideslip coefficient, sideslip stiffness
$C_{\alpha s}$	Tire lateral force drop factor
$C_{\alpha\gamma}$	Tire lateral-camber force drop factor
$C_{\alpha_1}, C_{\alpha_2}$	Velocity drop coefficient in lateral force
C_γ	Camber coefficient, camber stiffness
d	Displacement
$\mathbf{d}$	Location vector
d_F	No slip tire travel
d_A	Actual tire travel
D	Tire diameter
f	Function
$F, \mathbf{F}$	Force
F_x	Longitudinal force, forward force
F_{xM}	Maximum longitudinal force
F_y	Lateral force

F_{y_M}	Maximum lateral force
F_z	Normal force, vertical force, wheel load
g $\mathbf{g}$	Gravitational acceleration
h	Height of mass center from the ground
H	Heaviside function
I	Mass moment
$\mathbf{I}$	Unity matrix
k	Stiffness
k_x	Tire stiffness in the x-direction
k_y	Tire stiffness in the y-direction
k_z	Tire stiffness in the z-direction
K	Radial and non-radial tires parameter in $\mu_r = \mu_r(p, v_x)$
m	Mass
m_{vf}	Virtual front mass
m_{vr}	Virtual rear mass
M_r $\mathbf{M}_r$	Rolling resistance moment
M_x, $\mathbf{M}_x$	Roll moment, bank moment, tilting torque
M_y	Pitch moment, rolling resistance torque
M_z	Yaw moment, aligning moment, self-aligning moment
n	Exponent for shape and stress distribution of A_P
n_1	Number of tire rotations
o, O	Origin of a coordinate frame
O	Order of magnitude
p	Tire inflation pressure
P	Point
r	Radial position of tire periphery
$\mathbf{r}$	Position vector
R_g	Geometric tire radius
R_h	Loaded tire height
R_e	Equivalent tire radius
R	Tire radius, rolling radius, effective radius
	Rotation transformation matrix
s	Longitudinal slip
s_s	Saturation value of longitudinal slip
s_y	Lateral slip
S	Saturation function
T	Wheel torque, tire coordinate frame
$v \equiv \dot{x}$, $\mathbf{v}$	Velocity, tread velocity in tireprint
v_g	Velocity of the ground
v_{rel}	Relative velocity of tire tread with respect to ground
x, y, z, $\mathbf{x}$	Displacement

x, y, z	Coordinate axes
x_g	Displacement of the ground
x_{rel}	Relative displacement of tread with respect to ground
W	Wheel coordinate frame
$\mathbf{0}$	Zero vector
α	Tire sideslip angle, tire angular acceleration
α_s	Saturation sideslip angle
α_w	Tire angular acceleration
β	Tire-body sideslip, wheel-body sideslip, vehicle sideslip
	Transversal slope, attitude angle
γ	Camber angle
δ	Steer angle
θ	Tire angular rotation
μ	Friction coefficient
μ_x	Longitudinal friction coefficient
μ_y	Lateral friction coefficient
μ_{dp}	Friction coefficient driving peak value
μ_{ds}	Friction coefficient steady-state value
τ	Shear stress
τ_x, τ_y	Shear stresses over the tireprint
τ_{xM}, τ_{yM}	Maximum shear stresses
φ	Contact angle, angular length of A
ϕ	Slope angle
ω	Angular velocity, frequency
ω_{eq}	Equivalent tire angular velocity
ω_w	Angular velocity of a wheel, actual tire angular velocity

Exercises

1. Tire-wheel coordinate frames.
   Assume an equivalent rigid tire disc with the following equation in T-coordinate.

$$(z - R)^2 + x^2 = R^2 \tag{1.468}$$

   (a) Assuming a positive γ, determine the equation of the tire in W-frame.
   (b) Assuming a positive δ, determine the equation of the tire in C-frame.
   (c) Assuming a positive γ and δ, determine the equation of the tire in C-frame.
   (d) Assuming a positive γ and δ, determine the equation of the tire in B-frame.

2. Time-dependent transformation.
Consider the point P in Fig. 1.4 as the heading point of tire in T-frame. The coordinates of P in T-frame are:

$$ {}^{T}\mathbf{r}_{P} = \begin{bmatrix} R & 0 & R \end{bmatrix}^{T} \tag{1.469} $$

Assume the steer angle δ is varying with time as

$$ \delta = \frac{\pi}{10}\sin\frac{\pi}{10}t \tag{1.470} $$

(a) Determine the coordinates of P in W-frame.
(b) Determine the coordinates of P in C-frame.
(c) Determine the coordinates of P in B-frame.

3. T-frame vector to W-frame.

(a) Consider a vector quantity ${}^{T}\mathbf{F}$ in the tire T-frame as:

$$ {}^{T}\mathbf{F} = \begin{bmatrix} F_1 & F_2 & F_3 \end{bmatrix}^{T} \tag{1.471} $$

and knowing that:

$$ {}^{W}T_{T} = \begin{bmatrix} {}^{W}R_{T} & {}^{W}\mathbf{d}_{T} \\ 0 & 1 \end{bmatrix} = \begin{bmatrix} 1 & 0 & 0 & 0 \\ 0 & \cos\gamma & \sin\gamma & 0 \\ 0 & -\sin\gamma & \cos\gamma & -R \\ 0 & 0 & 0 & 1 \end{bmatrix} \tag{1.472} $$

$$ {}^{C}T_{T} = \begin{bmatrix} {}^{C}R_{T} & {}^{C}\mathbf{d}_{T} \\ 0 & 1 \end{bmatrix} = \begin{bmatrix} \cos\delta & -\sin\delta & 0 & 0 \\ \sin\delta & \cos\delta & 0 & 0 \\ 0 & 0 & 1 & -R \\ 0 & 0 & 0 & 1 \end{bmatrix} \tag{1.473} $$

(b) transform the vector to W-frame.
(c) transform the vector to C-frame.
(d) transform the vector to B-frame if:

$$ \begin{aligned} {}^{B}R_{W} &= {}^{B}R_{T}\,{}^{T}R_{W} \\ &= \begin{bmatrix} \cos\delta_1 & -\cos\gamma\sin\delta_1 & \sin\gamma\sin\delta_1 \\ \sin\delta_1 & \cos\gamma\cos\delta_1 & -\cos\delta_1\sin\gamma \\ 0 & \sin\gamma & \cos\gamma \end{bmatrix} \end{aligned} \tag{1.474} $$

$$ {}^{T}\mathbf{d}_{W} = \begin{bmatrix} 0 & -R\sin\gamma & R\cos\gamma \end{bmatrix}^{T} \tag{1.475} $$

4. Phase portrait for longitudinal force.
   Assume

$$\frac{F_x}{F_z} = C_s s - \frac{1}{10} C_s s^3 \tag{1.476}$$

   (a) Plot F_x/F_z as a function of s for $C_s = 10.01001$.
   (b) How much slip ratio is needed to have $F_x = F_z$ and how much to have $F_x = 2F_z$.
   (c) Plot $d(F_x/F_z)/dt$ versus F_x/F_z for $s = 0.01$, $s = 0.05$, $s = 0.075$, $s = 0.1$, $s = 0.01$, $s = 0.15$ and make a conclusion.

5. Tire sideslip, steer, wheel sideslip angles relationship.
   The relationship of $\alpha = \beta - \delta$ is correct from top view at zero camber $\gamma = 0$. Determine the relationship when $\gamma \neq 0$.
6. *Sigmoid* or *Logistic* function derivative.
   Prove that if

$$f(x) = \text{Sig}(x) = \frac{1}{1 + e^{-x}} \tag{1.477}$$

   then

$$f'(x) = f(x)(1 - f(x)) \tag{1.478}$$

7. *Sigmoid* and *Saturation* function.
   Determine the coefficient k in the Sigmoid function

$$f(x) = \text{Sig}(x) = \frac{1}{1 + e^{-kx}} \tag{1.479}$$

   such that the difference between Saturation and Sigmoid function to be minimum.
8. Activation functions.
   Show that

   (a) If $f(x) = x/(1 + |x|)$, then

$$f'(x) = \frac{x}{(1 + |x|)^2} \tag{1.480}$$

   (b) If $f(x) = \arctan(x) = \tan^{-1}(x)$, then

$$f'(x) = \frac{x}{1 + x^2} \tag{1.481}$$

(c) If $f(x) = x/\sqrt{1+kx^2}$, then

$$f'(x) = \left(\frac{1}{\sqrt{1+kx^2}}\right)^3 \tag{1.482}$$

(d) If $f(x) = \tanh(x)$, then

$$f'(x) = 1 - f(x)^2 \tag{1.483}$$

(e) If $f(x) = \tanh(x) = 2/\left(1+e^{-2x}\right) - 1$, then

$$\tanh(x) = 2\,\text{Sig}(2x) - 1 \tag{1.484}$$

Therefore, *Sigmoid* function is a scaled tanh function.

9. Graphical view of tire force.
Consider a vehicle with the following data.

$$\begin{array}{llll} C_{\alpha f} = 8.5 & C_{\alpha r} = 8.5 & \alpha_s = 5\,\text{deg} & s_s = 0.1 \\ C_{sf} = 7.5 & C_{sr} = 7.5 & C_{\alpha s} = 0.5 & C_{s\alpha} = 0.5 \end{array} \tag{1.485}$$

and the tire force equations:

$$\frac{F_x}{F_z} = C_s s \sqrt{1 - C_{s\alpha}\left(\frac{\alpha}{\alpha_s}\right)^2} \qquad |\alpha| < \alpha_s \qquad |s| < s_s \tag{1.486}$$

$$\frac{F_y}{F_z} = -C_\alpha \alpha \sqrt{1 - C_{\alpha s}\left(\frac{s}{s_s}\right)^2} \qquad |\alpha| < \alpha_s \qquad |s| < s_s \tag{1.487}$$

(a) Plot F_x/F_z as functions of α and s for $-2\alpha_s < \alpha < 2\alpha_s$ and $-2s_s < s < 2s_s$. Replot F_x/F_z for $C_{\alpha s} = 0$ and $C_{s\alpha} = 0$ and compare the plots.
(b) Plot F_z/F_z as functions of α and s for $-2\alpha_s < \alpha < 2\alpha_s$ and $-2s_s < s < 2s_s$. Replot F_y/F_z for $C_{\alpha s} = 0$ and $C_{s\alpha} = 0$ and compare the plots.
(c) If there is any sharp edges in any of the above plots, explain what would happen to the tire forces before and after the sharp edges.

10. Tire camber and steer angles connection.
Practically, we design a suspension and steering mechanism such that the camber angle γ is proportional to the steer angle δ.

$$\gamma = C_{\gamma\delta}\delta \tag{1.488}$$

Rewrite the lateral force equation to be a function of steer angle and sideslip angle.

11. Slope of limit ellipse tire force.
    (a) Determine the slope of both sides of the limit curve at the point that F_x/F_z and F_y/F_z reach each other.
    (b) Determine the required condition to have the same slope of both sides of the limit curve at the point that F_x/F_z and F_z/F_z reach each other.

12. Jump in limit ellipse tire force.
    Determine if it is possible the limit elliptical model curve at the point where both F_x/F_z and F_y/F_z are saturated do not reach each other and show a jump?
13. Approximate elliptic tire model.
    Use the approximate elliptical tire force model (1.266) and (1.267),

$$\frac{F_x}{F_z} \simeq C_s\, s \left(1 - \frac{1}{2} C_{s\alpha} \left(\frac{\alpha}{\alpha_s}\right)^2\right) \tag{1.489}$$

$$\frac{F_y}{F_z} \simeq -C_\alpha \alpha \left(1 - \frac{1}{2} C_{\alpha s} \left(\frac{s}{s_s}\right)^2\right) \tag{1.490}$$

    and use the following date

$$\begin{array}{lll} s_s = 0.1 & C_s = 10 & C_\alpha = 0.24 \\ C_{s\alpha} = 0.5 & C_{\alpha s} = 0.5 & \alpha_s = 5\,\text{deg} \end{array} \tag{1.491}$$

    and

    (a) Plot F_x/F_z as a function α for $-\alpha_s < \alpha < \alpha_s$ and for $s = 0.01$, $s = 0.02$, $s = 0.03$, $s = 0.04$.
    (b) Plot F_y/F_z as a function s for $-s_s < s < s_s$ and for $\alpha = 0.5$ deg, $\alpha = 1$ deg, $\alpha = 1.5$ deg, $\alpha = 2$ deg.
    (c) Plot F_x/F_z versus F_y/F_z for $s = 0.01$, $s = 0.02$, $s = 0.03$, $s = 0.04$ and $\alpha = 0.5$ deg, $\alpha = 1$ deg, $\alpha = 1.5$ deg, $\alpha = 2$ deg.
    (d) Plot F_x/F_z versus F_y/F_z for $s = s_s$ and for $-\alpha_s < \alpha < \alpha_s$, and also plot F_x/F_z versus F_y/F_z for $\alpha = \alpha_s$ and for $-s_s < s < s_s$. The closed curve will show the possible limit zone of tire slips.
    (e) Compare plots of 11(a) with elliptical tire model in the book and explain the differences.
    (f) Compare plots of 11(b) with elliptical tire model in the book and explain the differences.
    (g) Compare the limit of 11(c) with limit shape of elliptical tire model and explain the differences.

14. Jump in limit diamond tire force.
    Determine if it is possible the limit diamond model curve at the point where both F_x/F_z and F_y/F_z are saturated do not reach each other and show a jump?

15. Tires' loads order of magnitude.
Assuming that for most passenger cars the longitudinal and lateral accelerations are $a_x/g < 1$ and $a_y/g < 0.05$, determine the order of magnitude of the term with $a_x a_y$ compared to other terms in equations for vertical force on tires.
16. Vertical force examination
Use Eqs. (1.431)–(1.434) and show that all Eq. (1.459) are correct.
17. Longitudinal and lateral accelerating vehicle.
Reprove the equations for tires vertical force calculation starting with lateral acceleration. Then define a left and right virtual masses, m_{vl} m_{vr}, and use the longitudinal analysis. Compare your results with Eqs. (1.431)–(1.434).
18. Fourth equation.
The set of Eq. (1.459) provides three equations among four unknown forces, F_{z_1}, F_{z_2}, F_{z_3}, F_{z_4}. We have a set of solutions (1.431)–(1.434) that satisfy the set of equations. What can be the fourth equation to be added to Eq. (1.459) to provide us with the solutions (1.431)–(1.434).

$$\begin{bmatrix} F_{z_1} + F_{z_2} + F_{z_3} + F_{z_4} \\ a_2 F_{z_3} - a_1 F_{z_2} - a_1 F_{z_1} + a_2 F_{z_4} \\ F_{z_1} b_f + F_{z_3} b_r - F_{z_2} b_{2f} - F_{z_4} b_{2r} \end{bmatrix} = \begin{bmatrix} mg \\ ma_x h \\ -ma_y h \end{bmatrix} \tag{1.492}$$

Is it possible to rewrite Eq. (1.459) as

$$\begin{bmatrix} 1 & 1 & 1 & 1 \\ -a_1 & -a_1 & a_2 & a_2 \\ b_{1f} & -b_{2f} & b_{1r} & -b_{2r} \\ C_1 & C_2 & C_2 & C_3 \end{bmatrix} \begin{bmatrix} F_{z1} \\ F_{z2} \\ F_{z3} \\ F_{z4} \end{bmatrix} = \begin{bmatrix} mg \\ ma_x h \\ -ma_y h \\ A \end{bmatrix} \tag{1.493}$$

and solve the equations to find F_{z_1}, F_{z_2}, F_{z_3}, F_{z_4}

$$\begin{bmatrix} 1 & 1 & 1 & 1 \\ -a_1 & -a_1 & a_2 & a_2 \\ b_{1f} & -b_{2f} & b_{1r} & -b_{2r} \\ C_1 & C_2 & C_2 & C_3 \end{bmatrix}^{-1} \begin{bmatrix} mg \\ ma_x h \\ -ma_y h \\ A \end{bmatrix} = \begin{bmatrix} F_{z1} \\ F_{z2} \\ F_{z3} \\ F_{z4} \end{bmatrix} \tag{1.494}$$

and then expand the equations and pick one of the coefficients C_1, C_2, C_3, C_4 equal to one, say $C_1 = 1$, and find C_2, C_3, C_4, A?
19. Independent longitudinal and lateral forces.
In case the longitudinal and lateral forces are independent and not been affected when the other force is developed, then the following questions will be applied.

$$\frac{F_x}{F_z} = F_x(\alpha, s, \gamma) = C_s\, s \tag{1.495}$$

$$\frac{F_y}{F_z} = F_y(\alpha, s, \gamma) = -C_\alpha\, \alpha - C_\gamma\, \gamma \tag{1.496}$$

(a) Use $C_s = 10$, $\alpha_s = 5$ deg, $C_\alpha = 0.24$, $s_s = 0.1$ and plot force relations similar to Figs. 1.39 and 1.40.
(b) Use $C_{\alpha s} = 0.5$, and $C_{s\alpha} = 0.5$, and redraw the elliptical and diamond and independent models on one plot similar to Fig. 1.40.

20. ★Elliptical camber equation.
Consider the elliptical relationships (1.200) and (1.201) and include an elliptical camber equation to the lateral force.

$$\frac{F_y}{F_z} = -C_\alpha \alpha \sqrt{1 - C_{\alpha s}\left(\frac{s}{s_s}\right)^2 - C_{\alpha\gamma}\left(\frac{\gamma}{\gamma_s}\right)^2}$$
$$-C_\gamma \gamma \sqrt{1 - C_{\gamma s}\left(\frac{s}{s_s}\right)^2 - C_{\gamma\alpha}\left(\frac{\alpha}{\alpha_s}\right)^2} \tag{1.497}$$
$$\alpha < \alpha_s \qquad s < s_s \qquad \gamma < \gamma_s$$

(a) Plot F_y/F_z as a function of α and γ at constant values of s for $C_s = 10$, $\alpha_s = 5$ deg, $C_\alpha = 0.24$, $s_s = 0.1$, $C_\gamma = 0.05$, $\gamma_s = 20$ deg, $C_{\alpha s} = 0.5$, $C_{s\alpha} = 0.5$, $C_{\alpha\gamma} = 0.5$, and $C_{\gamma\alpha} = 0.5$.

21. Path of motion.
Assume

$$r = 0.01\ \text{rad/s} \qquad {}^G\mathbf{v} = \begin{bmatrix} \dot{Y} \\ \dot{X} \end{bmatrix} = \begin{bmatrix} 0.1 \\ 15 \end{bmatrix}\ \text{m/s} \tag{1.498}$$

Determine the path of motion of the vehicle.

22. Path of alternative motion.
Assume

$$r = 0.01\ \text{rad/s} \qquad {}^G\mathbf{v} = \begin{bmatrix} \dot{Y} \\ \dot{X} \end{bmatrix} = \begin{bmatrix} 0.1 \sin \frac{t}{100} \\ 15 \end{bmatrix}\ \text{m/s} \tag{1.499}$$

Determine the path of motion of the vehicle.

23. Path of yaw rate related motion.
Assume

$$r = 0.01 \sin \frac{t}{100}\ \text{rad/s} \qquad {}^G\mathbf{v} = \begin{bmatrix} \dot{Y} \\ \dot{X} \end{bmatrix} = \begin{bmatrix} 10lr \\ 100lr \end{bmatrix}\ \text{m/s} \tag{1.500}$$
$$l = 3\ \text{m} \tag{1.501}$$

Determine the path of motion of the vehicle.

Chapter 2
Vehicle Planar Dynamics

The planar model of vehicles is mathematically the simplest model to determine dynamic characteristics of vehicles and it still predicts the dynamic behavior of vehicles very well. In this chapter we study this principal model to examine maneuvering of vehicles by steering as well as the wheel torque control. The wheel torque and steer angle are the inputs and the longitudinal velocity, lateral velocity, and yaw rate are the main output variables of the planar vehicle dynamics model.

2.1 Vehicle Dynamics Equations

The planar vehicle dynamic model is the simplest applied modeling in which we assume the vehicle remains parallel to the ground and has no roll, no pitch, and no bounce motions. Figure 2.1 illustrates the variables of the planar vehicle dynamic model. A vehicle body coordinate $B\,(C, x, y, z)$ is attached to the mass center C of the vehicle. The planar motion of vehicles has three degrees of freedom: translation in the x and y directions, and a rotation about the z-axis. The longitudinal velocity v_x along the x-axis, the lateral velocity v_y along the y-axis, and the yaw rate $r = \dot{\psi}$ about the z-axis are the outputs of the dynamic equations of motion (Jazar 2017, 2011; MacMillan 1936). The *Newton–Euler equations of motion* for the vehicle in the body coordinate frame B are:

$$F_x = m\,\dot{v}_x - mr\,v_y \tag{2.1}$$

$$F_y = m\,\dot{v}_y + mr\,v_x \tag{2.2}$$

$$M_z = I_z\,\dot{r} \tag{2.3}$$

$$T_i = I_{w_i}\,\dot{\omega}_{w_i} + R_w\,F_{x_i} \tag{2.4}$$

R. N. Jazar, *Advanced Vehicle Dynamics*,
https://doi.org/10.1007/978-3-030-13062-6_2

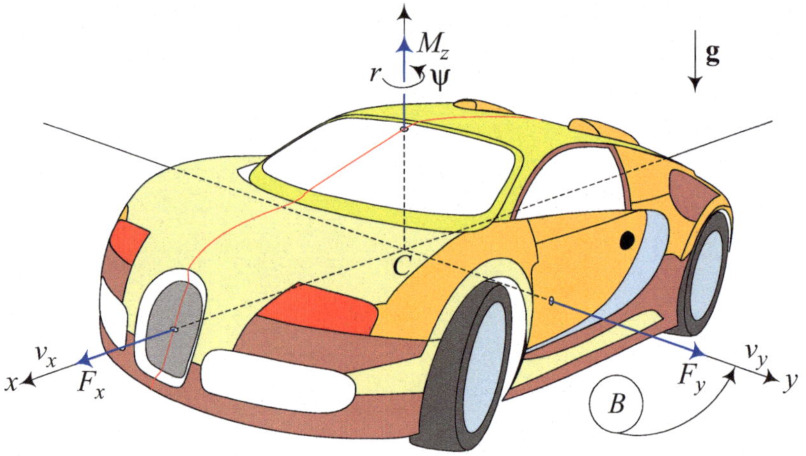

Fig. 2.1 Vehicle body coordinate frame $B(Cxyz)$, dynamic variables, and force system of the planar dynamic model

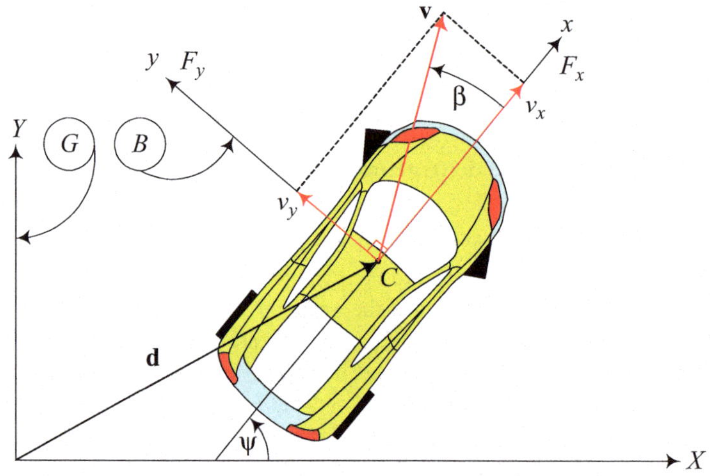

Fig. 2.2 A rigid vehicle in a planar motion in a globally fixed coordinate frame $G\ (X, Y, Z)$

Proof Figure 2.2 illustrates a vehicle in a planar motion. We attach a body coordinate frame $B\ (x, y, z)$ to the vehicle at its mass center C. The vehicle is moving in a global coordinate frame $G\ (X, Y, Z)$ that is attached to the ground at a given fixed point. The Z and z axes are always parallel. The orientation of the frame B in G is indicated by the heading angle ψ between the X and x axes, measured from X. The global position vector of the mass center of the vehicle is denoted by the location vector ${}^G\mathbf{d}$.

The equations of motion of a rigid body in the body coordinate frame are:

$$\begin{aligned} {}^B\mathbf{F} &= {}^BR_G\,{}^G\mathbf{F} = {}^BR_G\left(m\,{}^G\mathbf{a}_B\right) = m\,{}^B_G\mathbf{a}_B \\ &= m\,{}^B\dot{\mathbf{v}}_B + m\,{}^B_G\boldsymbol{\omega}_B \times {}^B\mathbf{v}_B \end{aligned} \tag{2.5}$$

$$\begin{aligned} {}^B\mathbf{M} &= {}^BR_G\,{}^G\mathbf{M} = \frac{{}^Gd}{dt}\,{}^B\mathbf{L} = {}^B_G\dot{\mathbf{L}}_B = {}^B\dot{\mathbf{L}} + {}^B_G\boldsymbol{\omega}_B \times {}^B\mathbf{L} \\ &= {}^BI\,{}^B_G\dot{\boldsymbol{\omega}}_B + {}^B_G\boldsymbol{\omega}_B \times \left({}^BI\,{}^B_G\boldsymbol{\omega}_B\right) \end{aligned} \tag{2.6}$$

The force, moment, and kinematic vectors for the vehicle are:

$$ {}^B\mathbf{F} = \begin{bmatrix} F_x & F_y & 0 \end{bmatrix}^T \qquad {}^B\mathbf{M} = \begin{bmatrix} 0 & 0 & M_z \end{bmatrix}^T \tag{2.7}$$

$$ {}^B_G\boldsymbol{\omega}_B = \begin{bmatrix} 0 & 0 & r \end{bmatrix}^T \qquad {}^B_G\dot{\boldsymbol{\omega}}_B = \begin{bmatrix} 0 & 0 & \dot{r} \end{bmatrix}^T \tag{2.8}$$

$$ {}^B\mathbf{v} = \begin{bmatrix} v_x & v_y & 0 \end{bmatrix}^T \qquad {}^B\dot{\mathbf{v}} = \begin{bmatrix} \dot{v}_x & \dot{v}_y & 0 \end{bmatrix}^T \tag{2.9}$$

The v_x is the forward component and the v_y is the lateral component of the velocity vector ${}^B\mathbf{v}$. The $r = \dot{\psi} = \omega_z$ is the yaw rate of the vehicle.

We assume that the body coordinate is the principal coordinate frame of the vehicle, and therefore we have a diagonal mass moment matrix (Beatty 1986; Bottema and Roth 1979).

$$ {}^BI = \begin{bmatrix} I_1 & 0 & 0 \\ 0 & I_2 & 0 \\ 0 & 0 & I_3 \end{bmatrix} = \begin{bmatrix} I_x & 0 & 0 \\ 0 & I_y & 0 \\ 0 & 0 & I_z \end{bmatrix} \tag{2.10}$$

Substituting the above vectors and matrices in the equations of motion (2.5)–(2.6) provides us with the following equations:

$$\begin{aligned} {}^B\mathbf{F} &= m\,{}^B\dot{\mathbf{v}}_B + m\,{}^B_G\boldsymbol{\omega}_B \times {}^B\mathbf{v}_B \\ &= m\begin{bmatrix} \dot{v}_x \\ \dot{v}_y \\ 0 \end{bmatrix} + m\begin{bmatrix} 0 \\ 0 \\ r \end{bmatrix} \times \begin{bmatrix} v_x \\ v_y \\ 0 \end{bmatrix} = \begin{bmatrix} m\dot{v}_x - mrv_y \\ m\dot{v}_y + mrv_x \\ 0 \end{bmatrix} \end{aligned} \tag{2.11}$$

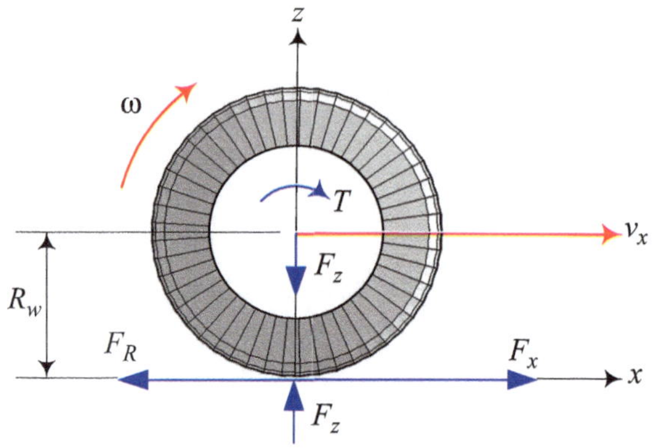

Fig. 2.3 Free-body-diagram of a wheel

$$
\begin{aligned}
{}^B\mathbf{M} &= {}^B I \, {}^B_G\dot{\boldsymbol{\omega}}_B + {}^B_G\boldsymbol{\omega}_B \times \left({}^B I \, {}^B_G\boldsymbol{\omega}_B \right) \\
&= \begin{bmatrix} I_x & 0 & 0 \\ 0 & I_y & 0 \\ 0 & 0 & I_z \end{bmatrix} \begin{bmatrix} 0 \\ 0 \\ \dot{r} \end{bmatrix} \\
&+ \begin{bmatrix} 0 \\ 0 \\ r \end{bmatrix} \times \left(\begin{bmatrix} I_x & 0 & 0 \\ 0 & I_y & 0 \\ 0 & 0 & I_z \end{bmatrix} \begin{bmatrix} 0 \\ 0 \\ r \end{bmatrix} \right) = \begin{bmatrix} 0 \\ 0 \\ I_z \dot{r} \end{bmatrix}
\end{aligned} \tag{2.12}
$$

The first two Equations of (2.11) and the third Equation of (2.12) make the set of equations of motion (2.1)–(2.3) for the planar vehicle dynamics.

$$F_x = m\,\dot{v}_x - mr\,v_y \tag{2.13}$$

$$F_y = m\,\dot{v}_y + mr\,v_x \tag{2.14}$$

$$M_z = \dot{r}\,I_z \tag{2.15}$$

The right-hand side of the equations of motion are the resultant of the kinematics of the rigid vehicle acceleration ${}^B_G\mathbf{a}$ expressed in B-frame. The left-hand side is the resultant of the external force system on the vehicle, expressed in body coordinate frame, ${}^B\mathbf{F}$, ${}^B\mathbf{M}$.

The fourth equation (2.4) is a result of dynamic analysis of individual wheels of the vehicle. Consider the free-body-diagram of the wheel number i as shown in Fig. 2.3. There is a traction force F_x and a roll resistance force F_R in the x-direction according to the tire coordinate frame. The equal action and reaction forces of wheel load and ground reaction are also applied to the tire in the z-direction by ignoring the mass of the wheel. The in-wheel torque T_i will provide us with the resultant traction force F_{x_i} according to:

$$T_i = I_{w_i}\,\dot{\omega}_{w_i} + R_{w_i}\,F_{x_i} \tag{2.16}$$

where I_w is the mass moment of the wheel about its spin axis, ω_w is the angular velocity of the wheel about its spin axis, and R_w is the equivalent tire radius. We may replace R_w with tire geometric radius R_g considering that in normal conditions they are very close to each other.

Besides these main components of the force system on a wheel, there are several other less important forces as aerodynamic resistance torque and force, aligning moments and forces, etc. (Schiehlen 1982; Milliken and Milliken 1995, 2002). ■

Example 56 Global equations of motion.

The equation of motion of a planar vehicle, expressed in the global coordinate frame, is called the G-expression.

$$F_X = m\frac{d}{dt}\dot{X} = m\dot{v}_X \tag{2.17}$$

$$F_Y = m\frac{d}{dt}\dot{Y} = m\dot{v}_Y \tag{2.18}$$

$$M_Z = I_z\frac{d}{dt}\dot{\psi} = I_z\,\dot{\omega}_Z \tag{2.19}$$

$$\dot{\omega}_Z = \dot{r}\,\hat{K} \tag{2.20}$$

Although these equations look simpler than (2.1)–(2.3), they are not practical because the forces F_X, F_Y are dependent on the orientation of the vehicle, and they are well expressed in the B-frame. The orientation of the vehicle is indicated by the heading angle ψ.

To recover the vehicle's equations of motion in B-frame we need to transform the global equations of motion (2.17)–(2.19) to the vehicle's body coordinate frame B, using the transformation matrix ${}^GR_B$ to go from B-frame to G-frame.

$${}^GR_B = \begin{bmatrix} \cos\psi & -\sin\psi & 0 \\ \sin\psi & \cos\psi & 0 \\ 0 & 0 & 1 \end{bmatrix} \tag{2.21}$$

The G-expression of the velocity vector is

$${}^G\mathbf{v}_C = {}^GR_B\,{}^B\mathbf{v}_C \tag{2.22}$$

$$\begin{bmatrix} v_X \\ v_Y \\ 0 \end{bmatrix} = \begin{bmatrix} \cos\psi & -\sin\psi & 0 \\ \sin\psi & \cos\psi & 0 \\ 0 & 0 & 1 \end{bmatrix} \begin{bmatrix} v_x \\ v_y \\ 0 \end{bmatrix}$$

$$= \begin{bmatrix} v_x\cos\psi - v_y\sin\psi \\ v_y\cos\psi + v_x\sin\psi \\ 0 \end{bmatrix} \tag{2.23}$$

where ${}^{B}\mathbf{v}_C$ is the velocity of vehicle at C expressed in B-frame. Therefore, the global acceleration components are

$$ {}^{G}\dot{\mathbf{v}}_C = \frac{{}^{G}d}{dt}\,{}^{G}\mathbf{v}_C \tag{2.24} $$

$$ \begin{bmatrix} \dot{v}_X \\ \dot{v}_Y \\ 0 \end{bmatrix} = \begin{bmatrix} \left(\dot{v}_x - \dot{\psi}\, v_y\right)\cos\psi - \left(\dot{v}_y + \dot{\psi}\, v_x\right)\sin\psi \\ \left(\dot{v}_y + \dot{\psi}\, v_x\right)\cos\psi + \left(\dot{v}_x - \dot{\psi}\, v_y\right)\sin\psi \\ 0 \end{bmatrix} \tag{2.25} $$

The B-expression of the angular velocity vector is

$$ {}_{G}\boldsymbol{\omega}_B = {}^{G}R_B\,{}^{B}_{G}\boldsymbol{\omega}_B $$

$$ \begin{bmatrix} 0 \\ 0 \\ \omega_Z \end{bmatrix} = \begin{bmatrix} \cos\psi & -\sin\psi & 0 \\ \sin\psi & \cos\psi & 0 \\ 0 & 0 & 1 \end{bmatrix} \begin{bmatrix} 0 \\ 0 \\ r \end{bmatrix} = \begin{bmatrix} 0 \\ 0 \\ \omega_z \end{bmatrix} \tag{2.26} $$

where ${}^{B}_{G}\boldsymbol{\omega}_B$ is the angular velocity of B-frame with respect to G-frame expressed in B-frame. The global equations of motion are

$$ {}^{G}\mathbf{F}_C = m\,{}^{G}\dot{\mathbf{v}}_C \qquad {}^{G}\mathbf{M}_C = {}^{G}I\,{}_{G}\dot{\boldsymbol{\omega}}_B \tag{2.27} $$

where the force system vector transformation is

$$ {}^{G}\mathbf{F}_C = {}^{G}R_B\,{}^{B}\mathbf{F}_C \qquad {}^{G}\mathbf{M}_C = {}^{G}R_B\,{}^{B}\mathbf{M}_C \tag{2.28} $$

therefore, the B-expression of the equations of motion is

$$ {}^{B}\mathbf{F}_C = {}^{G}R_B^T\,{}^{G}\mathbf{F}_C = m\,{}^{G}R_B^T\,{}^{G}\dot{\mathbf{v}}_C \tag{2.29} $$

$$ {}^{B}\mathbf{M}_C = {}^{G}R_B^T\,{}^{G}\mathbf{M}_C = {}^{G}R_B\,{}^{B}I\,{}^{G}R_B^T\,{}^{B}_{G}\dot{\boldsymbol{\omega}}_B \tag{2.30} $$

Substituting the associated vectors generates the Newton's equations of motion in the body coordinate frame.

$$ \begin{bmatrix} F_x \\ F_y \\ 0 \end{bmatrix} = {}^{G}R_B^T \begin{bmatrix} F_X \\ F_Y \\ 0 \end{bmatrix} $$

$$ = m\,{}^{G}R_B^T \begin{bmatrix} \left(\dot{v}_x - \dot{\psi}\, v_y\right)\cos\psi - \left(\dot{v}_y + \dot{\psi}\, v_x\right)\sin\psi \\ \left(\dot{v}_y + \dot{\psi}\, v_x\right)\cos\psi + \left(\dot{v}_x - \dot{\psi}\, v_y\right)\sin\psi \\ 0 \end{bmatrix} $$

$$ = m \begin{bmatrix} \dot{v}_x - \dot{\psi}\, v_y \\ \dot{v}_y + \dot{\psi}\, v_x \\ 0 \end{bmatrix} \tag{2.31} $$

Applying the same method for moment transformation,

$$\begin{bmatrix} 0 \\ 0 \\ M_Z \end{bmatrix} = \begin{bmatrix} \cos\psi & -\sin\psi & 0 \\ \sin\psi & \cos\psi & 0 \\ 0 & 0 & 1 \end{bmatrix} \begin{bmatrix} 0 \\ 0 \\ M_z \end{bmatrix} = \begin{bmatrix} 0 \\ 0 \\ M_z \end{bmatrix} \tag{2.32}$$

we find the Euler equation in the body coordinate frame.

$$M_z = {}^G R_B \, {}^B I \, {}^G R_B^T \, {}^B_G \dot{\boldsymbol{\omega}}_B = \dot{r} \, I_z \tag{2.33}$$

These are the same equations we found in (2.13)–(2.15).

Example 57 Lagrange method and equations of motion.

We may use the Lagrange method to determine the equations of motion of planar vehicle model. The kinetic energy K of a vehicle in a planar motion is,

$$\begin{aligned} K &= \frac{1}{2} {}^G\mathbf{v}_B^T \, m \, {}^G\mathbf{v}_B + \frac{1}{2} {}_G\boldsymbol{\omega}_B^T \, {}^G I \, {}_G\boldsymbol{\omega}_B \\ &= \frac{1}{2} \begin{bmatrix} v_X \\ v_Y \\ 0 \end{bmatrix}^T m \begin{bmatrix} v_X \\ v_Y \\ 0 \end{bmatrix} + \frac{1}{2} \begin{bmatrix} 0 \\ 0 \\ \omega_Z \end{bmatrix}^T {}^G I \begin{bmatrix} 0 \\ 0 \\ \omega_Z \end{bmatrix} \\ &= \frac{1}{2} m v_X^2 + \frac{1}{2} m v_Y^2 + \frac{1}{2} I_z \omega_Z^2 = \frac{1}{2} m \left(\dot{X}^2 + \dot{Y}^2 \right) + \frac{1}{2} I_z \, \dot{\psi}^2 \end{aligned} \tag{2.34}$$

where its mass moment matrix in global coordinate is:

$$\begin{aligned} {}^G I &= {}^G R_B \, {}^B I \, {}^G R_B^T \\ &= \begin{bmatrix} \cos\psi & -\sin\psi & 0 \\ \sin\psi & \cos\psi & 0 \\ 0 & 0 & 1 \end{bmatrix} \begin{bmatrix} I_x & 0 & 0 \\ 0 & I_y & 0 \\ 0 & 0 & I_z \end{bmatrix} \begin{bmatrix} \cos\psi & -\sin\psi & 0 \\ \sin\psi & \cos\psi & 0 \\ 0 & 0 & 1 \end{bmatrix}^T \\ &= \begin{bmatrix} I_x \cos^2\psi + I_y \sin^2\psi & \left(I_x - I_y\right) \sin\psi \cos\psi & 0 \\ \left(I_x - I_y\right) \sin\psi \cos\psi & I_y \cos^2\psi + I_x \sin^2\psi & 0 \\ 0 & 0 & I_z \end{bmatrix} \end{aligned} \tag{2.35}$$

and

$${}^G\mathbf{v}_B = \begin{bmatrix} v_X & v_Y & 0 \end{bmatrix}^T = \begin{bmatrix} \dot{X} & \dot{Y} & 0 \end{bmatrix}^T \tag{2.36}$$

$${}_G\boldsymbol{\omega}_B = \begin{bmatrix} 0 \\ 0 \\ \omega_Z \end{bmatrix} = \begin{bmatrix} 0 \\ 0 \\ r \end{bmatrix} = \begin{bmatrix} 0 \\ 0 \\ \dot{\psi} \end{bmatrix} \tag{2.37}$$

The resultant external force system in G-frame is:

$$
{}^{G}\mathbf{F}_{C} = \begin{bmatrix} F_X & F_Y & 0 \end{bmatrix}^T \qquad {}^{G}\mathbf{M}_{C} = \begin{bmatrix} 0 & 0 & M_Z \end{bmatrix}^T \tag{2.38}
$$

Employing the Lagrange method

$$
\frac{d}{dt}\left(\frac{\partial K}{\partial \dot{q}_i}\right) - \frac{\partial K}{\partial q_i} = F_i \qquad i = 1, 2, \cdots n \tag{2.39}
$$

and using the coordinates X, Y, and ψ as generalized coordinates q_i provide us with the following equations of motion in the global coordinate frame:

$$
F_X = m \frac{d}{dt}\dot{X} = m\dot{v}_X \tag{2.40}
$$

$$
F_Y = m \frac{d}{dt}\dot{Y} = m\dot{v}_Y \tag{2.41}
$$

$$
M_Z = I_z \frac{d}{dt}\dot{\psi} = I_z\,\dot{\omega}_Z \tag{2.42}
$$

These are the same equations as (2.17)–(2.19) (Jazar 2011; Goldstein et al. 2002).

Example 58 Comments on dynamic equations of motion.

Equations of motion of a moving rigid body such as a vehicle should be expressed in its principal body coordinate frame at its mass center. This is the only coordinate frame in which the mass moment matrix of the rigid body is diagonal and is constant. Furthermore, in study of vehicles, the external forces are resultant of the generated forces under the tires. Such forces have their simplest expression in the vehicle body frame. The traction force in x-direction and the lateral force in the y-direction are both independent of the orientation of the vehicle in the global frame. Similarly, the equations of motion of spacecraft, aircraft, helicopter, bicycle, motorcycle, hovercraft, and any other vehicle should be expressed in body coordinate frame. Majority of external forces have simpler expression in B-frame and the mass moments are constant in B-frame (Jazar 2011; Yang et al. 2015).

Example 59 Aerodynamic force.

The left-hand side of Eqs. (2.1) and (2.2) indicates all external forces applied on vehicles. Main reason of external forces are the resultant of tire road interactions. However, we may also consider other external forces such as aerodynamic and tilted gravitational forces.

$$
F_x = m\,\dot{v}_x - mr\,v_y \tag{2.43}
$$

$$
F_y = m\,\dot{v}_y + mr\,v_x \tag{2.44}
$$

Aerodynamics force F_A is considered to be proportional to relative velocity between the vehicle and ambient air. Therefore, the aerodynamic force may be a result of wind as well as vehicle speed. In this book we only consider F_A to be proportional to the square of the relative velocity v^2.

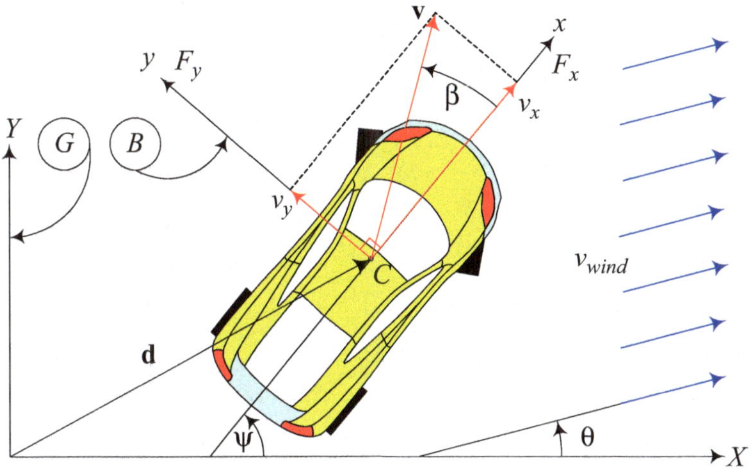

Fig. 2.4 A vehicle is moving with attitude angle ψ with velocity $\mathbf{v}$ while ambient wind is moving at an angle θ with respect to the global X-axis with velocity $\mathbf{v}_{wind}$

Assuming stationary air, the aerodynamic force F_A is only in the x-direction. We separate the tire traction force and air resistance force and write the equations of motion as:

$$F_x - F_A = m\,\dot{v}_x - mr\,v_y \tag{2.45}$$

$$F_y = m\,\dot{v}_y + mr\,v_x \tag{2.46}$$

The force F_A is the air resistance aerodynamic forces

$$F_A = \frac{1}{2}\rho\,C_D\,A_f\,v_x^2 = C_A\,v_x^2 \tag{2.47}$$

where ρ is the air density in $\left[kg/m^3\right]$, C_D is the drag coefficient, and A_f is the frontal area or projected area of vehicle in x-direction in $\left[\mathrm{m}^2\right]$. We may consider ρ, C_D, A_f to be constant for a vehicle; then, we combine the coefficients into a single aerodynamic coefficient C_A.

Example 60 Global force, wind.

Assume there exists a wind at the area the vehicle is moving. The wind force on the vehicle is an example of external force which is well expressed in the G-frame instead of B-frame. Figure 2.4 illustrates a vehicle facing a side wind at an angle θ with respect to the global X-axis.

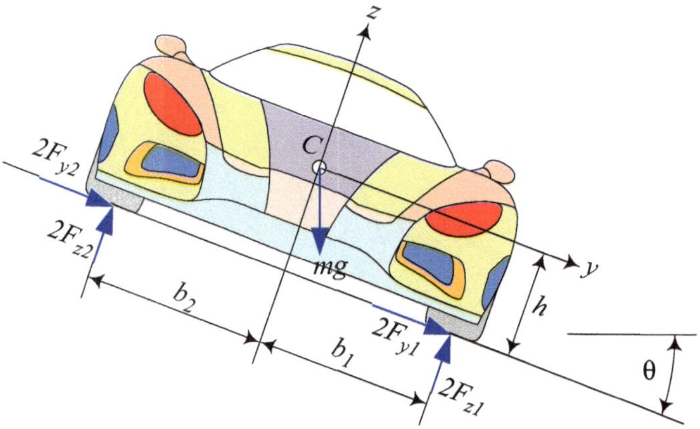

Fig. 2.5 A vehicle on a laterally tilted road to the angle $-\theta$ about the x-axis

$$\begin{aligned} {}^G\mathbf{F}_A &= C_A \left({}^G\mathbf{v} - \mathbf{v}_{wind}\right)^2 \qquad (2.48) \\ &= C_A \left({}^G R_B \begin{bmatrix} v_x \\ v_y \\ 0 \end{bmatrix} - \begin{bmatrix} v_{wind}\cos\theta \\ v_{wind}\sin\theta \\ 0 \end{bmatrix}\right)^2 \\ &= C_A \begin{bmatrix} \left(v_x\cos\psi - v_y\sin\psi - v_{wind}\cos\theta\right)^2 \\ \left(v_y\cos\psi + v_x\sin\psi - v_{wind}\sin\theta\right)^2 \\ 0 \end{bmatrix} \end{aligned}$$

To include the wind force into the vehicle equations of motion, we need to express ${}^G\mathbf{F}_A$ in B-frame.

$$\begin{aligned} {}^B\mathbf{F}_A &= {}^G R_B^T\, {}^G\mathbf{F}_A = C_A \left({}^B\mathbf{v} - {}^G R_B^T\, \mathbf{v}_{wind}\right)^2 \qquad (2.49) \\ &= C_A \begin{bmatrix} \left(v_x - v_{wind}\cos(\theta - \psi)\right)^2 \\ \left(v_y - v_{wind}\sin(\theta - \psi)\right)^2 \\ 0 \end{bmatrix} \end{aligned}$$

Example 61 Global force, gravitation.

Gravitational force is also an example of external forces that is well expressed in the G-frame instead of B-frame; however, it is easy to express it in the B-frame. Figure 2.5 illustrates a vehicle on a laterally tilted road to the angle $-\theta$ about the x-axis. Due to the tilting angle we have a lateral external force.

$$F_{y_{mg}} = mg\sin\theta \qquad (2.50)$$

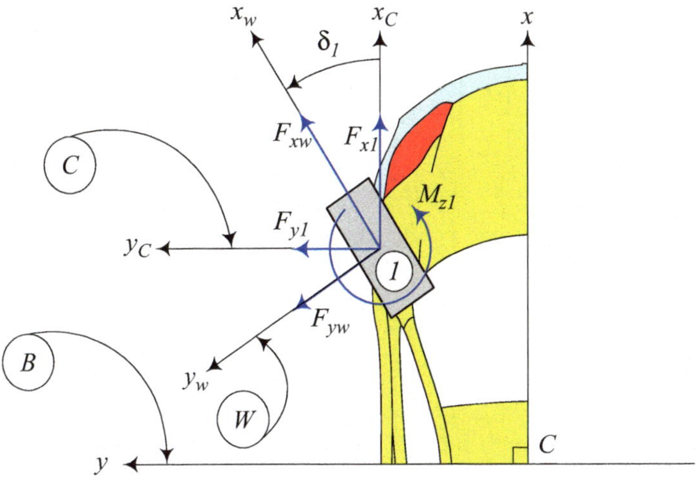

Fig. 2.6 The force system of tire number 1

Therefore the equation of motion of the vehicle on a laterally tilted road will be

$$m\,\dot{v}_x - mr\,v_y = F_x \tag{2.51}$$

$$\begin{aligned} m\,\dot{v}_y + mr\,v_x &= F_y + F_{y_{mg}} \\ &= F_y + mg\sin\theta \end{aligned} \tag{2.52}$$

Road angle about the y-axis provides us with similar equation in the x-component of the equations of motion.

2.2 Tire Force System

Figure 2.6 illustrates wheel number 1 of a vehicle and its force system $\left(F_{xW_1}, F_{yW_1}, M_{zW_1}\right)$ in the wheel coordinate frame W, as well as the force system $\left(F_{x_1}, F_{y_1}, M_{z_1}\right)$ in the wheel-body coordinate frame C (Jazar 2017).

If the force system at the tireprint in the tire coordinate frame T is:

$${}^T\mathbf{F}_w = \left[\begin{array}{ccc} F_{xT_i} & F_{yT_i} & F_{zT_i} \end{array}\right]^T \tag{2.53}$$

$${}^T\mathbf{M}_w = \left[\begin{array}{ccc} M_{xT_i} & M_{yT_i} & M_{zT_i} \end{array}\right]^T \tag{2.54}$$

then the force system at the center of the wheel in W-frame is:

$$ {}^{W}\mathbf{F}_{w} = \begin{bmatrix} F_{xW_i} \\ F_{yW_i} \\ F_{zW_i} \end{bmatrix} = \begin{bmatrix} F_{xT_i} \\ F_{yT_i} \cos\gamma_i + F_{zT_i} \sin\gamma_i \\ F_{zT_i} \cos\gamma_i - F_{yT_i} \sin\gamma_i \end{bmatrix} \tag{2.55} $$

$$ {}^{W}\mathbf{M}_{w} = \begin{bmatrix} M_{xW_i} \\ M_{yW_i} \\ M_{zW_i} \end{bmatrix} = \begin{bmatrix} M_{xT_i} + RF_{yT_i} \cos\gamma + RF_{zT_i} \sin\gamma \\ M_{yT_i} \cos\gamma - RF_{xT_i} + M_{zT_i} \sin\gamma \\ M_{zT_i} \cos\gamma - M_{yT_i} \sin\gamma \end{bmatrix} \tag{2.56} $$

where R is the tire radius. The force system at the center of the wheel number i in the wheel-body coordinate frame C is:

$$ {}^{C}\mathbf{F}_{w} = \begin{bmatrix} F_{x_i} \\ F_{y_i} \\ F_{z_i} \end{bmatrix} = \begin{bmatrix} F_{xW_i} \cos\delta_1 - F_{yW_i} \sin\delta_1 \\ F_{yW_i} \cos\delta_1 + F_{xW_i} \sin\delta_1 \\ F_{zW_i} \end{bmatrix} \tag{2.57} $$

$$ {}^{C}\mathbf{M}_{w} = \begin{bmatrix} M_{x_i} \\ M_{y_i} \\ M_{z_i} \end{bmatrix} = \begin{bmatrix} M_{xW_i} \cos\delta_1 - M_{yW_i} \sin\delta_1 \\ M_{yW_i} \cos\delta_1 + M_{xW_i} \sin\delta_1 \\ M_{zW_i} \end{bmatrix} \tag{2.58} $$

Therefore, the total planar force system on the vehicle in the body coordinate frame B is:

$$ {}^{B}F_{x} = \sum_i F_{x_i} = \sum_i F_{xW_i} \cos\delta_i - \sum_i F_{yW_i} \sin\delta_i \tag{2.59} $$

$$ {}^{B}F_{y} = \sum_i F_{y_i} = \sum_i F_{yW_i} \cos\delta_i + \sum_i F_{xW_i} \sin\delta_i \tag{2.60} $$

$$ {}^{B}M_{z} = \sum_i M_{z_i} + \sum_i x_i F_{y_i} - \sum_i y_i F_{x_i} \tag{2.61} $$

Proof There exists a tire coordinate frame T at the center of the tireprint at the intersection of tire-plane and the ground. The z_T-axis is always perpendicular to the ground and upward. The T-frame does not follow the spin and camber rotations of the tire; however, it follows the steer angle rotation about the z_T-axis. The W-frame that is attached to the center of the wheel follows every motion of the wheel except the spin. The C-frame at the center of neutral wheel is parallel to the body coordinate frame B. When the wheel is at the rest position, then W-frame and C-frame become coincident. The C-frame is motionless with respect to the vehicle and does not follow any motion of the wheel. The vehicle body coordinate frame $B\,(x, y, z)$ is attached to the vehicle at its mass center. The wheel force system is generated in T-frame and must be transformed to the C-frame and then B-frame to develop the vehicle equities of motion.

Let us assume that the force system in the T-frame at the tireprint of the wheel number i is

$$ {}^{T}\mathbf{F}_{w}=\begin{bmatrix} F_{x_{T_i}} & F_{y_{T_i}} & F_{z_{T_i}} \end{bmatrix}^{T} \tag{2.62}$$

$$ {}^{T}\mathbf{M}_{w}=\begin{bmatrix} M_{x_{T_i}} & M_{y_{T_i}} & M_{z_{T_i}} \end{bmatrix}^{T} \tag{2.63}$$

then the force system in the W-frame at the center of the wheel will be

$$ {}^{W}\mathbf{F}_{w}={}^{W}R_{T}\,{}^{T}\mathbf{F}_{w}=\begin{bmatrix} F_{x_{T_i}} \\ F_{y_{T_i}}\cos\gamma_i+F_{z_{T_i}}\sin\gamma_i \\ F_{z_{T_i}}\cos\gamma_i-F_{y_{T_i}}\sin\gamma_i \end{bmatrix}=\begin{bmatrix} F_{x_{W_i}} \\ F_{y_{W_i}} \\ F_{z_{W_i}} \end{bmatrix} \tag{2.64}$$

$$\begin{aligned} {}^{W}\mathbf{M}_{w} &= {}^{W}R_{T}\,{}^{T}\mathbf{M}_{w}+{}^{W}\mathbf{R}\times{}^{W}_{T}\mathbf{F}_{w} \\ &= {}^{W}R_{T}\,{}^{T}\mathbf{M}_{w}+(-R)\,{}^{W}\hat{k}\times{}^{W}R_{T}\,{}^{T}\mathbf{F}_{w} \\ &= {}^{W}R_{T}\begin{bmatrix} M_{x_{T_i}} \\ M_{y_{T_i}} \\ M_{z_{T_i}} \end{bmatrix}+\begin{bmatrix} 0 \\ 0 \\ -R \end{bmatrix}\times{}^{W}R_{T}\begin{bmatrix} F_{x_{T_i}} \\ F_{y_{T_i}} \\ F_{z_{T_i}} \end{bmatrix} \\ &= \begin{bmatrix} M_{x_{T_i}}+RF_{y_{T_i}}\cos\gamma+RF_{z_{T_i}}\sin\gamma \\ M_{y_{T_i}}\cos\gamma-RF_{x_{T_i}}+M_{z_{T_i}}\sin\gamma \\ M_{z_{T_i}}\cos\gamma-M_{y_{T_i}}\sin\gamma \end{bmatrix}=\begin{bmatrix} M_{x_{W_i}} \\ M_{y_{W_i}} \\ M_{z_{W_i}} \end{bmatrix} \end{aligned} \tag{2.65}$$

where ${}^{W}\mathbf{R}$ is the position vector of the T-frame in W-frame which is equal to radius of the wheel.

$$ {}^{W}\mathbf{R}=-R\,{}^{W}\hat{k} \tag{2.66}$$

The transformation ${}^{W}R_{T}$ from T-frame in W-frame is:

$$ {}^{W}R_{T}=\begin{bmatrix} 1 & 0 & 0 \\ 0 & \cos\gamma & \sin\gamma \\ 0 & -\sin\gamma & \cos\gamma \end{bmatrix} \tag{2.67}$$

Assuming W-frame and C-frame have a common origin, the transformation matrix between the W-frame and the C-frame is

$$ {}^{C}R_{W}=\begin{bmatrix} \cos\delta_1 & -\sin\delta_1 & 0 \\ \sin\delta_1 & \cos\delta_1 & 0 \\ 0 & 0 & 1 \end{bmatrix} \tag{2.68}$$

and therefore, the force system at the center of the wheel, parallel to the vehicle coordinate frame, is

$$^{C}\mathbf{F}_{w} = {}^{C}R_{W}\,{}^{W}\mathbf{F}_{w} \tag{2.69}$$

$$\begin{bmatrix} F_{x_1} \\ F_{y_1} \\ F_{z_1} \end{bmatrix} = \begin{bmatrix} \cos\delta_1 & -\sin\delta_1 & 0 \\ \sin\delta_1 & \cos\delta_1 & 0 \\ 0 & 0 & 1 \end{bmatrix} \begin{bmatrix} F_{xW_i} \\ F_{yW_i} \\ F_{zW_i} \end{bmatrix}$$

$$= \begin{bmatrix} F_{xW_i}\cos\delta_1 - F_{yW_i}\sin\delta_1 \\ F_{yW_i}\cos\delta_1 + F_{xW_i}\sin\delta_1 \\ F_{zW_i} \end{bmatrix} \tag{2.70}$$

$$^{C}\mathbf{M}_{w} = {}^{C}R_{W}\,{}^{W}\mathbf{M}_{w}$$

$$\begin{bmatrix} M_{x_1} \\ M_{y_1} \\ M_{z_1} \end{bmatrix} = \begin{bmatrix} \cos\delta_1 & -\sin\delta_1 & 0 \\ \sin\delta_1 & \cos\delta_1 & 0 \\ 0 & 0 & 1 \end{bmatrix} \begin{bmatrix} M_{xW_i} \\ M_{yW_i} \\ M_{zW_i} \end{bmatrix}$$

$$= \begin{bmatrix} M_{xW_i}\cos\delta_1 - M_{yW_i}\sin\delta_1 \\ M_{yW_i}\cos\delta_1 + M_{xW_i}\sin\delta_1 \\ M_{zW_i} \end{bmatrix} \tag{2.71}$$

Transforming the force system of each tire to the body coordinate frame B, located at the body mass center C, generates the total force system applied on the vehicle

$$^{B}\mathbf{F} = \sum_i {}^{C}\mathbf{F}_{w} = \sum_i F_{x_i}\,\hat{\imath} + \sum_i F_{y_i}\,\hat{\jmath} + \sum_i F_{z_i}\,\hat{k} \tag{2.72}$$

$$^{B}\mathbf{M} = \sum_i {}^{C}\mathbf{M}_{w} + \sum_i {}^{B}\mathbf{r}_i \times {}^{B}\mathbf{F}_{w_i} \tag{2.73}$$

$$= \sum_i M_{x_i}\,\hat{\imath} + \sum_i M_{y_i}\,\hat{\jmath} + \sum_i M_{z_i}\,\hat{k} + \sum_i \begin{bmatrix} y_i F_{z_i} - z_i F_{y_i} \\ z_i F_{x_i} - x_i F_{z_i} \\ x_i F_{y_i} - y_i F_{x_i} \end{bmatrix}$$

where ${}^{B}\mathbf{r}_i$ is the position vector of the wheel number i.

$$^{B}\mathbf{r}_i = \begin{bmatrix} x_i & y_i & z_i \end{bmatrix}^T = \begin{bmatrix} a_i & b_i & z_i \end{bmatrix}^T \tag{2.74}$$

$$z_i \simeq 0 \tag{2.75}$$

Expanding Eqs. (2.72) and (2.73) and assuming that

$$ {}^{T}\mathbf{F}_w \simeq \begin{bmatrix} F_{xT_i} & F_{yT_i} & 0 \end{bmatrix}^T \tag{2.76} $$

$$ {}^{T}\mathbf{M}_w \simeq \begin{bmatrix} 0 & 0 & M_{zT_i} \end{bmatrix}^T \tag{2.77} $$

provide us with the total planar force system.

$$ {}^{B}F_x = \sum_i F_{xW_i} \cos\delta_i - \sum_i F_{yW_i} \sin\delta_i \tag{2.78} $$

$$ {}^{B}F_y = \sum_i F_{yW_i} \cos\delta_i + \sum_i F_{xW_i} \sin\delta_i \tag{2.79} $$

$$ {}^{B}M_z = \sum_i M_{z_i} + \sum_i x_i F_{y_i} - \sum_i y_i F_{x_i} \tag{2.80} $$

■

Example 62 Tire force system in W-frame.

If the force system at the tireprint is ${}^{T}\mathbf{F}_w$ and ${}^{T}\mathbf{M}_w = M_{zT_i}\,{}^{T}\hat{k}$, then the force system in the W-frame at the center of the wheel would be

$$ {}^{W}\mathbf{F}_w = {}^{W}R_T\,{}^{T}\mathbf{F}_w = {}^{T}R_W^T\,{}^{T}\mathbf{F}_w $$

$$ \begin{bmatrix} F_{x_{w_i}} \\ F_{y_{w_i}} \\ F_{z_{w_i}} \end{bmatrix} = \begin{bmatrix} 1 & 0 & 0 \\ 0 & \cos\gamma & -\sin\gamma \\ 0 & \sin\gamma & \cos\gamma \end{bmatrix}^T \begin{bmatrix} F_{xT_i} \\ F_{yT_i} \\ F_{zT_i} \end{bmatrix} = \begin{bmatrix} F_{xT_i} \\ F_{yT_i} \cos\gamma + F_{zT_i} \sin\gamma \\ F_{zT_i} \cos\gamma - F_{yT_i} \sin\gamma \end{bmatrix} \tag{2.81} $$

$$ {}^{W}\mathbf{M}_w = {}^{T}R_W^T \left({}^{T}\mathbf{M}_w - {}^{T}\mathbf{r}_o \times {}^{T}\mathbf{F}_w \right) = \begin{bmatrix} RF_{y_{w_i}} \cos\gamma + RF_{z_{w_i}} \sin\gamma \\ M_{z_{w_i}} \sin\gamma - RF_{x_{w_i}} \\ M_{z_{w_i}} \cos\gamma \end{bmatrix} \tag{2.82} $$

where

$$
{}^{T}\mathbf{r}_{o} = \begin{bmatrix} 0 \\ -R\sin\gamma \\ R\cos\gamma \end{bmatrix} \qquad {}^{T}\mathbf{M}_{w} = \begin{bmatrix} 0 \\ 0 \\ M_{z_{w_i}} \end{bmatrix} \tag{2.83}
$$

The wheel force system at zero camber, $\gamma = 0$, reduces to:

$$
{}^{W}\mathbf{F}_{w} = \begin{bmatrix} F_{x_{w_i}} \\ F_{y_{w_i}} \\ F_{z_{w_i}} \end{bmatrix} \qquad {}^{W}\mathbf{M}_{w} = \begin{bmatrix} RF_{y_{w_i}} \\ -RF_{x_{w_i}} \\ M_{z_{w_i}} \end{bmatrix} \tag{2.84}
$$

Example 63 Full force in C-frame.

Considering force $\mathbf{F}$ under the tire number i in the T-frame is:

$$
{}^{T}\mathbf{F}_{w} = \begin{bmatrix} F_{xT_i} & F_{yT_i} & F_{zT_i} \end{bmatrix}^{T} \tag{2.85}
$$

then the force in W-frame would be

$$
{}^{W}\mathbf{F}_{w} = {}^{W}R_{T}\,{}^{T}\mathbf{F}_{w} = \begin{bmatrix} F_{xT_i} \\ F_{yT_i}\cos\gamma_i + F_{zT_i}\sin\gamma_i \\ F_{zT_i}\cos\gamma_i - F_{yT_i}\sin\gamma_i \end{bmatrix} \tag{2.86}
$$

and the force in C-frame would be

$$
{}^{C}\mathbf{F}_{w} = {}^{C}R_{W}\,{}^{W}\mathbf{F}_{w} \tag{2.87}
$$

$$
\begin{bmatrix} F_{x_1} \\ F_{y_1} \\ F_{z_1} \end{bmatrix} = \begin{bmatrix} F_{xT_i}\cos\delta_1 - \left(F_{yT_i}\cos\gamma_i + F_{zT_i}\sin\gamma_i\right)\sin\delta_1 \\ F_{xT_i}\sin\delta_1 + \left(F_{yT_i}\cos\gamma_i + F_{zT_i}\sin\gamma_i\right)\cos\delta_1 \\ F_{zT_i}\cos\gamma_i - F_{yT_i}\sin\gamma_i \end{bmatrix}
$$

The moment $\mathbf{M}$ under the tire number i in the T-frame is:

$$
{}^{T}\mathbf{M}_{w} = \begin{bmatrix} M_{xT_i} & M_{yT_i} & M_{zT_i} \end{bmatrix}^{T} \tag{2.88}
$$

Therefore, the moment in W-frame is

$$
\begin{aligned}
{}^{W}\mathbf{M}_{w} &= {}^{W}R_{T}\,{}^{T}\mathbf{M}_{w} + {}^{W}\mathbf{R} \times {}^{W}_{T}\mathbf{F}_{w} \\
&= {}^{W}R_{T}\,{}^{T}\mathbf{M}_{w} - R\,{}^{W}\hat{k} \times {}^{W}R_{T}\,{}^{T}\mathbf{F}_{w} \\
&= \begin{bmatrix} M_{xT_i} + RF_{yT_i}\cos\gamma + RF_{zT_i}\sin\gamma \\ M_{yT_i}\cos\gamma - RF_{xT_i} + M_{zT_i}\sin\gamma \\ M_{zT_i}\cos\gamma - M_{yT_i}\sin\gamma \end{bmatrix}
\end{aligned} \tag{2.89}
$$

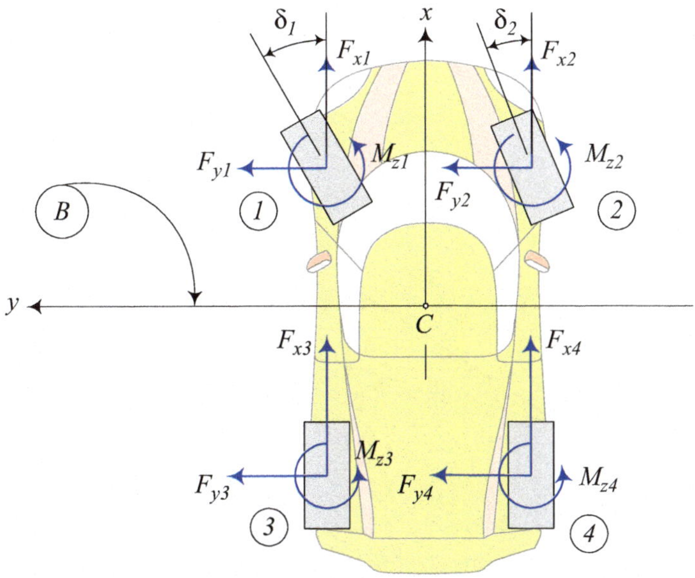

Fig. 2.7 The acting forces at the wheel center of a front-wheel-steering four-wheel vehicle

and the moment in C-frame would be

$$
\begin{aligned}
{}^{C}\mathbf{M}_w &= {}^{C}R_W\,{}^{W}\mathbf{M}_w \\
&= \begin{bmatrix}
\left(M_{xT_i} + RF_{yT_i}\cos\gamma + RF_{zT_i}\sin\gamma\right)\cos\delta_1 \\
\qquad -\left(M_{yT_i}\cos\gamma - RF_{xT_i} + M_{zT_i}\sin\gamma\right)\sin\delta_1 \\
\left(M_{xT_i} + RF_{yT_i}\cos\gamma + RF_{zT_i}\sin\gamma\right)\sin\delta_1 \\
\qquad +\left(M_{yT_i}\cos\gamma - RF_{xT_i} + M_{zT_i}\sin\gamma\right)\cos\delta_1 \\
M_{zT_i}\cos\gamma - M_{yT_i}\sin\gamma
\end{bmatrix}
\end{aligned}
\tag{2.90}
$$

2.3 Bicycle Planar Vehicle Force Components

Figure 2.7 illustrates the force system of each wheel in the wheel-body coordinate frame C. The forces are acting at the wheel center of a front-wheel-steering vehicle. When we ignore the roll and pitch motions of the vehicle, then the body z and global Z axes are parallel and the xy-plane remains parallel to the road's XY-plane.

Ignoring the roll motion as well as the lateral load transfer between left and right wheels, we may define a simplified *two-wheel model* for the vehicle. Figure 2.8

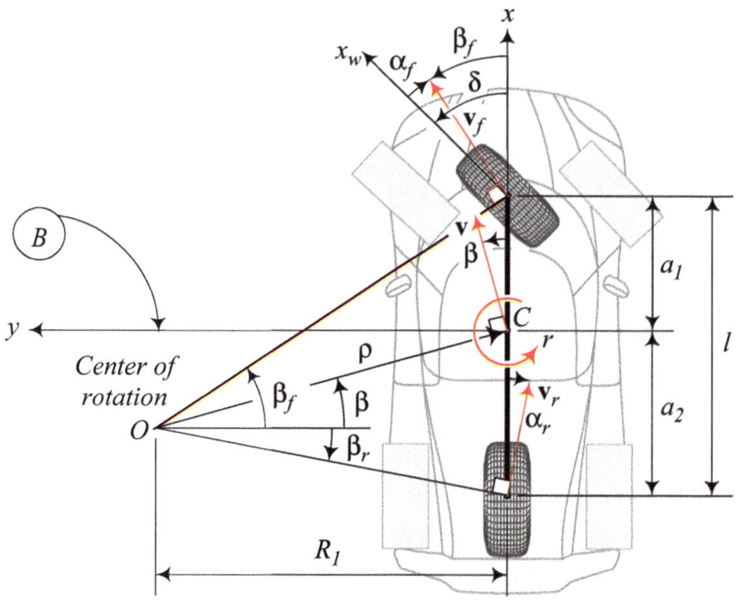

Fig. 2.8 A two-wheel model for a vehicle moving with no roll

illustrates a two-wheel model for a vehicle with no roll motion. The two-wheel model is also called the *bicycle model*, although a two-wheel vehicle model does not act similar to a traditional bicycle.

The force system applied on a bicycle model of vehicle at its mass center C is:

$$F_x = F_{x_f} \cos\delta + F_{x_r} - F_{y_f} \sin\delta \tag{2.91}$$

$$F_y = F_{y_f} \cos\delta + F_{y_r} + F_{x_f} \sin\delta \tag{2.92}$$

$$M_z = a_1 F_{y_f} \cos\delta + a_1 F_{x_f} \sin\delta - a_2 F_{y_r} \tag{2.93}$$

$$T_f = I_f\, \dot{\omega}_f + R_w\, F_{x_f} \tag{2.94}$$

$$T_r = I_r\, \dot{\omega}_r + R_w\, F_{x_r} \tag{2.95}$$

where tire forces based on elliptic combined tire forces are:

$$F_{x_f} = F_{z_f} C_{sf}\, S\left(s_f - s_s\right) \sqrt{1 - C_{s\alpha} \left(\frac{S\left(\alpha_f - \alpha_s\right)}{\alpha_s}\right)^2} \tag{2.96}$$

$$F_{x_r} = F_{z_r} C_{sr}\, S\left(s_r - s_s\right) \sqrt{1 - C_{s\alpha} \left(\frac{S\left(\alpha_r - \alpha_s\right)}{\alpha_s}\right)^2} \tag{2.97}$$

$$F_{y_f} = -F_{z_f} C_{\alpha f}\, S\left(\alpha_f - \alpha_s\right) \sqrt{1 - C_{\alpha s}\left(\frac{S\left(s_f - s_s\right)}{s_s}\right)^2} \tag{2.98}$$

$$F_{y_r} = -F_{z_r} C_{\alpha r}\, S\left(\alpha_r - \alpha_s\right) \sqrt{1 - C_{\alpha s}\left(\frac{S\left(s_r - s_s\right)}{s_s}\right)^2} \tag{2.99}$$

where S is the saturation function (1.60).

$$S(x - x_0) = \begin{cases} x_0 & x_0 < x \\ x & -x_0 < x < x_0 \\ -x_0 & x < -x_0 \end{cases} \tag{2.100}$$

The tire force characteristics are:

$$C_{\alpha f} = C_{\alpha_1} = C_{\alpha_2} \tag{2.101}$$
$$C_{\alpha r} = C_{\alpha_3} = C_{\alpha_4} \tag{2.102}$$
$$C_{sf} = C_{s_1} = C_{s_2} \tag{2.103}$$
$$C_{sr} = C_{s_3} = C_{s_4} \tag{2.104}$$

$$F_{z_f} = F_{z_1} + F_{z_2} = mg\frac{a_2}{l} - m\left(\dot{v}_x - r\, v_y\right)\frac{h}{l} \tag{2.105}$$
$$F_{z_r} = F_{z_3} + F_{z_4} = mg\frac{a_1}{l} + m\left(\dot{v}_x - r\, v_y\right)\frac{h}{l} \tag{2.106}$$
$$l = a_1 + a_2 \tag{2.107}$$

$$\alpha_f = \arctan\left(\frac{v_y}{v_x} + \frac{a_1}{v_x} r\right) - \delta \tag{2.108}$$
$$\alpha_r = \arctan\left(\frac{v_y}{v_x} - \frac{a_2}{v_x} r\right) \tag{2.109}$$
$$\beta = \arctan\frac{v_y}{v_x} \tag{2.110}$$

$$s_f = \frac{R_g\, \omega_f - v_{xT_f}}{R_g\, \omega_f\, H(R_g\, \omega_f - v_{xT_f}) + v_{xT_f}\, H(v_{xT_f} - R_g\, \omega_f)} \tag{2.111}$$

$$s_r = \frac{R_g\, \omega_r - v_x}{R_g\, \omega_r\, H(R_g\, \omega_r - v_x) + v_x\, H(v_x - R_g\, \omega_r)} \tag{2.112}$$

$$v_{xTf} = v_x \cos\delta + \left(v_y + a_1 r\right)\sin\delta \tag{2.113}$$

$$\dot{\omega}_f = \frac{T_f - R_{wf} F_{x_f}}{I_f} \tag{2.114}$$

$$\dot{\omega}_r = \frac{T_r - R_{wr} F_{x_r}}{I_r} \tag{2.115}$$

The T_f and T_r are the applied torques on front and rear wheels, R_{wf} and R_{wr} are equivalent tire radii of front and rear tires, and I_f and I_r are the front and rear wheels' mass moments about their spin axes. The forces $\left(F_{x_f}, F_{y_f}\right)$ and $\left(F_{x_r}, F_{y_r}\right)$ are the planar forces at the tireprint of the front and rear wheels and we consider them to be at the wheel center.

Proof For the bicycle vehicle model, we use the cot-average δ of the outer δ_o and inner δ_i steer angles as the only steer angle for the model

$$\cot\delta = \frac{\cot\delta_o + \cot\delta_i}{2} \tag{2.116}$$

and we define sideslip coefficients $C_{\alpha f}$ and $C_{\alpha r}$ as well as slip ratio coefficients C_{sf} and C_{sr} for the front and rear tires.

$$C_{\alpha f} = \left(C_{\alpha_1} + C_{\alpha_2}\right)/2 \tag{2.117}$$

$$C_{\alpha r} = \left(C_{\alpha_3} + C_{\alpha_4}\right)/2 \tag{2.118}$$

$$C_{sf} = \left(C_{s_1} + C_{s_2}\right)/2 \tag{2.119}$$

$$C_{sr} = \left(C_{s_3} + C_{s_4}\right)/2 \tag{2.120}$$

Assuming the left and right tires are identical, then we have:

$$C_{\alpha f} = C_{\alpha_1} = C_{\alpha_2} \tag{2.121}$$

$$C_{\alpha r} = C_{\alpha_3} = C_{\alpha_4} \tag{2.122}$$

$$C_{sf} = C_{s_1} = C_{s_2} \tag{2.123}$$

$$C_{sr} = C_{s_3} = C_{s_4} \tag{2.124}$$

Assuming similar force components to be equal on the left and right tires and adding them up, we will have longitudinal and lateral forces $\left(F_{x_f}, F_{y_f}\right)$ and $\left(F_{x_r}, F_{y_r}\right)$ applied on front and rear tires.

$$F_{x_f} = F_{x_1} + F_{x_2} \qquad F_{x_1} = F_{x_2} \tag{2.125}$$

$$F_{x_r} = F_{x_3} + F_{x_4} \qquad F_{x_3} = F_{x_4} \tag{2.126}$$

$$F_{y_f} = F_{y_1} + F_{y_2} \qquad F_{y_1} = F_{y_2} \tag{2.127}$$

$$F_{y_r} = F_{y_3} + F_{y_4} \qquad F_{y_3} = F_{y_4} \tag{2.128}$$

$$F_{z_f} = F_{z_1} + F_{z_2} \qquad F_{z_1} = F_{z_2} \tag{2.129}$$

$$F_{z_r} = F_{z_3} + F_{z_4} \qquad F_{z_3} = F_{z_4} \tag{2.130}$$

Employing the elliptical combined tire force models (1.202) and (1.203), and proportional-saturation tire force behaviors (1.67) and (1.152), the longitudinal and lateral forces on front and rear wheels will be

$$\begin{aligned} F_{x_f} &= F_{x_1} + F_{x_2} \\ &= F_{z_f} C_{sf}\, S\left(s_f - s_s\right) \sqrt{1 - C_{s\alpha} \left(\frac{S\left(\alpha_f - \alpha_s\right)}{\alpha_s} \right)^2} \end{aligned} \tag{2.131}$$

$$\begin{aligned} F_{x_r} &= F_{x_3} + F_{x_4} \\ &= F_{z_r} C_{sr}\, S\left(s_r - s_s\right) \sqrt{1 - C_{s\alpha} \left(\frac{S\left(\alpha_r - \alpha_s\right)}{\alpha_s} \right)^2} \end{aligned} \tag{2.132}$$

$$\begin{aligned} F_{y_f} &= F_{y_1} + F_{y_2} \\ &= -F_{z_f} C_{\alpha f}\, S\left(\alpha_f - \alpha_s\right) \sqrt{1 - C_{\alpha s} \left(\frac{S\left(s_f - s_s\right)}{s_s} \right)^2} \end{aligned} \tag{2.133}$$

$$\begin{aligned} F_{y_r} &= F_{y_3} + F_{y_4} \\ &= -F_{z_r} C_{\alpha r}\, S\left(\alpha_r - \alpha_s\right) \sqrt{1 - C_{\alpha s} \left(\frac{S\left(s_r - s_s\right)}{s_s} \right)^2} \end{aligned} \tag{2.134}$$

Using Eqs. (2.59)–(2.61) and ignoring the aligning moments M_{z_i}, the applied forces on the two-wheel vehicle are:

$$F_x = F_{x_f} \cos\delta - F_{y_f} \sin\delta + F_{x_r} \tag{2.135}$$

$$F_y = F_{y_f} \cos\delta + F_{x_f} \sin\delta + F_{y_r} \tag{2.136}$$

$$M_z = a_1 F_{y_f} \cos\delta + a_1 F_{x_f} \sin\delta - a_2 F_{y_r} \tag{2.137}$$

Assuming

$$F_{z_1} = F_{z_2} \qquad F_{z_3} = F_{z_4} \tag{2.138}$$

and using the weight transfer equations in body frame (1.462)–(1.465), the vertical load on front and rear tires is:

$$F_{z_f} = F_{z_1} + F_{z_2} = m\left(\frac{a_2}{l}g - \frac{h}{l}\left(\dot{v}_x - r\,v_y\right)\right) \tag{2.139}$$

$$F_{z_r} = F_{z_3} + F_{z_4} = m\left(\frac{a_1}{l}g + \frac{h}{l}\left(\dot{v}_x - r\,v_y\right)\right) \tag{2.140}$$

$$l = a_1 + a_2 \tag{2.141}$$

If we assume δ to be small, then the force equations may be approximated by the following equations.

$$F_x \approx F_{x_f} + F_{x_r} \tag{2.142}$$

$$F_y \approx F_{y_f} + F_{y_r} \tag{2.143}$$

$$M_z \approx a_1 F_{y_f} - a_2 F_{y_r} \tag{2.144}$$

Assume the wheel number i of the vehicle is located at (x_i, y_i) in the body coordinate frame B. The velocity of the wheel number i in the B-frame is

$${}^B\mathbf{v}_i = {}^B\mathbf{v} + {}^B_G\boldsymbol{\omega}_B \times {}^B\mathbf{r}_i \tag{2.145}$$

$${}^B_G\boldsymbol{\omega}_B = \omega_z\,{}^B\hat{k} = r\,{}^B\hat{k} = \dot{\psi}\,{}^B\hat{k} \tag{2.146}$$

in which ${}^B\mathbf{r}_i$ is the position vector of the wheel number i,

$${}^B\mathbf{r}_1 = \begin{bmatrix} a_1 & 0 & 0 \end{bmatrix}^T \tag{2.147}$$

$${}^B\mathbf{r}_2 = \begin{bmatrix} -a_2 & 0 & 0 \end{bmatrix}^T \tag{2.148}$$

${}^B\mathbf{v}$ is the velocity vector of the vehicle at its mass center C, and $\omega_z = r$ is the yaw rate of the vehicle as shown in Fig. 2.8. Expanding Eq. (2.145) provides the following velocity vector for the wheel number i expressed in the B-frame at C.

$${}^B\mathbf{v}_i = \begin{bmatrix} v_{x_i} \\ v_{y_i} \\ 0 \end{bmatrix} = \begin{bmatrix} v_x \\ v_y \\ 0 \end{bmatrix} + \begin{bmatrix} 0 \\ 0 \\ r \end{bmatrix} \times \begin{bmatrix} x_i \\ 0 \\ 0 \end{bmatrix} = \begin{bmatrix} v_x \\ v_y + r x_i \\ 0 \end{bmatrix} \tag{2.149}$$

The wheel-body sideslip β_i for the wheel i is the angle between the vehicle body x-axis and the wheel velocity vector $\mathbf{v}_i$.

$$\beta_i = \arctan\left(\frac{v_{y_i}}{v_{x_i}}\right) = \arctan\left(\frac{v_y + x_i\,r}{v_x}\right) \tag{2.150}$$

The wheel sideslip angles β_i for the front and rear wheels of a two-wheel vehicle model, β_f and β_r, are

$$\beta_f = \arctan\left(\frac{v_{y_f}}{v_{x_f}}\right) = \arctan\left(\frac{v_y + a_1 r}{v_x}\right) \tag{2.151}$$

$$\beta_r = \arctan\left(\frac{v_{y_r}}{v_{x_r}}\right) = \arctan\left(\frac{v_y - a_2 r}{v_x}\right) \tag{2.152}$$

and the vehicle sideslip angle β is

$$\beta = \arctan\left(\frac{v_y}{v_x}\right) \tag{2.153}$$

Having the steer angle δ of the front wheel, the tire sideslip angles will be

$$\alpha_f = \beta_f - \delta = \arctan\left(\frac{v_y + a_1 r}{v_x}\right) - \delta \tag{2.154}$$

$$\alpha_r = \beta_r = \arctan\left(\frac{v_y - a_2 r}{v_x}\right) \tag{2.155}$$

Assuming small angles for wheel and vehicle sideslip β_f, β_r, and β, the tire sideslip angles for the front and rear wheels, α_f and α_r, may be approximated as

$$\alpha_f = \beta_f - \delta \simeq \frac{1}{v_x}\left(v_y + a_1 r\right) - \delta = \beta + \frac{a_1}{v_x} r - \delta \tag{2.156}$$

$$\alpha_r = \beta_r \simeq \frac{1}{v_x}\left(v_y - a_2 r\right) = \beta - \frac{a_2}{v_x} r \tag{2.157}$$

Also we substitute s from (1.57),

$$s = \frac{R_g\,\omega_w - v_x}{R_g\,\omega_w\,H(R_g\,\omega_w - v_x) + v_x\,H(v_x - R_g\,\omega_w)} \tag{2.158}$$

however, the slip ratio of the front tire need to be adapted as the steer angle will change velocity of the tire in the x_T-direction as is shown in Fig. 1.13. The velocity of the front wheel center in the x-direction of its local C-frame is

$$v_{x_{Tf}} = v_x \cos\delta + \left(v_y + a_1 r\right)\sin\delta$$

Therefore, the longitudinal slip ratios of the front and rear tire are:

$$s_f = \frac{R_g\,\omega_f - v_{x_{Tf}}}{R_g\,\omega_f\,H(R_g\,\omega_f - v_{x_{Tf}}) + v_{x_{Tf}}\,H(v_{x_{Tf}} - R_g\,\omega_f)} \tag{2.159}$$

$$s_r = \frac{R_g\,\omega_r - v_x}{R_g\,\omega_r\,H(R_g\,\omega_r - v_x) + v_x\,H(v_x - R_g\,\omega_r)} \tag{2.160}$$

To derive Eq. (2.93), we calculate the resultant yaw moment M_z because of tire forces.

$$\begin{aligned} {}^B\mathbf{M} &= \sum_{i=1}^{2} \mathbf{r}_i \times {}^B\mathbf{F}_i = \sum_{i=1}^{2} \mathbf{r}_i \times {}^C R_T{}^T\mathbf{F}_i \\ &= \begin{bmatrix} a_1 \\ 0 \\ 0 \end{bmatrix} \times {}^C R_T \begin{bmatrix} F_{x_f} \\ F_{y_f} \\ 0 \end{bmatrix} + \begin{bmatrix} -a_2 \\ 0 \\ 0 \end{bmatrix} \times \begin{bmatrix} F_{x_r} \\ F_{y_r} \\ 0 \end{bmatrix} \\ &= \begin{bmatrix} 0 \\ 0 \\ a_1\left(F_{y_f}\cos\delta + F_{x_f}\sin\delta\right) - a_2 F_{y_r} \end{bmatrix} \end{aligned} \tag{2.161}$$

where ${}^C R_T$ for the front wheel is

$$ {}^C R_T = \begin{bmatrix} \cos\delta & -\sin\delta & 0 \\ \sin\delta & \cos\delta & 0 \\ 0 & 0 & 1 \end{bmatrix} \tag{2.162}$$

and ${}^C R_T$ is an identity matrix for the rear ${}^C R_T = [\mathbf{I}]$ because of $\delta_r = 0$. Therefore, the force equations (2.91)–(2.93) are completed. ■

Example 64 Kinematic steering of a two-wheel vehicle.

For the two-wheel vehicle shown in Fig. 2.9, we use the cot-average (2.116) of the outer and inner steer angles as the input steer angle,

$$\cot\delta = \frac{\cot\delta_o + \cot\delta_i}{2} \tag{2.163}$$

Using the geometry of the vehicle of Fig. 2.9 and assuming same track in front and rear, $w = w_f = w_r$, we have

$$\tan\delta_i = \frac{l}{R_1 - \frac{w}{2}} \qquad \tan\delta_o = \frac{l}{R_1 + \frac{w}{2}} \tag{2.164}$$

The radius of rotation ρ for the two-wheel vehicle will be

$$\rho = \sqrt{a_2^2 + l^2\cot^2\delta} \tag{2.165}$$

when the vehicle is moving very slowly such that the rotation center remains on the axis of the rear axle (Fenton 1996; Karnopp 2013).

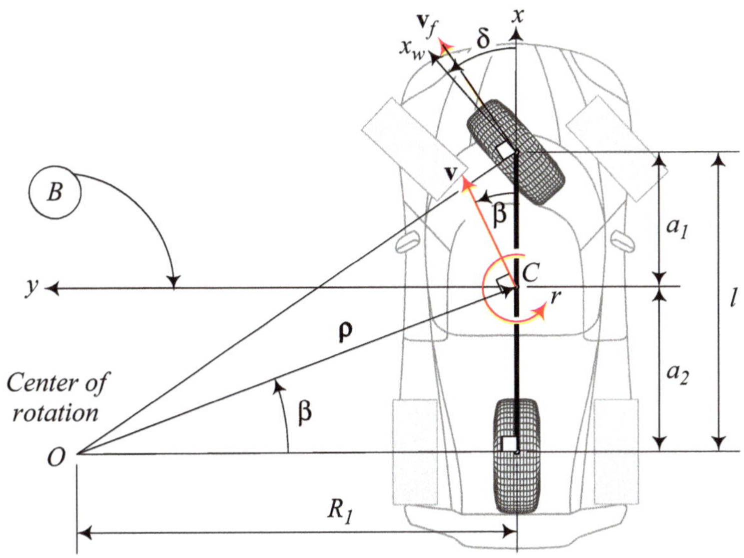

Fig. 2.9 Steer angle and rotation center of a two-wheel model for a vehicle moving with no roll at low speed

Example 65 Unsaturated force system.

Let us assume δ is very small steer angle and

$$s < s_s \qquad \alpha < \alpha_s \tag{2.166}$$

so the tire forces never reach their limit of saturation. In this case, the force system of the vehicle bicycle model will be simplified to:

$$F_x = F_{x_f} + F_{x_r} \tag{2.167}$$

$$F_y = F_{y_f} + F_{y_r} \tag{2.168}$$

$$M_z = a_1 F_{y_f} - a_2 F_{y_r} \tag{2.169}$$

$$F_{x_f} = F_{z_f} C_{sf}\, s_f \sqrt{1 - C_{s\alpha}\left(\frac{\alpha_f}{\alpha_s}\right)^2} \tag{2.170}$$

$$F_{x_r} = F_{z_r} C_{sr}\, s_r \sqrt{1 - C_{s\alpha}\left(\frac{\alpha_r}{\alpha_s}\right)^2} \tag{2.171}$$

$$F_{y_f} = -F_{z_f} C_{\alpha f}\, \alpha_f \sqrt{1 - C_{\alpha s}\left(\frac{s_f}{s_s}\right)^2} \tag{2.172}$$

$$F_{y_r} = -F_{z_r} C_{\alpha r}\, \alpha_r \sqrt{1 - C_{\alpha s}\left(\frac{s_r}{s_s}\right)^2} \tag{2.173}$$

$$F_{z_f} = F_{z_1} + F_{z_2} = mg\frac{a_2}{l} - m\left(\dot{v}_x - r\, v_y\right)\frac{h}{l} \tag{2.174}$$

$$F_{z_r} = F_{z_3} + F_{z_4} = mg\frac{a_1}{l} + m\left(\dot{v}_x - r\, v_y\right)\frac{h}{l} \tag{2.175}$$

$$l = a_1 + a_2 \tag{2.176}$$

$$\alpha_f = \beta + \frac{a_1}{v_x} r - \delta \qquad \alpha_r = \beta - \frac{a_2}{v_x} r \tag{2.177}$$

$$s_f = \frac{R_g\, \omega_f - v_x}{R_g\, \omega_f} \qquad s_r = \frac{R_g\, \omega_r - v_x}{R_g\, \omega_r} \tag{2.178}$$

$$\dot{\omega}_f = \frac{T_f - R_{wf} F_{x_f}}{I_f} \qquad \dot{\omega}_r = \frac{T_r - R_{wr} F_{x_r}}{I_r} \tag{2.179}$$

The vehicle assumed to be in acceleration.

Example 66 Four-wheel rigid planar vehicle.

The force on each wheel of a planar model of vehicles is

$$\mathbf{F}_1 = \begin{bmatrix} F_{x_1}\cos\delta_1 - F_{y_1}\sin\delta_1 \\ F_{x_1}\sin\delta_1 + F_{y_1}\cos\delta_1 \\ 0 \end{bmatrix} \qquad \mathbf{F}_3 = \begin{bmatrix} F_{x_3} \\ F_{y_3} \\ 0 \end{bmatrix} \tag{2.180}$$

$$\mathbf{F}_2 = \begin{bmatrix} F_{x_2}\cos\delta_2 - F_{y_2}\sin\delta_2 \\ F_{x_2}\sin\delta_2 + F_{y_2}\cos\delta_2 \\ 0 \end{bmatrix} \qquad \mathbf{F}_4 = \begin{bmatrix} F_{x_4} \\ F_{y_4} \\ 0 \end{bmatrix} \tag{2.181}$$

The position vector of the wheels is

$$\mathbf{r}_1 = \begin{bmatrix} a_1 \\ b_1 \\ 0 \end{bmatrix} \qquad \mathbf{r}_2 = \begin{bmatrix} a_1 \\ -b_2 \\ 0 \end{bmatrix} \tag{2.182}$$

$$\mathbf{r}_3 = \begin{bmatrix} -a_2 \\ b_1 \\ 0 \end{bmatrix} \qquad \mathbf{r}_4 = \begin{bmatrix} -a_2 \\ -b_2 \\ 0 \end{bmatrix} \tag{2.183}$$

and therefore the force system on a planar vehicles model is

$$\mathbf{F} = \sum \mathbf{F}_i \tag{2.184}$$

$$= \begin{bmatrix} F_{x_1}\cos\delta_1 + F_{x_2}\cos\delta_2 - F_{y_1}\sin\delta_1 - F_{y_2}\sin\delta_2 + F_{x_3} + F_{x_4} \\ F_{y_1}\cos\delta_1 + F_{y_2}\cos\delta_2 + F_{x_1}\sin\delta_1 + F_{x_2}\sin\delta_2 + F_{y_3} + F_{y_4} \\ 0 \end{bmatrix}$$

$$\mathbf{M} = \sum \mathbf{r}_i \times \mathbf{F}_i = \begin{bmatrix} 0 \\ 0 \\ M_z \end{bmatrix} \tag{2.185}$$

$$\begin{aligned} M_z = {} & a_1 F_{y_1}\cos\delta_1 + a_1 F_{y_2}\cos\delta_2 - a_2 F_{y_3} - a_2 F_{y_4} \\ & + b_1 F_{y_1}\sin\delta_1 - b_2 F_{y_2}\sin\delta_2 + b_2 F_{x_2}\cos\delta_2 - b_1 F_{x_1}\cos\delta_1 \\ & + a_1 F_{x_1}\sin\delta_1 + a_1 F_{x_2}\sin\delta_2 + b_2 F_{x_4} - b_1 F_{x_3} \end{aligned} \tag{2.186}$$

Let us assume a planar dynamic model of vehicles with four wheels such that its equation to be approximated by

$$F_x \approx F_{x_1} + F_{x_2} + F_{x_3} + F_{x_4} \tag{2.187}$$

$$F_y \approx F_{y_1} + F_{y_2} + F_{y_3} + F_{y_4} \tag{2.188}$$

$$M_z \approx a_1 F_{y_1} + a_1 F_{y_2} - a_2 F_{y_3} - a_2 F_{y_4} \tag{2.189}$$

in which we assumed small steer angles and ignored the yaw moment caused by imbalance longitudinal forces, F_{x_i}. Knowing that

$$v_{x_i} = v_x - y_i\, r \qquad v_{y_i} = v_y + x_i\, r \tag{2.190}$$

and tire angles as

$$\beta_i = \arctan\left(\frac{v_{y_i}}{v_{x_i}}\right) = \arctan\left(\frac{v_y + x_i\, r}{v_x - y_i\, r}\right) \tag{2.191}$$

$$\alpha_i = \beta_i - \delta_i = \arctan\left(\frac{v_y + x_i\, r}{v_x - y_i\, r}\right) - \delta_i \tag{2.192}$$

we find that

$$\beta_1 = \arctan\left(\frac{v_{y1}}{v_{x1}}\right) = \arctan\left(\frac{v_y + a_1\, r}{v_x - b_1\, r}\right) \tag{2.193}$$

$$\beta_2 = \arctan\left(\frac{v_{y2}}{v_{x2}}\right) = \arctan\left(\frac{v_y + a_1\, r}{v_x + b_2\, r}\right) \tag{2.194}$$

$$\beta_3 = \arctan\left(\frac{v_{y3}}{v_{x3}}\right) = \arctan\left(\frac{v_y - a_2\, r}{v_x + b_2\, r}\right) \tag{2.195}$$

$$\beta_4 = \arctan\left(\frac{v_{y4}}{v_{x4}}\right) = \arctan\left(\frac{v_y - a_2\, r}{v_x - b_1\, r}\right) \tag{2.196}$$

Employing the vehicle sideslip angle $\beta = \arctan\left(v_y/v_x\right)$ and assuming small angles for β_f, β_r, and β, the tire sideslip angles would be:

$$\alpha_1 = \beta_1 - \delta_1 = \frac{v_y + a_1\, r}{v_x - b_1\, r} - \delta_1 = \frac{\beta + \dfrac{a_1}{v_x} r}{1 - \dfrac{b_1}{v_x} r} - \delta_1 \tag{2.197}$$

$$\alpha_2 = \beta_2 - \delta_2 = \frac{v_y + a_1\, r}{v_x + b_2\, r} - \delta_2 = \frac{\beta + \dfrac{a_1}{v_x} r}{1 + \dfrac{b_2}{v_x} r} - \delta_2 \tag{2.198}$$

$$\alpha_3 = \beta_3 = \frac{v_y - a_2\, r}{v_x + b_2\, r} = \frac{\beta - \dfrac{a_2}{v_x} r}{1 + \dfrac{b_2}{v_x} r} \tag{2.199}$$

$$\alpha_4 = \beta_4 = \frac{v_y - a_2\, r}{v_x - b_1\, r} = \frac{\beta - \dfrac{a_2}{v_x} r}{1 - \dfrac{b_1}{v_x} r} \tag{2.200}$$

Modeling the longitudinal and lateral forces as

$$F_{x_i} = F_{z_i} C_{si}\, S\left(s_i - s_s\right) \sqrt{1 - C_{s\alpha}\left(\frac{S\left(\alpha_i - \alpha_s\right)}{\alpha_s}\right)^2} \tag{2.201}$$

$$F_{y_i} = -F_{z_i} C_{\alpha i}\, S\left(\alpha_i - \alpha_s\right) \sqrt{1 - C_{\alpha s}\left(\frac{S\left(s_i - s_s\right)}{s_s}\right)^2} \tag{2.202}$$

and assuming that the left and right wheels have similar lateral and longitudinal slip coefficients we find the equations of motion (2.187)–(2.189).

2.4 Two-Wheel Planar Vehicle Dynamics

We combine the planar equations of motion of the bicycle vehicle model (2.1)–(2.4) with the force expressions (2.91)–(2.115) to derive the equations of motion of the two-wheel rigid bicycle vehicle:

$$\dot{v}_x = \frac{1}{m}(F_x - F_A) + r\, v_y \tag{2.203}$$

$$\dot{v}_y = \frac{1}{m}F_y - r\, v_x \tag{2.204}$$

$$\dot{r} = \frac{1}{I_z}M_z \tag{2.205}$$

$$\dot{\omega}_f = \frac{1}{I_f}T_f - \frac{R_g}{I_f}F_{x_f} \tag{2.206}$$

$$\dot{\omega}_r = \frac{1}{I_r}T_r - \frac{R_g}{I_r}F_{x_r} \tag{2.207}$$

$$F_x = F_{x_f}\cos\delta + F_{x_r} - F_{y_f}\sin\delta \tag{2.208}$$

$$F_y = F_{y_f}\cos\delta + F_{y_r} + F_{x_f}\sin\delta \tag{2.209}$$

$$F_A = C_A\, v_x^2 \tag{2.210}$$

$$M_z = a_1 F_{y_f}\cos\delta + a_1 F_{x_f}\sin\delta - a_2 F_{y_r} \tag{2.211}$$

$$F_{x_f} = F_{z_f} C_{sf}\, S\left(s_f - s_s\right)\sqrt{1 - C_{s\alpha}\left(\frac{S\left(\alpha_f - \alpha_s\right)}{\alpha_s}\right)^2} \tag{2.212}$$

$$F_{x_r} = F_{z_r} C_{sr}\, S\left(s_r - s_s\right)\sqrt{1 - C_{s\alpha}\left(\frac{S\left(\alpha_r - \alpha_s\right)}{\alpha_s}\right)^2} \tag{2.213}$$

$$F_{y_f} = -F_{z_f} C_{\alpha f}\, S\left(\alpha_f - \alpha_s\right)\sqrt{1 - C_{\alpha s}\left(\frac{S\left(s_f - s_s\right)}{s_s}\right)^2} \tag{2.214}$$

$$F_{y_r} = -F_{z_r} C_{\alpha r}\, S\left(\alpha_r - \alpha_s\right)\sqrt{1 - C_{\alpha s}\left(\frac{S\left(s_r - s_s\right)}{s_s}\right)^2} \tag{2.215}$$

$$F_{z_f} = F_{z_1} + F_{z_2} = mg\frac{a_2}{l} - m\left(\dot{v}_x - r\,v_y\right)\frac{h}{l} \tag{2.216}$$

$$F_{z_r} = F_{z_3} + F_{z_4} = mg\frac{a_1}{l} + m\left(\dot{v}_x - r\,v_y\right)\frac{h}{l} \tag{2.217}$$

$$l = a_1 + a_2 \tag{2.218}$$

$$\alpha_f = \arctan\left(\frac{v_y + a_1 r}{v_x}\right) - \delta \tag{2.219}$$

$$\alpha_r = \arctan\left(\frac{v_y - a_2 r}{v_x}\right) \tag{2.220}$$

$$\beta = \arctan\frac{v_y}{v_x} \tag{2.221}$$

$$s_f = \frac{R_g\,\omega_f - v_{x_{Tf}}}{R_g\,\omega_f\,H(R_g\,\omega_f - v_{x_{Tf}}) + v_{x_{Tf}}\,H(v_{x_{Tf}} - R_g\,\omega_f)} \tag{2.222}$$

$$s_r = \frac{R_g\,\omega_r - v_x}{R_g\,\omega_r\,H(R_g\,\omega_r - v_x) + v_x\,H(v_x - R_g\,\omega_r)} \tag{2.223}$$

$$v_{x_{Tf}} = v_x\cos\delta + \left(v_y + a_1 r\right)\sin\delta \tag{2.224}$$

Proof The Newton–Euler equations of motion for a rigid vehicle in the local coordinate frame B, attached to the vehicle at its mass center C, are given in Eqs. (2.1)–(2.4). Those equations for bicycle vehicle model will be:

$$F_x = m\,\dot{v}_x - mr\,v_y + F_A \tag{2.225}$$

$$F_y = m\,\dot{v}_y + mr\,v_x \tag{2.226}$$

$$M_z = \dot{r}\,I_z \tag{2.227}$$

$$T_f = I_{w_i}\,\dot{\omega}_f + R_w\,F_{x_f} \tag{2.228}$$

$$T_i = I_{w_i}\,\dot{\omega}_r + R_w\,F_{x_r} \tag{2.229}$$

The force F_A is the air resistance aerodynamic forces that we assume to be applied only in longitudinal direction.

$$F_A = \frac{1}{2}\rho\,C_D\,A_f\,v_x^2 = C_A\,v_x^2 \tag{2.230}$$

Considering ρ, C_D, and A_f to be constant for a vehicle, we may combine the coefficients of the aerodynamic force to be C_A.

From (2.91) to (2.93), the vehicle body forces are

$$F_x = F_{x_f} \cos\delta + F_{x_r} - F_{y_f} \sin\delta \tag{2.231}$$

$$F_y = F_{y_f} \cos\delta + F_{y_r} + F_{x_f} \sin\delta \tag{2.232}$$

$$M_z = a_1 F_{y_f} \cos\delta + a_1 F_{x_f} \sin\delta - a_2 F_{y_r} \tag{2.233}$$

The force components $F_{x_f}, F_{x_r}, F_{y_f}, F_{y_r}$ have been calculated in (2.96)–(2.99). The vertical forces on front and rear wheels F_{z_f}, F_{z_r} are also calculated in (2.139)–(2.140) as are shown in (2.212)–(2.217).

The complexity of the equations of motion are hidden in the forward and lateral forces $F_{x_f}, F_{x_r}, F_{y_f}, F_{y_r}$ in (2.212)–(2.215). These forces are functions of two new variables s and α. To be able to solve the equations of motions, we must be able to relate these variables to the inputs δ, F_x or output variables v_x, v_y, r, of the vehicle dynamic equations.

The tire side slip angle α_i has been shown in (1.121)–(1.123) to be a function of the input steer angle δ_i and the output *wheel-body sideslip angle* β_i of the tire.

$$\alpha_i = \beta_i - \delta_i \qquad \alpha_i = \arctan\frac{{}^{T}v_{y_i}}{{}^{T}v_{x_i}} \qquad \beta_i = \arctan\frac{{}^{C}v_{y_i}}{{}^{C}v_{x_i}} \tag{2.234}$$

The tire side slip angle α_i is defined based on the velocity components in T-frame while the wheel-body sideslip angle β_i is defined based on the velocity components in C-frame. The wheel-body sideslip β_i for the wheel i is shown in Eq. (2.150) to be a function of vehicle velocity components.

$$\beta_i = \arctan\left(\frac{v_{y_i}}{v_{x_i}}\right) = \arctan\left(\frac{v_y + x_i\, r}{v_x}\right) \tag{2.235}$$

$$\beta_1 = \arctan\left(\frac{v_y + a_1 r}{v_x}\right) \tag{2.236}$$

$$\beta_2 = \arctan\left(\frac{v_y - a_2 r}{v_x}\right) \tag{2.237}$$

Rewriting this equation for the front and rear wheels of the two-wheel vehicle model will be (2.219) and (2.220).

The longitudinal slip ratio s is shown in (1.57) to be a function of the tire forward velocity v_{x_T} and the angular velocity of tire ω_w.

$$s = \frac{R_g\,\omega_w - v_{x_T}}{R_g\,\omega_w\, H(R_g\,\omega_w - v_{x_T}) + v_{x_T}\, H(v_{x_T} - R_g\,\omega_w)} \tag{2.238}$$

According to (1.303), the longitudinal velocity of the vehicle at the wheels centers is:

$$ {}^{B}\mathbf{v}_i = \begin{bmatrix} v\cos\beta - ry_i \\ v\sin\beta + rx_i \end{bmatrix} = \begin{bmatrix} v_x - ry_i \\ v_y + rx_i \end{bmatrix} \tag{2.239} $$

therefore,

$$ {}^{B}\mathbf{v}_f = \begin{bmatrix} v_x \\ v_y + ra_1 \end{bmatrix} \qquad {}^{B}\mathbf{v}_r = \begin{bmatrix} v_x \\ v_y - ra_2 \end{bmatrix} \tag{2.240} $$

and hence the velocity v_{xT_f} at the center of the front wheel in the tire x-direction, ${}^{T}i$, is

$$ v_{xT_f} = v_x\cos\delta + \left(v_y + a_1 r\right)\sin\delta \tag{2.241} $$

because

$$ \begin{aligned} {}^{T}\mathbf{v}_f &= {}^{C}R_T^{T}\,{}^{B}\mathbf{v}_f = \begin{bmatrix} \cos\delta & -\sin\delta & 0 \\ \sin\delta & \cos\delta & 0 \\ 0 & 0 & 1 \end{bmatrix}^{T} \begin{bmatrix} v_x \\ v_y + ra_1 \\ 0 \end{bmatrix} \\ &= \begin{bmatrix} v_x\cos\delta + \left(v_y + ra_1\right)\sin\delta \\ \left(v_y + ra_1\right)\cos\delta - v_x\sin\delta \\ 0 \end{bmatrix} \end{aligned} \tag{2.242} $$

$$ \begin{aligned} {}^{T}\mathbf{v}_{rf} &= {}^{C}R_T^{T}\,{}^{B}\mathbf{v}_r = \begin{bmatrix} \cos\delta & -\sin\delta & 0 \\ \sin\delta & \cos\delta & 0 \\ 0 & 0 & 1 \end{bmatrix}^{T} \begin{bmatrix} v_x \\ v_y - ra_2 \\ 0 \end{bmatrix} \\ &= \begin{bmatrix} v_x\cos\delta + \left(v_y - ra_2\right)\sin\delta \\ \left(v_y - ra_2\right)\cos\delta - v_x\sin\delta \\ 0 \end{bmatrix} \end{aligned} \tag{2.243} $$

and therefore

$$ s_f = \frac{R_g\,\omega_w - v_{xT_f}}{R_g\,\omega_w\,H(R_g\,\omega_w - v_{xT_f}) + v_{xT_f}\,H(v_{xT_f} - R_g\,\omega_w)} \tag{2.244} $$

$$ s_r = \frac{R_g\,\omega_w - v_x}{R_g\,\omega_w\,H(R_g\,\omega_w - v_x) + v_x\,H(v_x - R_g\,\omega_w)} \tag{2.245} $$

Introducing ω_w as a new kinematic variable requires to have a new equation to determine ω_w. The free-body-diagram of a wheel provided us with the equation of motion of the wheel (1.76). Ignoring the tire resistance force, the rotational equation of motion of the wheel will be

$$I_w \dot{\omega}_w = T - R_w F_x \tag{2.246}$$

and therefore the tire longitudinal slip ratio s_i becomes a function of v_{x_i}, ω_i, F_{x_i}. This equation must be written for the front and rear wheels of the two-wheel vehicle model as shown in Eqs. (2.212) and (2.224).

These sets of equations of motion of bicycle vehicle model are nonlinear due to several reasons such as:

1. Mathematical nonlinearities: The equations of motion relate kinematic variables to applied forces. The force system also is a function of kinematic variables. For example, v_x is a variable and it appears in the denominators in Eqs. (2.234)–(2.245).
2. Model nonlinearities: The tire forces are functions of parameters which have saturation limits and therefore they show nonlinear characteristics. For example, The lateral force F_y is proportional to the sideslip angle α as long as $\alpha < \alpha_s$ and constant for $\alpha > \alpha_s$. Therefore, all proportional equations and their coefficients are modeled better by using nonlinear relationships. ■

Example 67 Aerodynamic effect.

Consider a vehicle with the following data.

$$\begin{aligned}
m &= 1000\,\text{kg} & Iz &= 2000\,\text{kg}\,\text{m}^2 & I_f &= I_r = 30\,\text{kg}\,\text{m}^2 \\
C_{\alpha f} &= 8.5 & C_{\alpha r} &= 8.5 & \alpha_s &= 5\,\text{deg} \\
C_{sf} &= 7.5 & C_{sr} &= 7.5 & s_s &= 0.1 \\
a_1 &= 1.35\,\text{m} & a_2 &= 1.5\,\text{m} & h &= 0.9\,\text{m} \\
R_g &= 0.35\,\text{m} & C_{\alpha s} &= 0.5 & C_{s\alpha} &= 0.5
\end{aligned} \tag{2.247}$$

Assume the car initially is moving at

$$v_x = 20\,\text{m/s} \qquad \delta = 0 \qquad T_f = T_r = 0 \tag{2.248}$$

and

$$\omega_f = \omega_r = \frac{v_x}{R_g} = \frac{20}{0.35} = 57.143\,\text{rad/s} \tag{2.249}$$

To show the effect of the aerodynamic resistance force, we determine the forward velocity of the vehicle for a C_D. Figure 2.10 illustrates how the forward velocity v_x will reduce with

$$C_A = 0.8 \tag{2.250}$$

Figure 2.11 depicts the angular velocities ω_f and ω_r of the front and rear wheels.

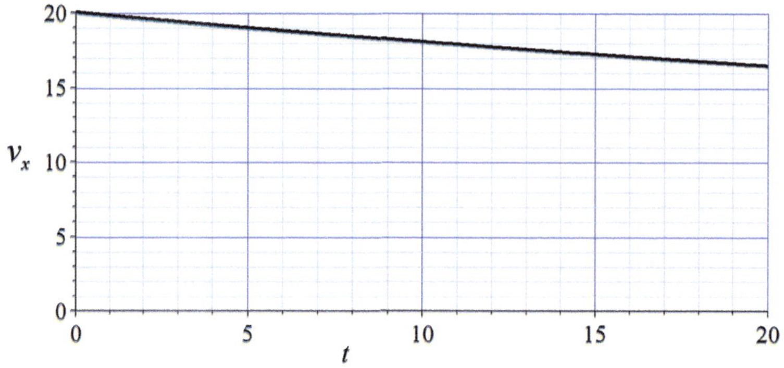

Fig. 2.10 The effect of aerodynamic drag force $F_A = C_A v_x^2$ on forward velocity of a vehicle in free straight motion

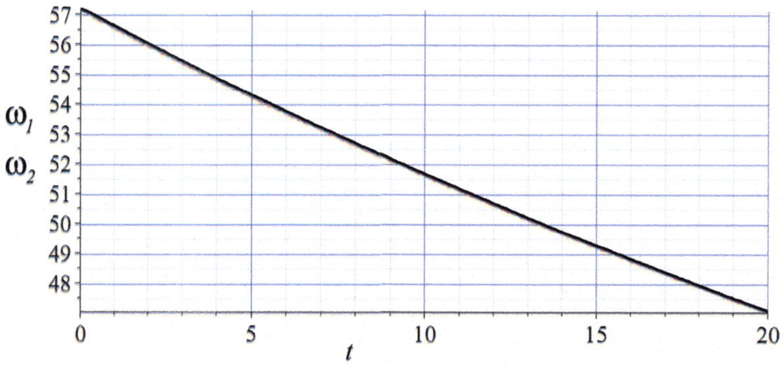

Fig. 2.11 The angular velocities ω_f and ω_r of the front and rear wheels

Example 68 Increasing rear torque, straight driving.

Consider a vehicle with the data given in (2.251).

$$\begin{aligned} m &= 1000\,\text{kg} & Iz &= 2000\,\text{kg m}^2 & I_f &= I_r = 30\,\text{kg m}^2 \\ C_{\alpha f} &= C_{\alpha r} = 8.5 & \alpha_s &= 5\,\text{deg} & C_A &= 0.8 \\ C_{sf} &= C_{sr} = 7.5 & s_s &= 0.1 \\ a_1 &= 1.35\,\text{m} & a_2 &= 1.5\,\text{m} & h &= 0.9\,\text{m} \\ R_g &= 0.35\,\text{m} & C_{\alpha s} &= 0 & C_{s\alpha} &= 0 \end{aligned} \tag{2.251}$$

Assume the vehicle is moving slowly straight with

$$v_x = 2\,\text{m/s} \qquad \omega_f = \omega_r = \frac{v_x}{R_g} = \frac{2}{0.35} = 5.714\,\text{rad/s} \tag{2.252}$$

$$\delta = 0 \tag{2.253}$$

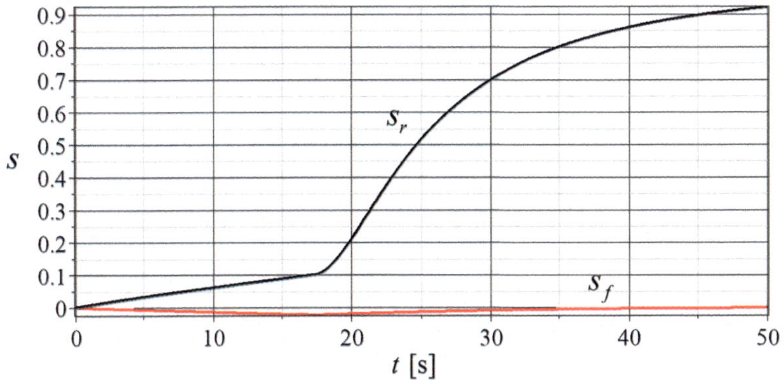

Fig. 2.12 The longitudinal slip ratios of the front and rear tires

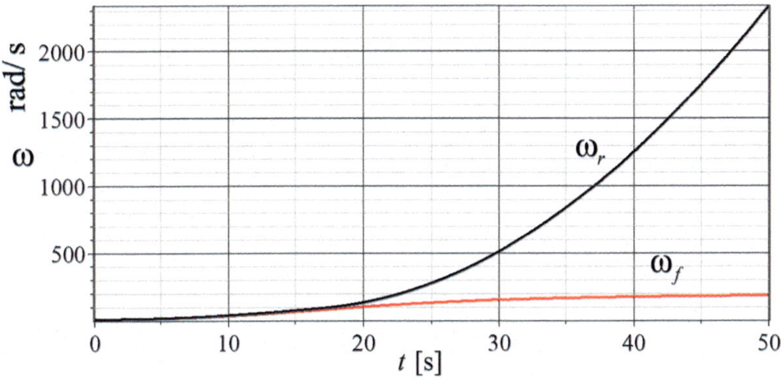

Fig. 2.13 The angular velocities of the front and rear wheels ω_f and ω_r

At time $t = 0$, we apply an increasing torque on the rear wheel.

$$T_f = 0 \qquad T_r = 100t \text{ N m} \tag{2.254}$$

This example shows how the rear longitudinal slip ratio will increase and reaches its saturation, and how the rear wheel starts spinning after that. Figure 2.12 depicts the longitudinal slip ratios. Considering that the saturation limit of the front and rear tires are at $s_s = 0.1$, we see that the slip ratio of the rear tire s_r increases linearly from zero up to the point that reaches $s_s = 0.1$. At that point, tire starts sliding on the ground and the applied torque increases the wheels' angular velocity. As a result, the slip ratio s_r increases rapidly. There is no torque on the front wheel and therefore the front wheel's slip ratio s_f remains unsaturated. Figure 2.13 illustrates the angular velocities of the front and rear wheels ω_f and ω_r. At the beginning of applying torque on the rear wheel, both angular velocities increase. At the point that the rear wheel starts sliding and spinning, the rear wheel angular velocity increases rapidly. The front angular velocity remains proportional to the vehicle velocity. Figure 2.14 depicts the forward velocity of the vehicle v_x.

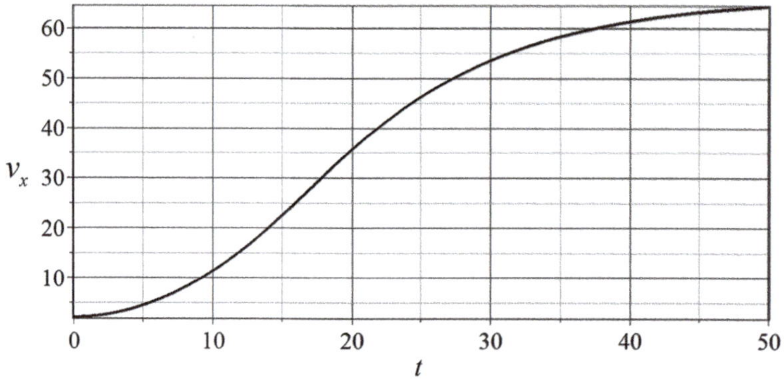

Fig. 2.14 The forward velocity of the vehicle v_x

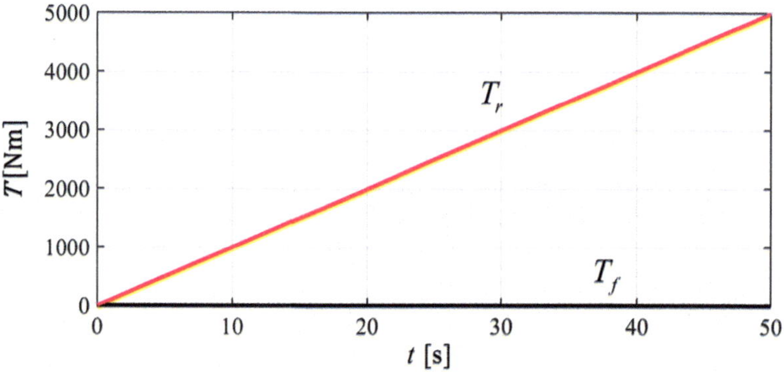

Fig. 2.15 The torques on front and rear wheels T_f and T_r

The torques on front and rear wheels T_f and T_r are shown in Fig. 2.15. Because of variable acceleration of the vehicle, the vertical load on the wheels are not constant. Figure 2.16 illustrates how the vertical loads F_{zf} and F_{zr} will change, and the traction forces F_{xf} and F_{xr} are shown in Fig. 2.17.

Example 69 Increasing front torque, straight driving.

Consider a vehicle with the data given in (2.251) and assume the vehicle is moving slowly straight with

$$v_x = 2\,\text{m/s} \qquad \omega_f = \omega_r = \frac{v_x}{R_g} = \frac{2}{0.35} = 5.714\,\text{rad/s} \tag{2.255}$$

$$\delta = 0 \tag{2.256}$$

At time $t = 0$, we apply an increasing torque on the front wheel

$$T_f = 100t\,\text{N m} \qquad T_r = 0 \tag{2.257}$$

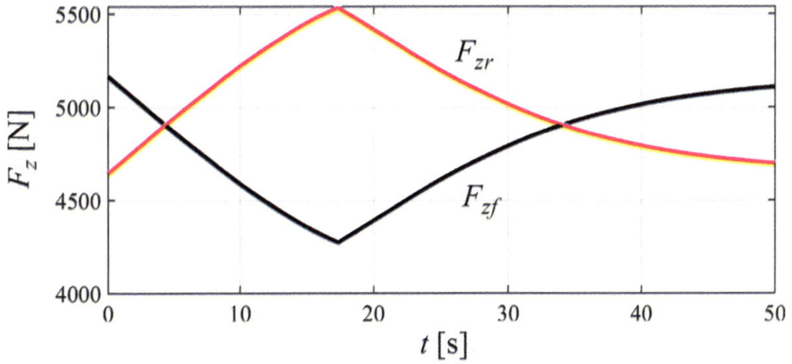

Fig. 2.16 The vertical load F_{zf} and F_{zr} on front and rear wheels

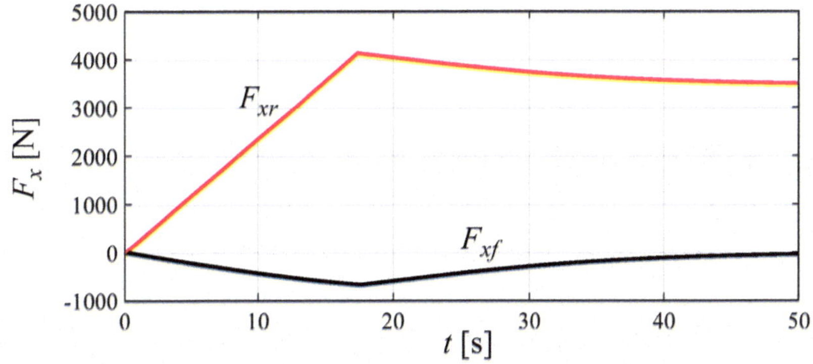

Fig. 2.17 The traction forces F_{xf} and F_{xr} on front and rear wheels

to investigate the differences of vehicle response to front or rear wheel drive. Figure 2.18 depicts the longitudinal slip ratios s_f and s_r for the front and rear tires. Figure 2.19 illustrates the angular velocities of the front and rear wheels ω_f and ω_r. Figure 2.20 illustrates variation of the vertical loads F_{zf} and F_{zr}. Figure 2.21 depicts the forces F_x and F_y on the vehicle at its mass center. The force components on front and rear wheels are slightly different. Figure 2.22 illustrates the longitudinal forces F_{xf} and F_{xr} on front and rear tires. The acceleration components a_x and a_y of the vehicle are plotted in Fig. 2.23.

Example 70 Increasing steer angle and front torque, slip saturation.

Consider a vehicle with the following data

$$\begin{array}{lll} m = 1000\,\text{kg} & Iz = 2000\,\text{kg m}^2 & I_f = I_r = 30\,\text{kg m}^2 \\ C_{\alpha f} = C_{\alpha r} = 8.5 & \alpha_s = 5\,\text{deg} & C_A = 0.8 \\ C_{sf} = C_{sr} = 7.5 & s_s = 0.1 & \end{array} \tag{2.258}$$

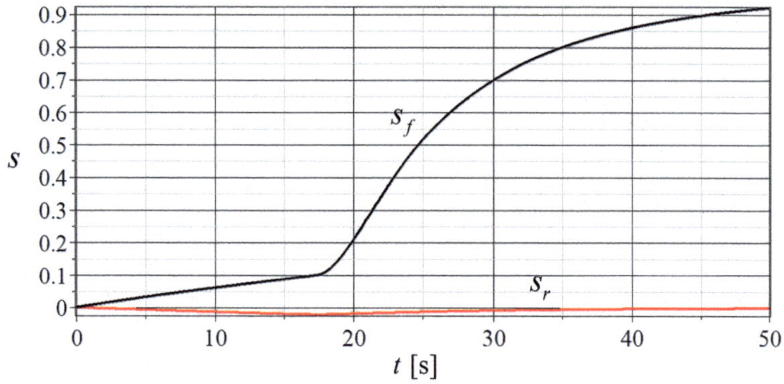

Fig. 2.18 The longitudinal slip ratios s_f and s_r for the front and rear tires

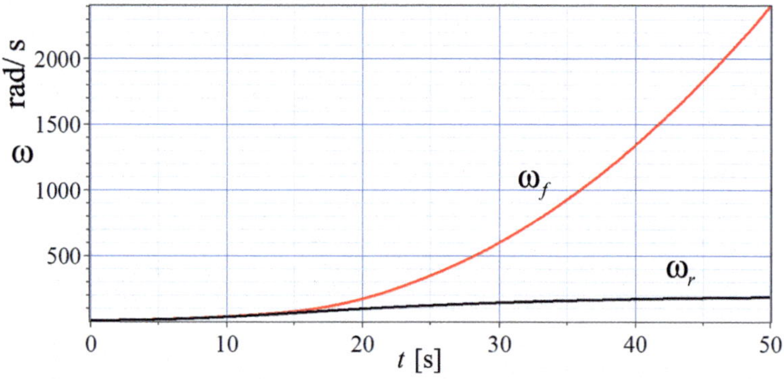

Fig. 2.19 The angular velocities of the front and rear wheels ω_f and ω_r

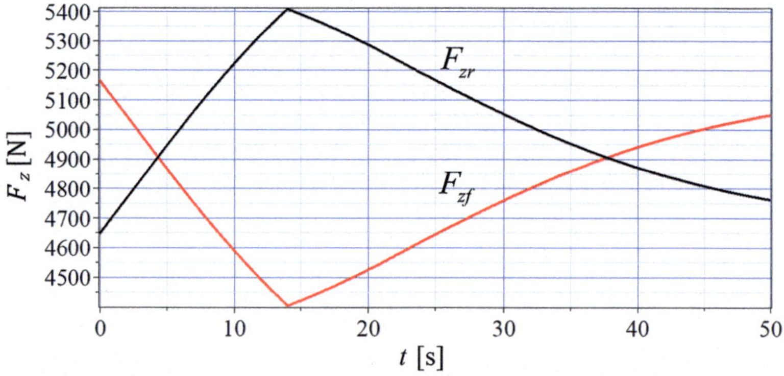

Fig. 2.20 The vertical load F_{zf} and F_{zr} on front and rear wheels

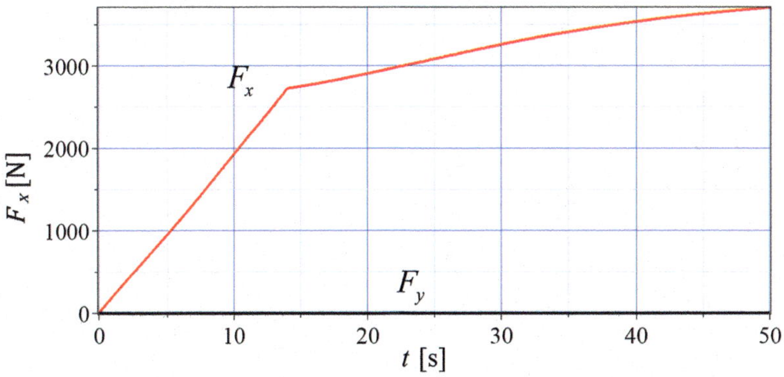

Fig. 2.21 The traction forces F_x and F_y on the vehicle at its mass center

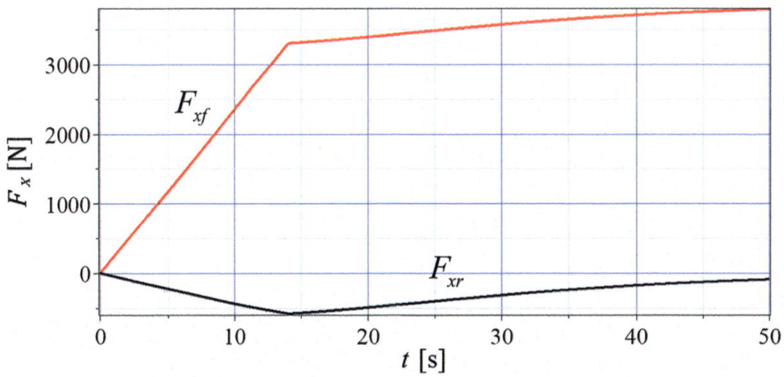

Fig. 2.22 The traction forces F_{xf} and F_{xr} on front and rear tires

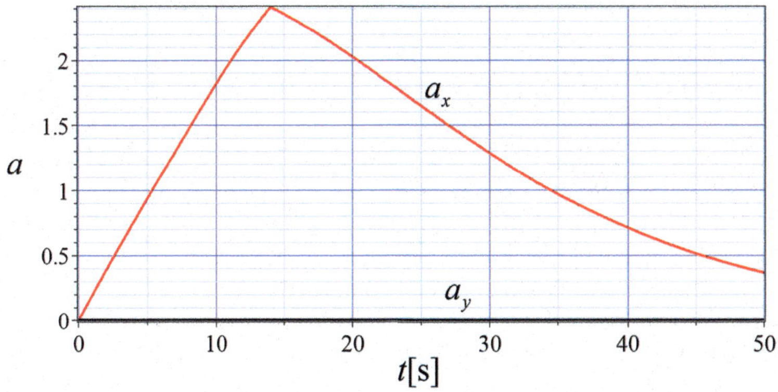

Fig. 2.23 The acceleration components a_x and a_y/g of the vehicle

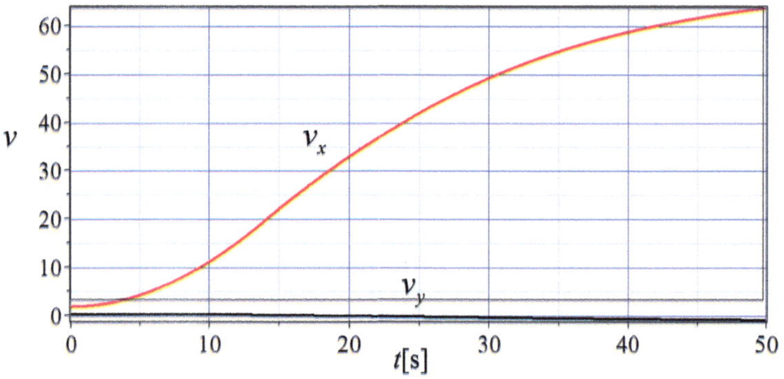

Fig. 2.24 The forward velocity components of the vehicle v_x and v_y

$$a_1 = 1.35\,\text{m} \qquad a_2 = 1.5\,\text{m} \qquad h = 0.9\,\text{m}$$

$$R_g = 0.35\,\text{m} \qquad C_{\alpha s} = C_{s\alpha} = 0.5$$

and assume the vehicle is moving slowly straight.

$$v_x = 2\,\text{m/s} \qquad \omega_f = \omega_r = \frac{v_x}{R_g} = \frac{2}{0.35} = 5.714\,\text{rad/s} \tag{2.259}$$

At time $t = 0$, we apply a linearly increasing torque on the front wheel up to $T_f = 1500\,\text{N m}$ and keep constant after that. The steer angle is also linearly increasing at a very low rate up to $\delta = 0.5\,\text{deg}$ and remains constant after that.

$$T_f = \begin{cases} 100t\,\text{N m} & 0 < t < 15\,\text{s} \\ 1500\,\text{N m} & 15\,\text{s} < t \end{cases} \qquad T_r = 0 \tag{2.260}$$

$$\delta = \begin{cases} 0.05t\,\text{deg} = \frac{0.05\pi}{180}t\,\text{rad} & 0 < t < 10\,\text{s} \\ 0.5\,\text{deg} = \frac{0.5\pi}{180}\,\text{rad} & 10\,\text{s} < t \end{cases} \tag{2.261}$$

The front torque increases and goes beyond the limit of front wheel capability in producing traction force. As a result, the front tire slip ratio s_f and hence its traction force F_{xf} will become saturated.

Figure 2.24 depicts the velocity components of the vehicle v_x and v_y, measured in body coordinate frame B. Figure 2.25 illustrates the angular velocities of the front and rear wheels ω_f and ω_r. Figure 2.26 illustrates the sideslip angles of the front and rear wheels α_f and α_r. Figure 2.27 depicts the longitudinal slip ratios s_f and s_r for the front and rear tires. The acceleration components a_x and a_y of the vehicle are plotted in Fig. 2.28. Figure 2.29 depicts the forces F_x and F_y on the vehicle at its mass center. Figure 2.30 illustrates variation of the vertical loads F_{zf} and F_{zr}.

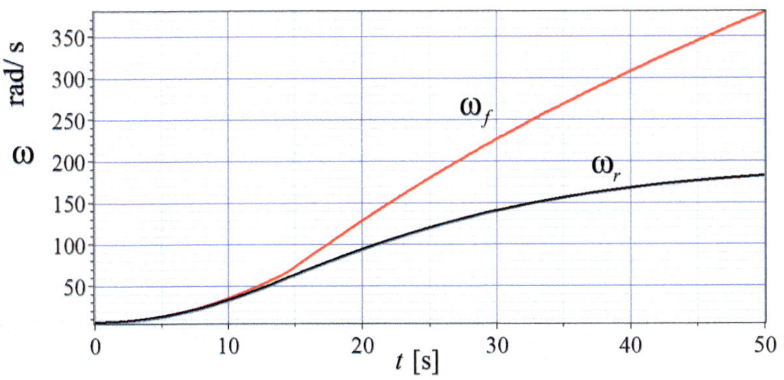

Fig. 2.25 The angular velocities of the front and rear wheels ω_f and ω_r

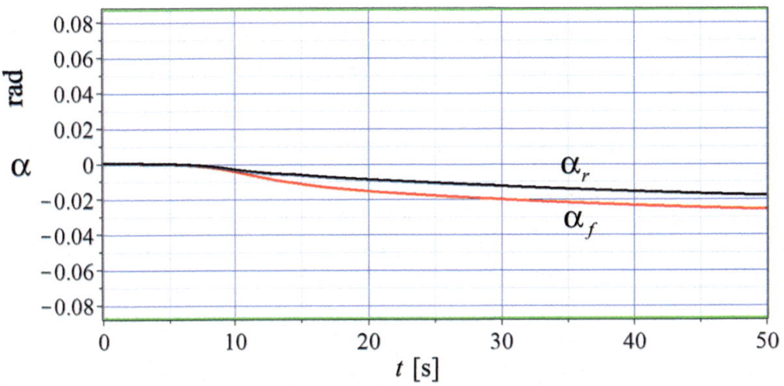

Fig. 2.26 The sideslip angles of the front and rear wheels α_f and α_r

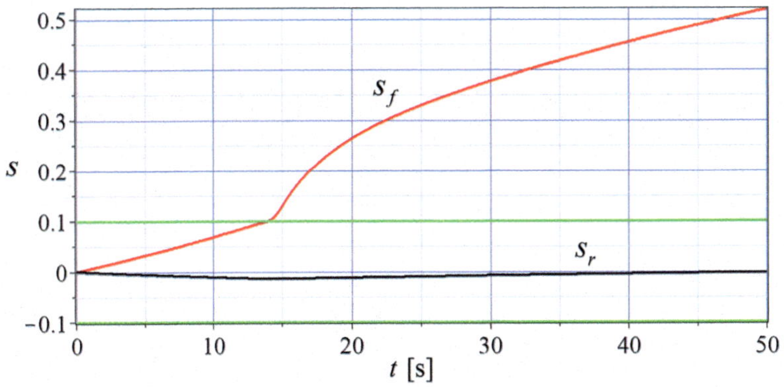

Fig. 2.27 The longitudinal slip ratios s_f and s_r for the front and rear tires

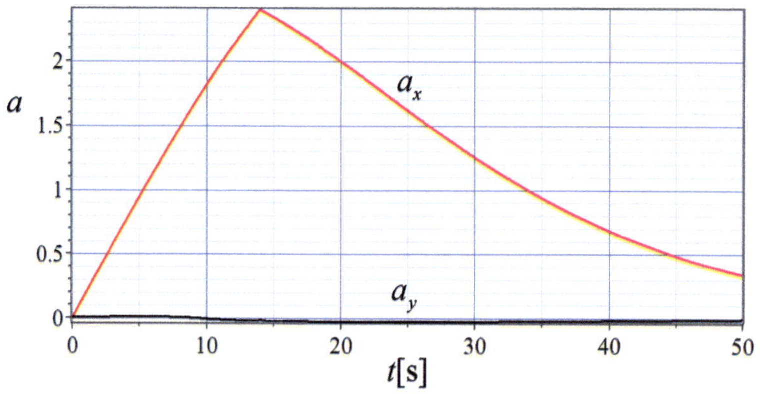

Fig. 2.28 The acceleration components a_x and a_y of the vehicle

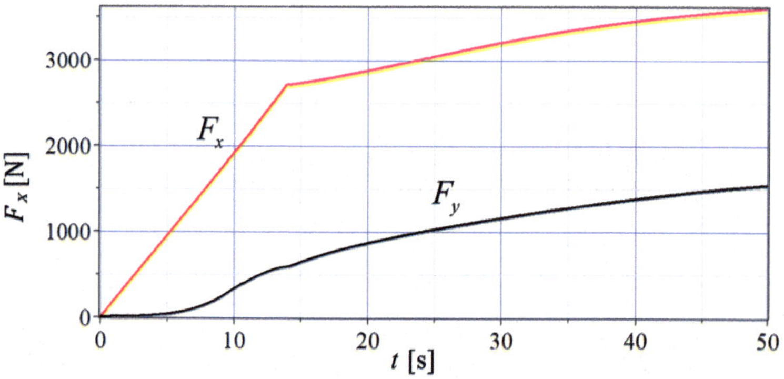

Fig. 2.29 The traction forces F_x and F_y on the vehicle at its mass center

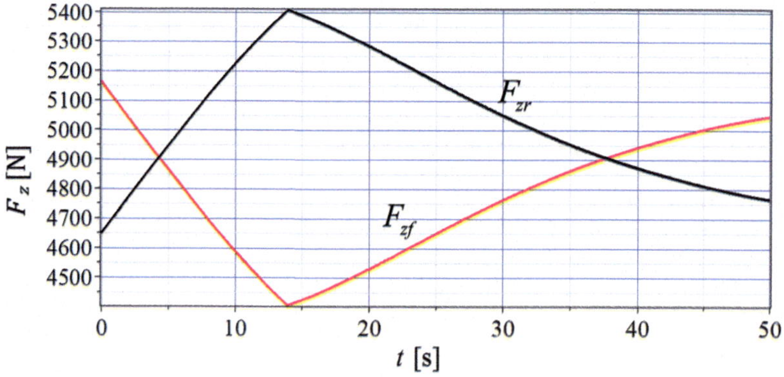

Fig. 2.30 The vertical load F_{zf} and F_{zr} on front and rear wheels

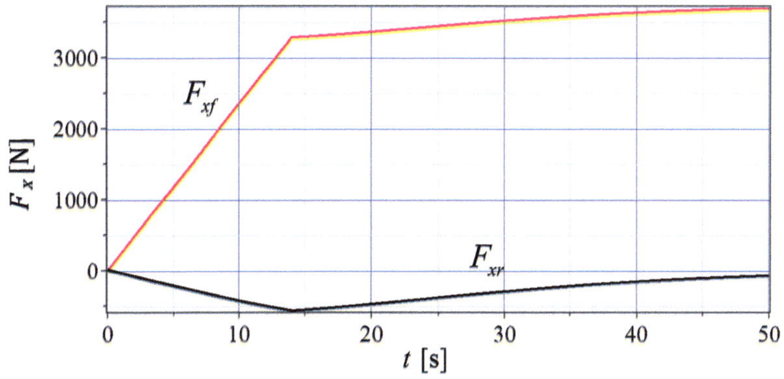

Fig. 2.31 The traction forces F_{xf} and F_{xr} on front and rear tires

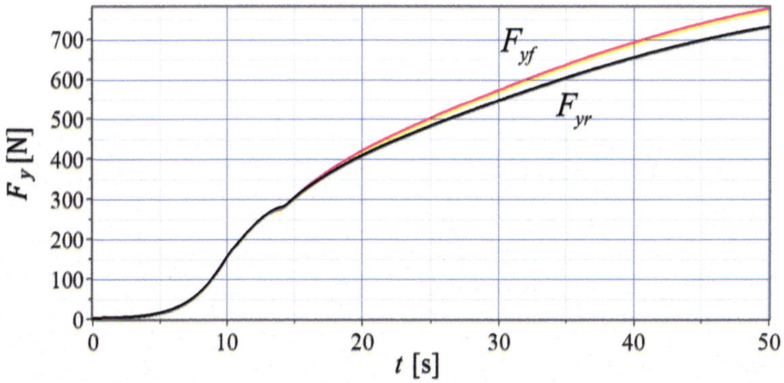

Fig. 2.32 The lateral forces F_{yf} and F_{yr} on front and rear tires

Figure 2.31 illustrates the longitudinal forces F_{xf} and F_{xr} on front and rear tires. Figure 2.32 shows the lateral forces F_{yf} and F_{yr} on front and rear tires.

The front wheel torque T_f is increasing from zero up to $T_f = 1500\,\mathrm{N\,m}$. Increasing torque will increase the slip ratio; however, before T_f reaches its maximum, the slip ratio reaches its saturation $s = s_s$ as it can be seen in Fig. 2.27. The front tire slip ratio increases rapidly after saturation. The sideslips α_f and α_r never get saturated in this maneuver. Hence the vehicle in this maneuver slips longitudinally and sticks laterally to the road. Figure 2.33 illustrates the path of the vehicle.

Example 71 Increasing steer angle and front torque, no combined forces.

Consider the vehicle in Example 70 with setting the tire slips not related by $C_{\alpha s} = 0$, $C_{s\alpha} = 0$. Ignoring the interaction between s and α makes both of them to be higher and their associated forces to be more than real saturation. Figures 2.34 and 2.35 illustrate the sideslip angles α and slip ratios s of the front and rear tires to

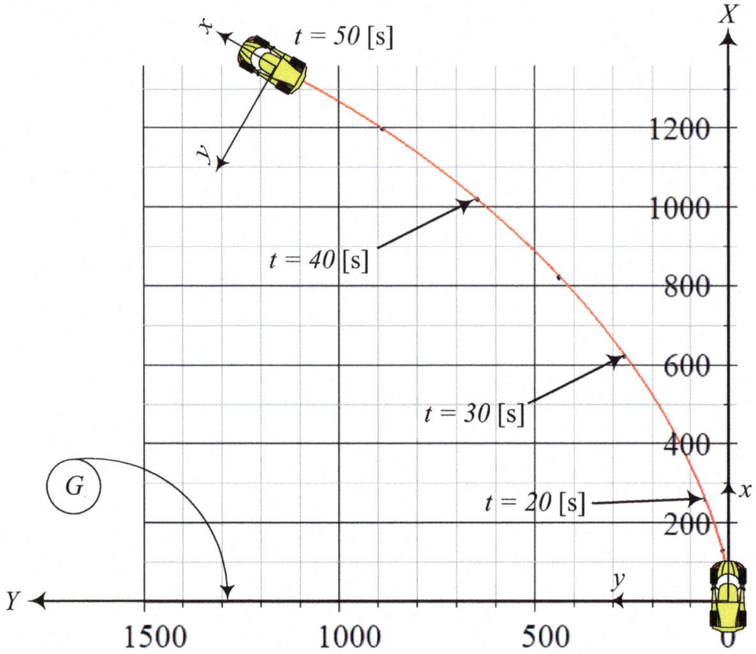

Fig. 2.33 Path of motion of the vehicle for $0 < t < 50$ s by increasing steer angle and applying front torque and having slip ratio saturation

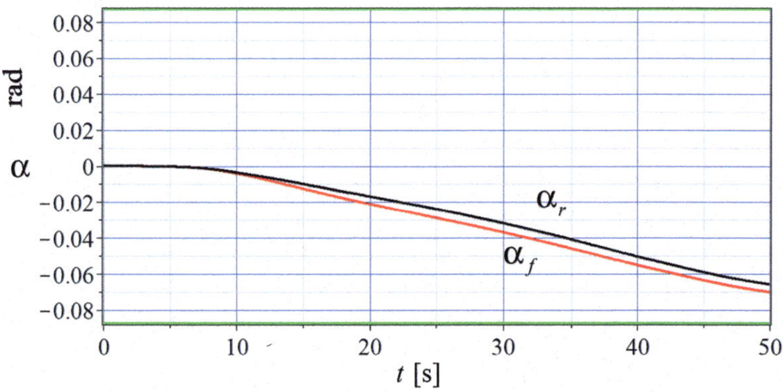

Fig. 2.34 The sideslip angles of the front and rear wheels α_f and α_r

be compared with Figs. 2.26 and 2.27. Figure 2.36 illustrates the path of the vehicle to be compared with Fig. 2.33. It can be seen that tire slip interaction has a big impact in the path of vehicle. The lateral force drops significantly and the rotation of the vehicle decreases.

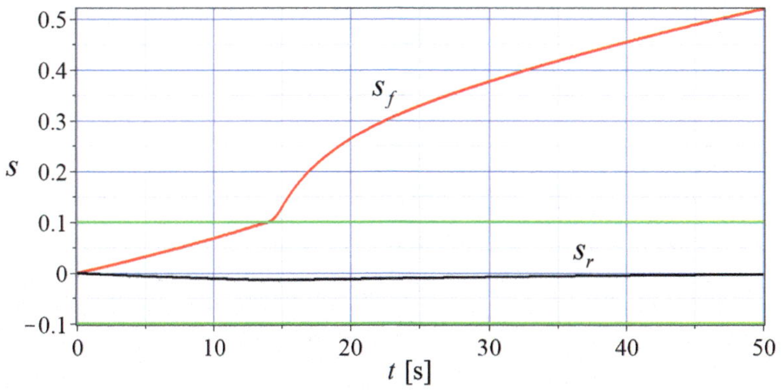

Fig. 2.35 The longitudinal slip ratios s_f and s_r for the front and rear tires

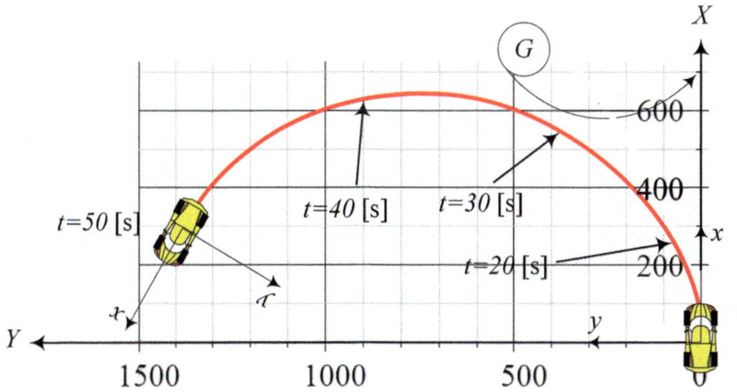

Fig. 2.36 Path of motion of the vehicle for $0 < t < 50$ s by increasing steer angle and applying front torque and having slip ratio saturation in case the sideslip and slip ratios were not connected

Example 72 Increasing steer angle and rear torque.

Consider a vehicle with the following parameters.

$$\begin{array}{lll} m = 1000\,\text{kg} & Iz = 2000\,\text{kg m}^2 & I_f = I_r = 30\,\text{kg m}^2 \\ C_{\alpha f} = C_{\alpha r} = 8.5 & \alpha_s = 5\,\text{deg} & \\ C_{sf} = C_{sr} = 7.5 & s_s = 0.1 & \\ a_1 = 1.35\,\text{m} & a_2 = 1.5\,\text{m} & h = 0.9\,\text{m} \\ R_g = 0.35\,\text{m} & C_{\alpha s} = C_{s\alpha} = 0.5 & \end{array} \tag{2.262}$$

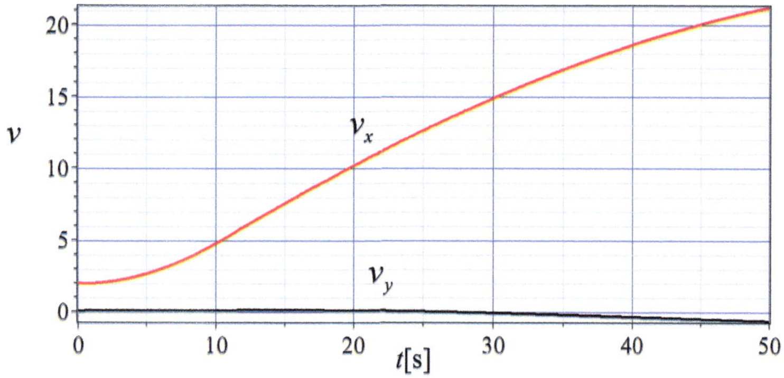

Fig. 2.37 The forward velocity components of the vehicle v_x and v_y

The vehicle is moving slowly straight with

$$\omega_f = \omega_r = \frac{v_x}{R_g} = \frac{2}{0.35} = 5.714 \,\mathrm{rad/\,s}$$
$$C_A = 0.8 \qquad v_x = 2\,\mathrm{m/s} \tag{2.263}$$

At time $t = 0$, we apply an increasing torque on the rear wheel as well as an increasing steer angle for limited times as expressed below.

$$T_r = \begin{cases} 30t\,\mathrm{N\,m} & 0 < t < 10\,\mathrm{s} \\ 300\,\mathrm{N\,m} & 10\,\mathrm{s} < t \end{cases} \qquad T_f = 0 \tag{2.264}$$

$$\delta = \begin{cases} 0.1t\,\deg = \frac{0.1\pi}{180}t\,\mathrm{rad} & 0 < t < 15\,\mathrm{s} \\ 1.5\,\deg = \frac{1.5\pi}{180}\,\mathrm{rad} & 15\,\mathrm{s} < t \end{cases} \tag{2.265}$$

Figure 2.37 depicts the velocity components of the vehicle v_x and v_y. Figure 2.38 illustrates the angular velocities of the front and rear wheels ω_f and ω_r. Figure 2.39 illustrates the sideslip angles of the front and rear wheels α_f and α_r. Figure 2.40 depicts the longitudinal slip ratios s_f and s_r for the front and rear tires. The angular accelerations of the front and rear wheels $\dot{\omega}_f$ and $\dot{\omega}_r$ are shown in Fig. 2.41. Figure 2.42 depicts the resultant forces F_x and F_y on the vehicle at its mass center. Figure 2.43 illustrates variation of the vertical loads F_{zf} and F_{zr}. Figure 2.44 illustrates the longitudinal forces F_{xf} and F_{xr} on front and rear tires. Figure 2.45 shows the lateral forces F_{yf} and F_{yr} on front and rear tires. Figure 2.46 illustrates the path of the vehicle.

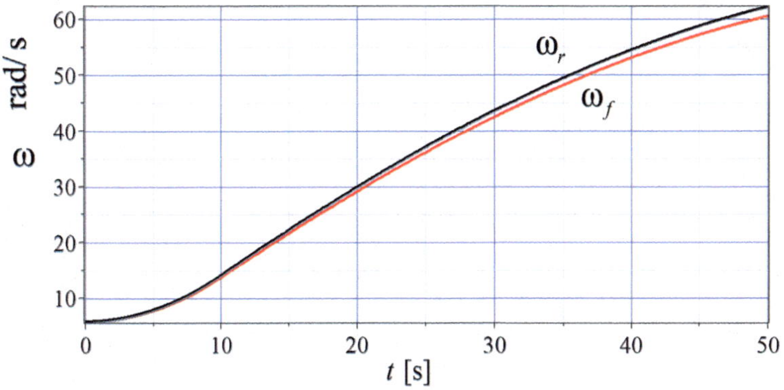

Fig. 2.38 The angular velocities of the front and rear wheels ω_f and ω_r

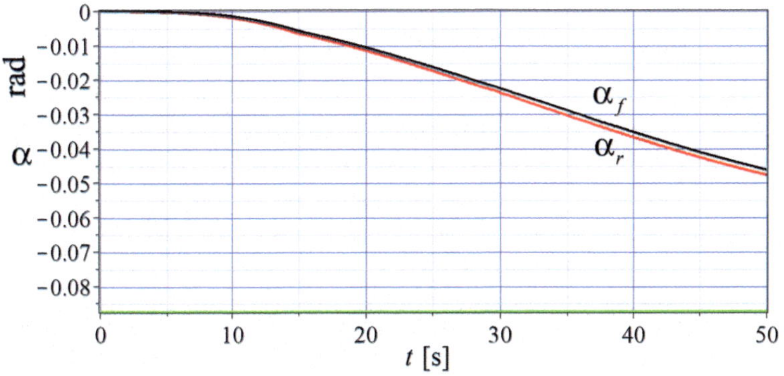

Fig. 2.39 The sideslip angles of the front and rear wheels α_f and α_r

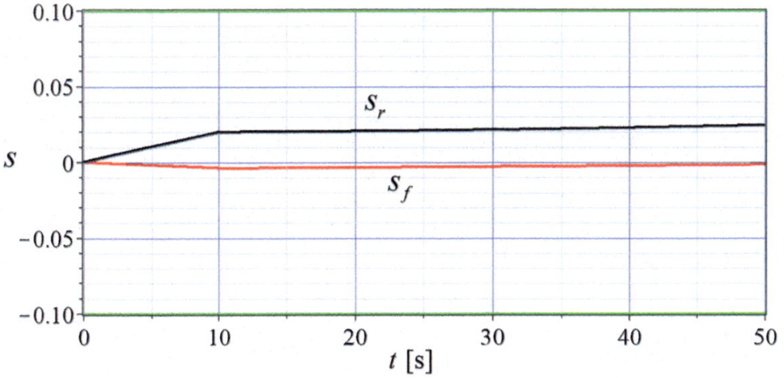

Fig. 2.40 The longitudinal slip ratios s_f and s_r for the front and rear tires

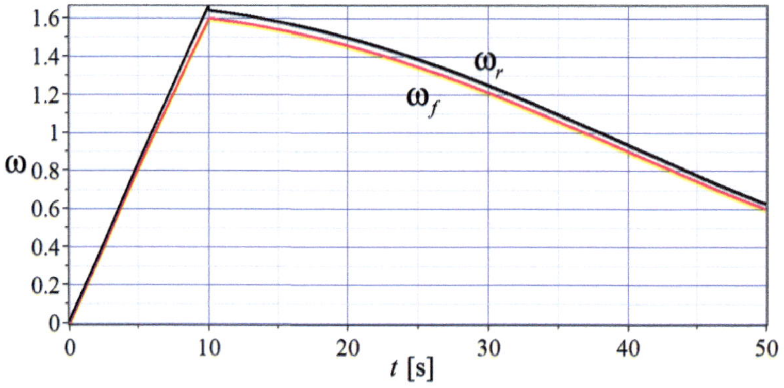

Fig. 2.41 The angular accelerations of the front and rear wheels $\dot{\omega}_f$ and $\dot{\omega}_r$

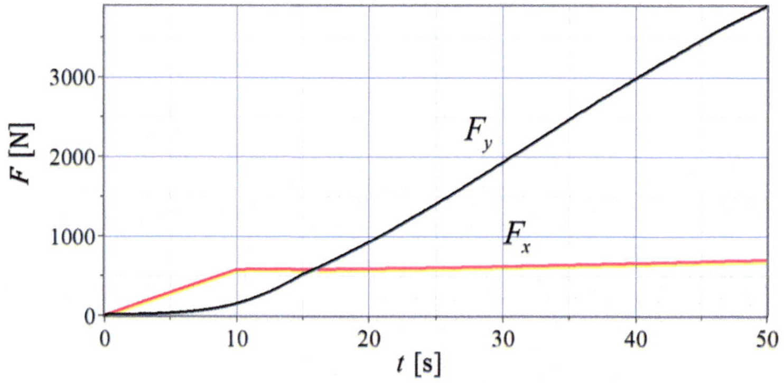

Fig. 2.42 The forward and lateral forces F_x and F_y on the vehicle at its mass center

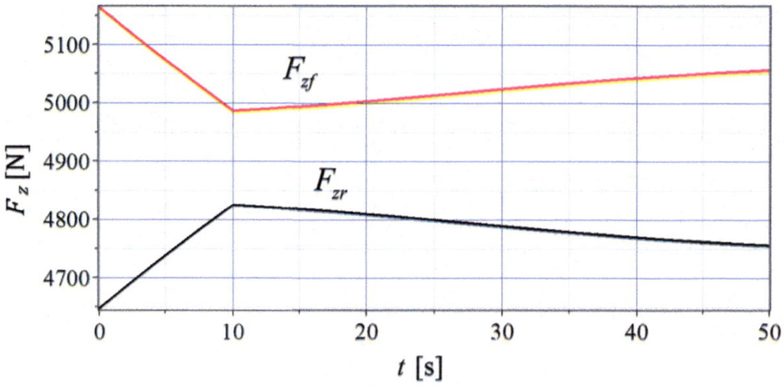

Fig. 2.43 The vertical load F_{zf} and F_{zr} on front and rear wheels

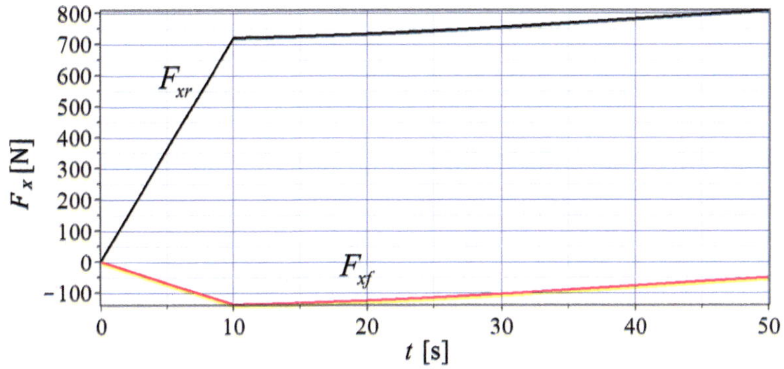

Fig. 2.44 The traction forces F_{xf} and F_{xr} on front and rear tires

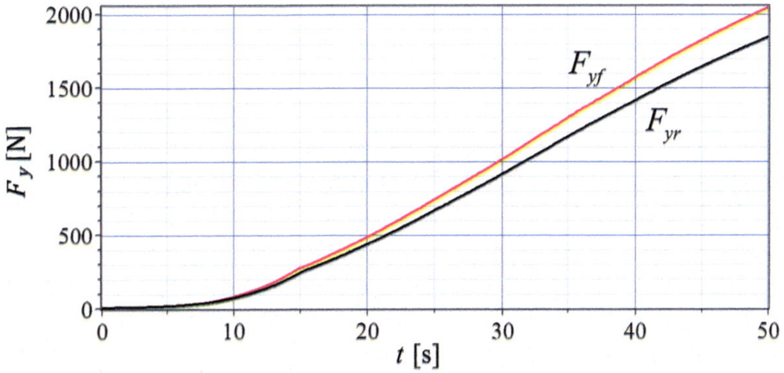

Fig. 2.45 The lateral forces F_{yf} and F_{yr} on front and rear tires

2.5 Steady-State Turning

The conditions at which all variables remain still and do not changes with time is called the steady-state conditions. To determine the steady-state relationships, it is enough to eliminate all time derivative terms from the equations of motion and solve the resultant algebraic equations for the steady-state values of the variables.

The steady-state output–input relationships of a front-wheel-steering two-wheel vehicle are defined by the following responses:

1. *Curvature response*, S_κ

$$S_\kappa = \frac{\kappa}{\delta} = \frac{1}{\rho\delta} = \frac{D_\delta C_\beta - D_\beta C_\delta}{l D_\beta v_x^2 + D_\kappa C_\beta - D_\beta C_\kappa} \tag{2.266}$$

$$= \frac{g C_{\alpha f} C_{\alpha r}}{l g C_{\alpha f} C_{\alpha r} + v_x^2 \left(C_{\alpha r} - C_{\alpha f} \right)} \tag{2.267}$$

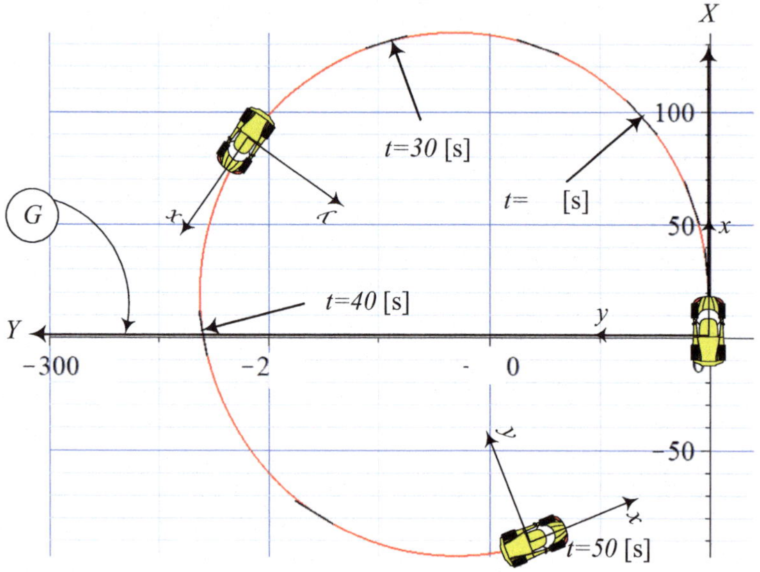

Fig. 2.46 Path of motion of the vehicle for $0 < t < 50$ s by increasing steer angle up to 1.5 deg and applying rear torque up to 300 N m

where

$$C_\beta = -ga_2 C_{\alpha f} - ga_1 C_{\alpha r} \tag{2.268}$$

$$C_\kappa = -ga_1 a_2 C_{\alpha f} + ga_1 a_2 C_{\alpha r} \tag{2.269}$$

$$C_\delta = ga_2 C_{\alpha f} \tag{2.270}$$

$$D_\beta = C_{\alpha r} - C_{\alpha f} \tag{2.271}$$

$$D_\kappa = -a_2 C_{\alpha r} - a_1 C_{\alpha f} \tag{2.272}$$

$$D_\delta = C_{\alpha f} \tag{2.273}$$

Curvature response S_κ determines the steady-state radius of rotation, $\rho = 1/\kappa$, of the vehicle for a given steer angle δ at a given v_x.

2. *Sideslip response*, S_β

$$S_\beta = \frac{\beta}{\delta} = \frac{D_\kappa C_\delta - D_\delta \left(C_\kappa - l v_x^2\right)}{l D_\beta v_x^2 + D_\kappa C_\beta - D_\beta C_\kappa} \tag{2.274}$$

$$= \frac{v_x^2 C_{\alpha f} - g a_2 C_{\alpha f} C_{\alpha r}}{l g C_{\alpha f} C_{\alpha r} + v_x^2 \left(C_{\alpha r} - C_{\alpha f}\right)} \tag{2.275}$$

Sideslip response S_β determines the steady-state angle of velocity vector, $\beta = v_y/v_x$, of the vehicle for a given steer angle δ at a given v_x (Marzbani and Jazar 2015).

Proof The equations of motion of two-wheel planar dynamics (2.203)–(2.207)

$$\dot{v}_x = \frac{1}{m}(F_x - F_A) + r\,v_y \tag{2.276}$$

$$\dot{v}_y = \frac{1}{m}F_y - r\,v_x \tag{2.277}$$

$$\dot{r} = \frac{1}{I_z}M_z \tag{2.278}$$

$$\dot{\omega}_f = \frac{1}{I_f}T_f - \frac{R_g}{I_f}F_{x_f} \tag{2.279}$$

$$\dot{\omega}_r = \frac{1}{I_r}T_r - \frac{R_g}{I_r}F_{x_r} \tag{2.280}$$

will be simplified to the following at steady-state conditions and ignoring the aerodynamic forces.

$$F_x = -mr\,v_y \tag{2.281}$$

$$F_y = mr\,v_x \tag{2.282}$$

$$M_z = 0 \tag{2.283}$$

$$F_{x_f} = \frac{1}{R_g}T_f \tag{2.284}$$

$$F_{x_r} = \frac{1}{R_g}T_r \tag{2.285}$$

The vehicle force system

$$F_x = F_{x_f}\cos\delta + F_{x_r} - F_{y_f}\sin\delta \tag{2.286}$$

$$F_y = F_{y_f}\cos\delta + F_{y_r} + F_{x_f}\sin\delta \tag{2.287}$$

$$M_z = a_1 F_{y_f}\cos\delta + a_1 F_{x_f}\sin\delta - a_2 F_{y_r} \tag{2.288}$$

will be approximated for small steer angles to:

$$F_x = F_{x_f} + F_{x_r} = \frac{1}{R_g}\left(T_f + T_r\right) \tag{2.289}$$

$$F_y = F_{y_f} + F_{y_r} \tag{2.290}$$

$$M_z = a_1 F_{y_f} - a_2 F_{y_r} \tag{2.291}$$

The tire forces with combined slips effect

$$F_{x_f} = F_{z_f} C_{sf}\, S\left(s_f - s_s\right)\sqrt{1 - C_{s\alpha}\left(\frac{S\left(\alpha_f - \alpha_s\right)}{\alpha_s}\right)^2} \tag{2.292}$$

$$F_{x_r} = F_{z_r} C_{sr}\, S\left(s_r - s_s\right)\sqrt{1 - C_{s\alpha}\left(\frac{S\left(\alpha_r - \alpha_s\right)}{\alpha_s}\right)^2} \tag{2.293}$$

$$F_{y_f} = -F_{z_f} C_{\alpha f}\, S\left(\alpha_f - \alpha_s\right)\sqrt{1 - C_{\alpha s}\left(\frac{S\left(s_f - s_s\right)}{s_s}\right)^2} \tag{2.294}$$

$$F_{y_r} = -F_{z_r} C_{\alpha r}\, S\left(\alpha_r - \alpha_s\right)\sqrt{1 - C_{\alpha s}\left(\frac{S\left(s_r - s_s\right)}{s_s}\right)^2} \tag{2.295}$$

may be well approximated to the following equations

$$F_{x_f} = F_{z_f} C_{sf}\, s_f \left(1 - \frac{1}{2} C_{s\alpha}\left(\frac{\alpha_f}{\alpha_s}\right)^2\right) \tag{2.296}$$

$$F_{x_r} = F_{z_r} C_{sr}\, s_r \left(1 - \frac{1}{2} C_{s\alpha}\left(\frac{\alpha_r}{\alpha_s}\right)^2\right) \tag{2.297}$$

$$F_{y_f} = -F_{z_f} C_{\alpha f}\, \alpha_f \left(1 - \frac{1}{2} C_{\alpha s}\left(\frac{s_f}{s_s}\right)^2\right) \tag{2.298}$$

$$F_{y_r} = -F_{z_r} C_{\alpha r}\, \alpha_r \left(1 - \frac{1}{2} C_{\alpha s}\left(\frac{s_r}{s_s}\right)^2\right) \tag{2.299}$$

and the vertical load equations

$$F_{z_f} = F_{z_1} + F_{z_2} = mg\frac{a_2}{l} - m\left(\dot{v}_x - r\, v_y\right)\frac{h}{l} \tag{2.300}$$

$$F_{z_r} = F_{z_3} + F_{z_4} = mg\frac{a_1}{l} + m\left(\dot{v}_x - r\, v_y\right)\frac{h}{l} \tag{2.301}$$

$$l = a_1 + a_2 \tag{2.302}$$

could be assumed to be equal to the following equation, ignoring weight transfer effects.

$$F_{z_f} = mg\frac{a_2}{l} \tag{2.303}$$

$$F_{z_r} = mg\frac{a_1}{l} \tag{2.304}$$

The lateral slip angles α_f and α_r are assumed to be small.

$$\alpha_f = \arctan\left(\frac{v_y + a_1 r}{v_x}\right) - \delta \tag{2.305}$$

$$\alpha_r = \arctan\left(\frac{v_y - a_2 r}{v_x}\right) \tag{2.306}$$

$$\beta = \arctan\frac{v_y}{v_x} \tag{2.307}$$

Introducing the curvature radius ρ and curvature $\kappa = 1/\rho$

$$\rho = \frac{1}{\kappa} = \frac{v_x}{r} \tag{2.308}$$

we may rewrite the lateral slips as:

$$\alpha_f = \frac{v_y + a_1 r}{v_x} - \delta = \beta + \frac{a_1}{\rho} - \delta = \beta + \kappa a_1 - \delta \tag{2.309}$$

$$\alpha_r = \frac{v_y - a_2 r}{v_x} = \beta - \frac{a_2}{\rho} = \beta - \kappa a_2 \tag{2.310}$$

$$\beta = \frac{v_y}{v_x} \tag{2.311}$$

Removing the braking case from the slip ratios s_f and s_r

$$s_f = \frac{R_g\,\omega_f - v_{xT_f}}{R_g\,\omega_f\, H(R_g\,\omega_f - v_{xT_f}) + v_{xT_f}\, H(v_{xT_f} - R_g\,\omega_f)} \tag{2.312}$$

$$s_r = \frac{R_g\,\omega_r - v_x}{R_g\,\omega_r\, H(R_g\,\omega_r - v_x) + v_x\, H(v_x - R_g\,\omega_r)} \tag{2.313}$$

$$v_{xT_f} = v_x \cos\delta + \left(v_y + a_1 r\right)\sin\delta \tag{2.314}$$

provides us with the primary definition of slip ratios in traction condition.

$$s_f = \frac{R_g\,\omega_f - v_x}{R_g\,\omega_f} \tag{2.315}$$

$$s_r = \frac{R_g\,\omega_r - v_x}{R_g\,\omega_r} \tag{2.316}$$

Substituting (2.289)–(2.291) in (2.281)–(2.285)

$$\frac{1}{R_g}\left(T_f + T_r\right) = F_{x_f} + F_{x_r} = -mr\,v_y \tag{2.317}$$

$$F_{y_f} + F_{y_r} = mr\, v_x \tag{2.318}$$

$$0 = a_1 F_{y_f} - a_2 F_{y_r} \tag{2.319}$$

$$T_f = R_g F_{x_f} \tag{2.320}$$

$$T_r = R_g\, F_{x_r} \tag{2.321}$$

and then using (2.296)–(2.299) and (2.303)–(2.304) provides us with the following equations.

$$\frac{1}{R_g}\left(T_f + T_r\right) = -mr\, v_y \tag{2.322}$$

$$-mg\frac{a_2}{l}C_{\alpha f}\,\alpha_f\left(1 - \frac{1}{2}C_{\alpha s}\left(\frac{s_f}{s_s}\right)^2\right)$$
$$-mg\frac{a_1}{l}C_{\alpha r}\,\alpha_r\left(1 - \frac{1}{2}C_{\alpha s}\left(\frac{s_r}{s_s}\right)^2\right) = mr\, v_x \tag{2.323}$$

$$-a_1 mg\frac{a_2}{l}C_{\alpha f}\,\alpha_f\left(1 - \frac{1}{2}C_{\alpha s}\left(\frac{s_f}{s_s}\right)^2\right)$$
$$+a_2 mg\frac{a_1}{l}C_{\alpha r}\,\alpha_r\left(1 - \frac{1}{2}C_{\alpha s}\left(\frac{s_r}{s_s}\right)^2\right) = 0 \tag{2.324}$$

$$R_g mg\frac{a_2}{l}C_{sf}\, s_f\left(1 - \frac{1}{2}C_{s\alpha}\left(\frac{\alpha_f}{\alpha_s}\right)^2\right) = R_g F_{x_f} = T_f \tag{2.325}$$

$$R_g\, mg\frac{a_1}{l}C_{sr}\, s_r\left(1 - \frac{1}{2}C_{s\alpha}\left(\frac{\alpha_r}{\alpha_s}\right)^2\right) = R_g\, F_{x_r} = T_r \tag{2.326}$$

Substituting lateral slips (2.309)–(2.311) makes the equations

$$\frac{1}{R_g}\left(T_f + T_r\right) = -mr\, v_y \tag{2.327}$$

$$-mg\frac{a_2}{l}C_{\alpha f}\,\left(\beta + \kappa a_1 - \delta\right)\left(1 - \frac{1}{2}C_{\alpha s}\left(\frac{s_f}{s_s}\right)^2\right)$$
$$-mg\frac{a_1}{l}C_{\alpha r}\,\left(\beta + \kappa a_2\right)\left(1 - \frac{1}{2}C_{\alpha s}\left(\frac{s_r}{s_s}\right)^2\right) = mr\, v_x \tag{2.328}$$

$$-a_1 mg\frac{a_2}{l}C_{\alpha f}\,\left(\beta + \kappa a_1 - \delta\right)\left(1 - \frac{1}{2}C_{\alpha s}\left(\frac{s_f}{s_s}\right)^2\right)$$

$$+a_2 mg\frac{a_1}{l}C_{\alpha r}\ (\beta+\kappa a_2)\left(1-\frac{1}{2}C_{\alpha s}\left(\frac{s_r}{s_s}\right)^2\right)=0 \tag{2.329}$$

$$R_g mg\frac{a_2}{l}C_{sf}\,s_f\left(1-\frac{1}{2}C_{s\alpha}\left(\frac{\beta+\kappa a_1-\delta}{\alpha_s}\right)^2\right)=R_g F_{x_f}=T_f \tag{2.330}$$

$$R_g\,mg\frac{a_1}{l}C_{sr}\,s_r\left(1-\frac{1}{2}C_{s\alpha}\left(\frac{\beta+\kappa a_2}{\alpha_s}\right)^2\right)=R_g\,F_{x_r}=T_r \tag{2.331}$$

Let us assume the slip ratios s_f and s_r to be very small such that the nonlinear terms of Eqs. (2.298)–(2.299) may be ignored and the slip ratios s_i will not affect sideslip angles α_i. The tire slip interaction will be only due to α_i reducing s_i. Therefore, the equations will become simpler as the following.

$$\frac{1}{R_g}\left(T_f+T_r\right)=-m\kappa v_x v_y \tag{2.332}$$

$$-ga_2C_{\alpha f}\ (\beta+\kappa a_1-\delta)-ga_1C_{\alpha r}\ (\beta-\kappa a_2)=\kappa l v_x^2 \tag{2.333}$$

$$-C_{\alpha f}\ (\beta+\kappa a_1-\delta)+C_{\alpha r}\ (\beta-\kappa a_2)=0 \tag{2.334}$$

$$R_g mg\frac{a_2}{l}C_{sf}\,s_f\left(1-\frac{1}{2}C_{s\alpha}\left(\frac{\beta+\kappa a_1-\delta}{\alpha_s}\right)^2\right)=R_g F_{x_f}=T_f \tag{2.335}$$

$$R_g\,mg\frac{a_1}{l}C_{sr}\,s_r\left(1-\frac{1}{2}C_{s\alpha}\left(\frac{\beta-\kappa a_2}{\alpha_s}\right)^2\right)=R_g\,F_{x_r}=T_r \tag{2.336}$$

Equations (2.333) and (2.334) are linear functions of β, κ, δ, and may be rearranged in the following forms

$$C_\beta\beta+C_\kappa\kappa+C_\delta\delta=\kappa l v_x^2 \tag{2.337}$$

$$D_\beta\beta+D_\kappa\kappa+D_\delta\delta=0 \tag{2.338}$$

where

$$C_\beta=-ga_2C_{\alpha f}-ga_1C_{\alpha r} \tag{2.339}$$

$$C_\kappa=-ga_1a_2C_{\alpha f}+ga_1a_2C_{\alpha r} \tag{2.340}$$

$$C_\delta=ga_2C_{\alpha f} \tag{2.341}$$

$$D_\beta=C_{\alpha r}-C_{\alpha f} \tag{2.342}$$

$$D_\kappa=-a_2C_{\alpha r}-a_1C_{\alpha f} \tag{2.343}$$

$$D_\delta=C_{\alpha f} \tag{2.344}$$

Equations (2.337) and (2.338) make a set of two linear algebraic equations

$$\begin{bmatrix} C_\beta & C_\kappa - l v_x^2 \\ D_\beta & D_\kappa \end{bmatrix} \begin{bmatrix} \beta \\ \kappa \end{bmatrix} = \begin{bmatrix} C_\delta \\ D_\delta \end{bmatrix} \delta$$

which provides us with the following solutions as the steady-state responses.

$$S_\kappa = \frac{\kappa}{\delta} = \frac{1}{\rho\delta} = \frac{D_\delta C_\beta - D_\beta C_\delta}{l D_\beta v_x^2 + D_\kappa C_\beta - D_\beta C_\kappa}$$

$$S_\beta = \frac{\beta}{\delta} = \frac{D_\kappa C_\delta - D_\delta \left(C_\kappa - l v_x^2\right)}{l D_\beta v_x^2 + D_\kappa C_\beta - D_\beta C_\kappa}$$

Substituting (2.339)–(2.344) determines the steady-state responses as functions of vehicle dynamic parameters.

$$S_\kappa = \frac{\kappa}{\delta} = \frac{g C_{\alpha f} C_{\alpha r}}{l g C_{\alpha f} C_{\alpha r} + v_x^2 \left(C_{\alpha r} - C_{\alpha f}\right)}$$

$$S_\beta = \frac{\beta}{\delta} = \frac{v_x^2 C_{\alpha f} - g a_2 C_{\alpha f} C_{\alpha r}}{l g C_{\alpha f} C_{\alpha r} + v_x^2 \left(C_{\alpha r} - C_{\alpha f}\right)}$$

■

Example 73 Under steering, over steering, neutral steering.

Curvature response S_κ indicates how the radius of turning will change with a change in steer angle. S_κ can be expressed as

$$S_\kappa = \frac{\kappa}{\delta} = \frac{1/\rho}{\delta} = \frac{1}{l} \frac{1}{1 + K v_x^2} \tag{2.345}$$

$$K = \frac{1}{g l} \left(\frac{1}{C_{\alpha f}} - \frac{1}{C_{\alpha r}} \right) \tag{2.346}$$

where K is called the **stability factor**. It determines if the vehicle is

$$\begin{array}{ll} Understeer & K > 0 \\ Neutral & K = 0 \\ Oversteer & K < 0 \end{array} \tag{2.347}$$

To find K we may rewrite S_κ as

$$S_\kappa = \frac{\kappa}{\delta} = \frac{1/\rho}{\delta} = \frac{g C_{\alpha f} C_{\alpha r}}{l g C_{\alpha f} C_{\alpha r} + v_x^2 \left(C_{\alpha r} - C_{\alpha f}\right)}$$

$$= \frac{1}{\dfrac{l g C_{\alpha f} C_{\alpha r}}{g C_{\alpha f} C_{\alpha r}} + \dfrac{\left(C_{\alpha r} - C_{\alpha f}\right) v_x^2}{g C_{\alpha f} C_{\alpha r}}}$$

$$= \frac{1}{l + \frac{1}{g}\left(\frac{1}{C_{\alpha f}} - \frac{1}{C_{\alpha r}}\right) v_x^2} = \frac{1}{l} \frac{1}{1 + \frac{1}{gl}\left(\frac{1}{C_{\alpha f}} - \frac{1}{C_{\alpha r}}\right) v_x^2}$$

$$= \frac{1}{l} \frac{1}{1 + K v_x^2} \tag{2.348}$$

Therefore,

$$K = \frac{1}{gl}\left(\frac{1}{C_{\alpha f}} - \frac{1}{C_{\alpha r}}\right) \tag{2.349}$$

The sign of stability factor K determines if S_κ is an increasing or decreasing function of velocity v_x. The sign of K depends on the relative weight of $1/C_{\alpha f}$ and $1/C_{\alpha r}$, which are dependent on the sideslip coefficients of the front and rear tires $C_{\alpha f}$, $C_{\alpha r}$.

If $K > 0$, then

$$\frac{1}{C_{\alpha f}} > \frac{1}{C_{\alpha r}} \tag{2.350}$$

and $S_\kappa = \kappa/\delta$ and $dS_\kappa/dv_x < 0$. Hence, the curvature of the path $\kappa = 1/\rho$ decreases with speed for a constant δ. Decreasing κ indicates that the radius of the steady-state circle of rotation, ρ, increases by increasing speed v_x. A positive stability factor is desirable. A vehicle with $K > 0$ is stable and is called **understeer**. For an understeer vehicle, we need to increase the steering angle if we increase the speed of the vehicle to keep the same turning circle.

If $K < 0$, then

$$\frac{1}{C_{\alpha f}} < \frac{1}{C_{\alpha r}} \tag{2.351}$$

and $S_\kappa = \kappa/\delta$ and $dS_\kappa/dv_x > 0$. Hence, the curvature of the path $\kappa = 1/\rho$ increases with speed for a constant δ. Increasing κ indicates that the radius of the steady-state circle of rotation, ρ, decreases by increasing speed v_x. A negative stability factor is undesirable. A vehicle with $K < 0$ is unstable and is called **oversteer**. For an oversteer vehicle, we need to decrease the steering angle when we increase the speed of the vehicle, to keep the same turning circle.

If $K = 0$, then

$$\frac{1}{C_{\alpha f}} = \frac{1}{C_{\alpha r}} \tag{2.352}$$

then $S_\kappa = \kappa/\delta$ is not a function of v_x because $dS_\kappa/dv_x = 0$. Hence, the curvature of the path $\kappa = 1/\rho$ remains constant for a constant δ regardless of v_x. Having a constant κ indicates that the radius of the steady-state circle, R, will not change by changing the speed v_x. A zero stability factor is neutral and a vehicle with $K = 0$ is on the border of stability and is called a **neutral steer**. When driving a neutral steer vehicle, we do not need to change the steering angle if we increase or decrease the speed of the vehicle, to keep the same turning circle.

2.6 Four-Wheel Planar Vehicle Dynamics

The four-wheel planar vehicle model is an extension to the two-wheel planar vehicle model to include the lateral weight transfer. The four-wheel planar model provides us with better simulation of drifting vehicles.

Let us assume that the front and rear tracks are different.

$$b_{1f} + b_{2f} = w_f \tag{2.353}$$

$$b_{1r} + b_{2r} = w_r \tag{2.354}$$

The equations of motion of the four-wheel rigid planar vehicle are:

$$\dot{v}_x = \frac{1}{m}(F_x - F_A) + r\, v_y \tag{2.355}$$

$$\dot{v}_y = \frac{1}{m}F_y - r\, v_x \tag{2.356}$$

$$\dot{r} = \frac{1}{I_z}M_z \tag{2.357}$$

$$\dot{\omega}_1 = \frac{1}{I_1}T_1 - \frac{R_g}{I_1}\, F_{x_1} \tag{2.358}$$

$$\dot{\omega}_2 = \frac{1}{I_2}T_2 - \frac{R_g}{I_2}\, F_{x_2} \tag{2.359}$$

$$\dot{\omega}_1 = \frac{1}{I_3}T_3 - \frac{R_g}{I_3}\, F_{x_3} \tag{2.360}$$

$$\dot{\omega}_1 = \frac{1}{I_4}T_4 - \frac{R_g}{I_4}\, F_{x_4} \tag{2.361}$$

$$F_x = F_{x_1}\cos\delta_1 + F_{x_2}\cos\delta_2 + F_{x_3} + F_{x_4} - F_{y_1}\sin\delta_1 - F_{y_2}\sin\delta_2 \tag{2.362}$$

$$F_y = F_{y_1}\cos\delta_1 + F_{y_2}\cos\delta_2 + F_{y_3} + F_{y_4} + F_{x_1}\sin\delta_1 + F_{x_2}\sin\delta_2 \tag{2.363}$$

$$F_A = C_A\, v_x^2 \tag{2.364}$$

$$\begin{aligned} M_z = & \, a_1 F_{x_1} \sin\delta_1 + a_1 F_{y_1} \cos\delta_1 + a_1 F_{x_2} \sin\delta_2 + a_1 F_{y_2} \cos\delta_2 \\ & -b_{1f} F_{x_1} \cos\delta_1 + b_{1f} F_{y_1} \sin\delta_1 - b_{2f} F_{y_2} \sin\delta_2 + b_{2f} F_{x_2} \cos\delta_2 \\ & -a_2 F_{y_3} - a_2 F_{y_4} - b_{1r} F_{x_3} + b_{2r} F_{x_4} \end{aligned} \tag{2.365}$$

$$F_{x_i} = F_{z_i} C_{si}\, S\left(s_i - s_s\right) \sqrt{1 - C_{s\alpha} \left(\frac{S\left(\alpha_i - \alpha_s\right)}{\alpha_s} \right)^2} \tag{2.366}$$

$$F_{y_i} = -F_{z_i} C_{\alpha i}\, S\left(\alpha_i - \alpha_s\right) \sqrt{1 - C_{\alpha s} \left(\frac{S\left(s_i - s_s\right)}{s_s} \right)^2} \tag{2.367}$$

$$F_{z_1} = \frac{m}{l w_f g} \left(a_2 g - h\left(\dot{v}_x - r\, v_y\right)\right)\left(b_{2f} g - h\left(\dot{v}_y + r\, v_x\right)\right) \tag{2.368}$$

$$F_{z_2} = \frac{m}{l w_f g} \left(a_2 g - h\left(\dot{v}_x - r\, v_y\right)\right)\left(b_{1f} g + h\left(\dot{v}_y + r\, v_x\right)\right) \tag{2.369}$$

$$F_{z_3} = \frac{m}{l w_r g} \left(a_1 g + h\left(\dot{v}_x - r\, v_y\right)\right)\left(b_{2r} g - h\left(\dot{v}_y + r\, v_x\right)\right) \tag{2.370}$$

$$F_{z_4} = \frac{m}{l w_r g} \left(a_1 g + h\left(\dot{v}_x - r\, v_y\right)\right)\left(b_{1r} g + h\left(\dot{v}_y + r\, v_x\right)\right) \tag{2.371}$$

and tire slips as

$$\alpha_i = \beta_i - \delta_i = \arctan\left(\frac{v_y + x_i\, r}{v_x - y_i\, r} \right) - \delta_i \tag{2.372}$$

$$\beta_i = \arctan\left(\frac{v_{y_i}}{v_{x_i}} \right) = \arctan\left(\frac{v_y + x_i\, r}{v_x - y_i\, r} \right) \tag{2.373}$$

$$\beta = \arctan \frac{v_y}{v_x} \tag{2.374}$$

$$x_1 = x_2 = a_1 \qquad y_1 = b_{1f} \qquad y_3 = b_{1r} \tag{2.375}$$

$$x_3 = x_4 = -a_2 \qquad y_2 = -b_{2f} \qquad y_4 = -b_{2r} \tag{2.376}$$

$$s_i = \frac{R_g\, \omega_i - v_{xTi}}{R_g\, \omega_i\, H(R_g\, \omega_i - v_{xT_i}) + v_{xTi}\, H(v_{xTi} - R_g\, \omega_i)} \tag{2.377}$$

$$v_{xT1} = \left(v_x - r b_{1f}\right) \cos\delta_1 + \left(v_y + r a_1\right) \sin\delta_1 \tag{2.378}$$

$$v_{xT2} = \left(v_x + rb_{1f}\right)\cos\delta_2 + \left(v_y + ra_1\right)\sin\delta_2 \tag{2.379}$$

$$v_{xT3} = v_x - rb_{2r} \tag{2.380}$$

$$v_{xT4} = v_x + rb_{2r} \tag{2.381}$$

The left and right steer angles are related due to steering mechanism.

$$\delta_2 = f\left(\delta_1\right) \tag{2.382}$$

Proof The equations of motion for a flat and rigid vehicle in the local coordinate frame B, at its mass center C, are given in Eqs. (2.1)–(2.4). The equations for a four-wheel vehicle model will be:

$$F_x = m\,\dot{v}_x - mr\,v_y + F_A \tag{2.383}$$

$$F_y = m\,\dot{v}_y + mr\,v_x \tag{2.384}$$

$$M_z = \dot{r}\,I_z \tag{2.385}$$

$$T_1 = I_{w_1}\,\dot{\omega}_1 + R_w\,F_{x_1} \tag{2.386}$$

$$T_2 = I_{w_2}\,\dot{\omega}_2 + R_w\,F_{x_2} \tag{2.387}$$

$$T_3 = I_{w_3}\,\dot{\omega}_r + R_w\,F_{x_3} \tag{2.388}$$

$$T_4 = I_{w_4}\,\dot{\omega}_r + R_w\,F_{x_4} \tag{2.389}$$

We add the aerodynamic force F_A as the only resistance forces on the vehicle. We assume F_A to be effective only in the x-direction

$$F_A = \frac{1}{2}\rho\,C_D\,A_f\,v_x^2 = C_A\,v_x^2 \tag{2.390}$$

where ρ is the air density, C_D is the drag coefficient, and A_f is the frontal area or projected area of vehicle in x-direction. We combine the coefficients ρ, C_D, and A_f into a single coefficient C_A.

From (2.59) to (2.61), the applied forces on the vehicle are:

$$\begin{aligned} F_x &= \sum_i F_{xw_i}\cos\delta_i - \sum_i F_{yw_i}\sin\delta_i \\ &= F_{x_1}\cos\delta_1 + F_{x_2}\cos\delta_2 + F_{x_3} + F_{x_4} \\ &\quad - F_{y_1}\sin\delta_1 - F_{y_2}\sin\delta_2 \end{aligned} \tag{2.391}$$

$$\begin{aligned}F_y &= \sum_i F_{yW_i} \cos\delta_i + \sum_i F_{xW_i} \sin\delta_i \\ &= F_{y_1} \cos\delta_1 + F_{y_2} \cos\delta_2 + F_{y_3} + F_{y_4} \\ &\quad + F_{x_1} \sin\delta_1 + F_{x_2} \sin\delta_2\end{aligned} \tag{2.392}$$

and the moment in the z-direction is:

$$\mathbf{M} = \sum \mathbf{r}_i \times \mathbf{F}_i = \begin{bmatrix} 0 \\ 0 \\ M_z \end{bmatrix} \tag{2.393}$$

$$\begin{aligned}M_z &= \sum_i x_i F_{y_i} - \sum_i y_i F_{x_i} \\ &= a_1 F_{x_1} \sin\delta_1 + a_1 F_{y_1} \cos\delta_1 + a_1 F_{x_2} \sin\delta_2 \\ &\quad + a_1 F_{y_2} \cos\delta_2 - a_2 F_{y_3} - a_2 F_{y_4} \\ &\quad - b_{1f} F_{x_1} \cos\delta_1 + b_{1f} F_{y_1} \sin\delta_1 - b_{2f} F_{y_2} \sin\delta_2 \\ &\quad + b_{2f} F_{x_2} \cos\delta_2 - b_{1r} F_{x_3} + b_{2r} F_{x_4}\end{aligned} \tag{2.394}$$

because the forces and location of each tire is:

$$\mathbf{F}_1 = \begin{bmatrix} F_{x_1} \cos\delta_1 - F_{y_1} \sin\delta_1 \\ F_{x_1} \sin\delta_1 + F_{y_1} \cos\delta_1 \\ 0 \end{bmatrix} \qquad \mathbf{F}_3 = \begin{bmatrix} F_{x3} \\ F_{y3} \\ 0 \end{bmatrix} \tag{2.395}$$

$$\mathbf{F}_2 = \begin{bmatrix} F_{x_2} \cos\delta_2 - F_{y_2} \sin\delta_2 \\ F_{x_2} \sin\delta_2 + F_{y_2} \cos\delta_2 \\ 0 \end{bmatrix} \qquad \mathbf{F}_4 = \begin{bmatrix} F_{x4} \\ F_{y4} \\ 0 \end{bmatrix} \tag{2.396}$$

$$\mathbf{r}_1 = \begin{bmatrix} a_1 \\ b_{1f} \\ 0 \end{bmatrix} \qquad \mathbf{r}_2 = \begin{bmatrix} a_1 \\ -b_{2f} \\ 0 \end{bmatrix} \tag{2.397}$$

$$\mathbf{r}_3 = \begin{bmatrix} -a_2 \\ b_{1r} \\ 0 \end{bmatrix} \qquad \mathbf{r}_4 = \begin{bmatrix} -a_2 \\ -b_{2r} \\ 0 \end{bmatrix} \tag{2.398}$$

Employing the combined slip elliptical model, the tire tangential forces F_{x_i} and F_{y_i} are:

$$F_{x_i} = F_{z_i} C_{si}\, S\left(s_i - s_s\right) \sqrt{1 - C_{s\alpha} \left(\frac{S\left(\alpha_i - \alpha_s\right)}{\alpha_s}\right)^2} \tag{2.399}$$

$$F_{y_i} = -F_{z_i} C_{\alpha i}\, S\left(\alpha_i - \alpha_s\right) \sqrt{1 - C_{\alpha s} \left(\frac{S\left(s_i - s_s\right)}{s_s}\right)^2} \tag{2.400}$$

The vertical forces on front and rear wheels F_{z_i} are calculated in (1.431)–(1.434) in which the acceleration components a_X and a_Y must be expressed in body coordinate B

$$a_X = \dot{v}_x - r\, v_y \tag{2.401}$$

$$a_Y = \dot{v}_y + r\, v_x \tag{2.402}$$

to be used in Eqs. (2.399) and (2.400). Rearrangement of Eqs. (1.431)–(1.434) is:

$$F_{z_1} = \frac{m}{l w_f g} \left(g a_2 - h \dot{v}_x + h r v_y\right) \left(h \dot{v}_y + g b_{2f} - h r v_x\right) \tag{2.403}$$

$$F_{z_2} = \frac{m}{l w_f g} \left(g a_2 - h \dot{v}_x + h r v_y\right) \left(h \dot{v}_y + g b_{1f} + h r v_x\right) \tag{2.404}$$

$$F_{z_3} = \frac{m}{l w_r g} \left(h \dot{v}_x + g a_1 - h r v_y\right) \left(h \dot{v}_y + g b_{2r} - h r v_x\right) \tag{2.405}$$

$$F_{z_4} = \frac{m}{l w_r g} \left(h \dot{v}_x + g a_1 - h r v_y\right) \left(h \dot{v}_y + g b_{1r} + h r v_x\right) \tag{2.406}$$

where

$$w_f = b_{1f} + b_{2f} \tag{2.407}$$

$$w_r = b_{1r} + b_{2r} \tag{2.408}$$

$$l = a_1 + a_2 \tag{2.409}$$

The tire side slip angle α_i has been calculated in (1.121)–(1.123)

$$\alpha_i = \beta_i - \delta_i = \arctan\left(\frac{v_y + x_i\, r}{v_x - y_i\, r}\right) - \delta_i$$

where

$$\beta_i = \arctan\left(\frac{v_{y_i}}{v_{x_i}}\right) = \arctan\left(\frac{v_y + x_i\, r}{v_x - y_i\, r}\right) \tag{2.410}$$

$$v_{x_i} = v_x - y_i\, r \qquad v_{y_i} = v_y + x_i\, r \tag{2.411}$$

and therefore,

$$\beta_1 = \arctan\left(\frac{v_{y_1}}{v_{x_1}}\right) = \arctan\left(\frac{v_y + a_1 r}{v_x - b_{1f} r}\right) \tag{2.412}$$

$$\beta_2 = \arctan\left(\frac{v_{y_2}}{v_{x_2}}\right) = \arctan\left(\frac{v_y + a_1 r}{v_x + b_{2f} r}\right) \tag{2.413}$$

$$\beta_3 = \arctan\left(\frac{v_{y_3}}{v_{x_3}}\right) = \arctan\left(\frac{v_y - a_2 r}{v_x + b_{2r} r}\right) \tag{2.414}$$

$$\beta_4 = \arctan\left(\frac{v_{y_4}}{v_{x_4}}\right) = \arctan\left(\frac{v_y - a_2 r}{v_x - b_{1r} r}\right) \tag{2.415}$$

The longitudinal slip ratio s_i is calculated in (1.57) as a function of the tire forward velocity $v_{x_{Ti}}$ and its angular velocity ω_i.

$$s_i = \frac{R_g\, \omega_i - v_{x_{Ti}}}{R_g\, \omega_i\, H(R_g\, \omega_w - v_{x_{Ti}}) + v_{x_{Ti}}\, H(v_{x_{Ti}} - R_g\, \omega_i)} \tag{2.416}$$

According to (1.303), the forward velocity of the vehicle at the wheels centers is:

$$ {}^B\mathbf{v}_i = \begin{bmatrix} v\cos\beta - r y_i \\ v\sin\beta + r x_i \end{bmatrix} = \begin{bmatrix} v_x - r y_i \\ v_y + r x_i \end{bmatrix} \tag{2.417}$$

because

$$\begin{aligned} {}^B\mathbf{v}_i &= {}^B\mathbf{v} + {}^B_G\boldsymbol{\omega}_B \times {}^B\mathbf{r}_i \\ &= \begin{bmatrix} v_x \\ v_y \\ 0 \end{bmatrix} + \begin{bmatrix} 0 \\ 0 \\ r \end{bmatrix} \times \begin{bmatrix} x_i \\ y_i \\ 0 \end{bmatrix} = \begin{bmatrix} v_x - r y_i \\ v_y + r x_i \\ 0 \end{bmatrix} \end{aligned} \tag{2.418}$$

and therefore,

$$ {}^B\mathbf{v}_1 = \begin{bmatrix} v_x - r b_{1f} \\ v_y + r a_1 \end{bmatrix} \qquad {}^B\mathbf{v}_2 = \begin{bmatrix} v_x + r b_{1f} \\ v_y + r a_1 \end{bmatrix} \tag{2.419}$$

$$ {}^B\mathbf{v}_3 = \begin{bmatrix} v_x - r b_{2r} \\ v_y - r a_2 \end{bmatrix} \qquad {}^B\mathbf{v}_4 = \begin{bmatrix} v_x + r b_{2r} \\ v_y - r a_2 \end{bmatrix} \tag{2.420}$$

Therefore, the velocity at the center of the front wheel in the tire x-direction, ${}^T\hat{\imath}$, will be

$$v_{x_{T1}} = \left(v_x - r b_{1f}\right)\cos\delta_1 + \left(v_y + r a_1\right)\sin\delta_1 \tag{2.421}$$

$$v_{xT_2} = \left(v_x + rb_{1f}\right)\cos\delta_2 + \left(v_y + ra_1\right)\sin\delta_2 \tag{2.422}$$

$$v_{xT_3} = v_x - rb_{2r} \tag{2.423}$$

$$v_{xT_4} = v_x + rb_{2r} \tag{2.424}$$

because

$$
\begin{aligned}
{}^T\mathbf{v}_i = {}^C R_T^T \, {}^B\mathbf{v}_i &= \begin{bmatrix} \cos\delta & -\sin\delta & 0 \\ \sin\delta & \cos\delta & 0 \\ 0 & 0 & 1 \end{bmatrix}^T \begin{bmatrix} v_x - ry_i \\ v_y + rx_i \\ 0 \end{bmatrix} \\
&= \begin{bmatrix} (v_x - ry_i)\cos\delta + \left(v_y + rx_i\right)\sin\delta \\ \left(v_y + rx_i\right)\cos\delta - (v_x - ry_i)\sin\delta \\ 0 \end{bmatrix}
\end{aligned} \tag{2.425}
$$

The tire angular velocity ω_i comes from the equation of moment balance on each wheel. Ignoring the tire resistance force, the rotational equation of motion of the wheels is:

$$I_{w_i}\,\dot{\omega}_i = T_i - R_i\,F_{x_i} \tag{2.426}$$

The traditional four-wheel vehicles with the front steerable wheels need to connect the left and right front wheels together such that only one steer angle is the driver's command and the other wheels steer according to the connection rule. The connection might be mechanical, as a passive steering mechanism does the job. Let us assume that the tire number 1 is the commander and all other steer angles are follower. The connection provides us with a mathematical equation to calculate δ_2 for a given δ_1.

$$\delta_2 = f\left(\delta_1\right) \tag{2.427}$$

The connection may also be virtual as steer by wire systems do the job. Such a smart system receives commands from a central system and activates individual actuators on each wheel to give them the required steer angles calculated by a computer system based on an objective function.

The simplest mathematical relation is the low speed neutral kinematic condition called the *Ackerman condition*

$$\cot\delta_o - \cot\delta_i = \frac{w}{l} \tag{2.428}$$

in which δ_i is the steer angle of the inner wheel, and δ_o is the steer angle of the outer wheel. The inner and outer wheels are defined based on the turning center of the vehicle. This equation is only a synthetic relation which theoretically works at zero speed and impossible to make a simple low bar mechanisms to provide the equation. However, it is a good starting point. ■

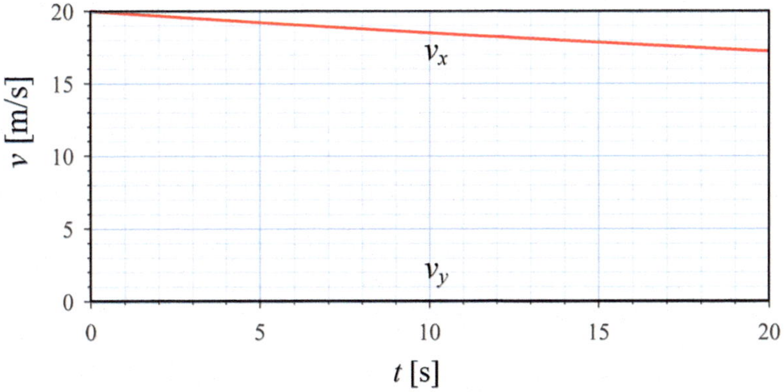

Fig. 2.47 The effect of aerodynamic drag force $F_A = C_A v_x^2$ on forward velocity of a four-wheel planar vehicle in free straight motion

Example 74 Aerodynamic effect.

Consider a vehicle with the following data.

$$\begin{aligned}
&m = 1000\,\text{kg} \qquad Iz = 2000\,\text{kg m}^2 \qquad I_1 = I_2 = I_3 = I_4 = 30\,\text{kg m}^2 \\
&C_{\alpha_1} = C_{\alpha_2} = C_{\alpha_3} = C_{\alpha_4} = 8.5 \qquad R_g = 0.35\,\text{m} \\
&C_{s_1} = C_{s_2} = C_{s_3} = C_{s_4} = 7.5 \\
&\alpha_s = 5\,\text{deg} \qquad s_s = 0.1 \qquad C_{\alpha s} = 0.5 \qquad C_{s\alpha} = 0.5 \\
&a_1 = 1.35\,\text{m} \qquad a_2 = 1.5\,\text{m} \qquad l = 2.85\,\text{m} \qquad h = 0.9\,\text{m} \\
&b_{1f} = b_{1r} = b_{2f} = b_{2r} = 0.9\,\text{m} \qquad w_f = w_r = 1.8\,\text{m}
\end{aligned} \tag{2.429}$$

Assume the car initially is moving at

$$v_x = 20\,\text{m/s} \qquad \delta = 0 \qquad T_i = 0 \tag{2.430}$$

and

$$\omega_i = \frac{v_x}{R_g} = \frac{20}{0.35} = 57.143\,\text{rad/s} \tag{2.431}$$

The effect of the aerodynamic resistance force in slowing down the vehicle can be seen in the time history of the forward velocity as is shown in Fig. 2.47 for:

$$C_A = 0.8 \tag{2.432}$$

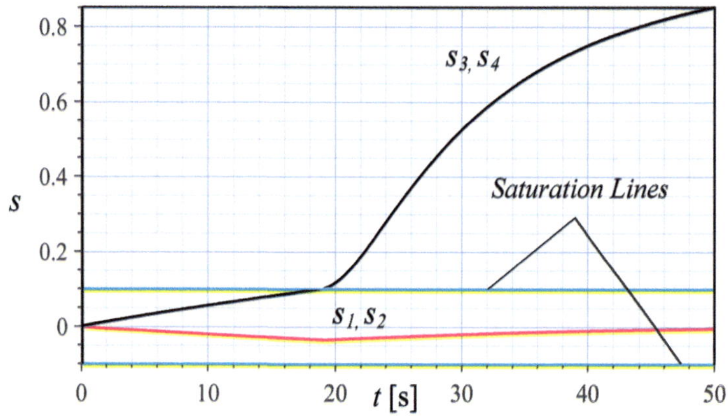

Fig. 2.48 The longitudinal slip ratios of the tires of a four-wheel planar vehicle

Example 75 Increasing rear torque, straight driving.

Consider a vehicle with the data given below.

$$
\begin{aligned}
m &= 1000\,\text{kg} \qquad Iz = 2000\,\text{kg m}^2 \qquad I_1 = I_2 = I_3 = I_4 = 30\,\text{kg m}^2 \\
C_{\alpha_1} &= C_{\alpha_2} = C_{\alpha_3} = C_{\alpha_4} = 8.5 \qquad C_{s_1} = C_{s_2} = C_{s_3} = C_{s_4} = 7.5 \\
\alpha_s &= 5\,\text{deg} \qquad s_s = 0.1 \qquad C_{\alpha s} = C_{s\alpha} = 0.5 \\
a_1 &= 1.35\,\text{m} \qquad a_2 = 1.5\,\text{m} \qquad l = 2.85\,\text{m} \qquad h = 0.9\,\text{m} \\
b_{1f} &= b_{1r} = b_{2f} = b_{2r} = 0.9\,\text{m} \qquad w_f = w_r = 1.8\,\text{m} \qquad R_g = 0.35\,\text{m}
\end{aligned}
\tag{2.433}
$$

Assume the vehicle is moving slowly straight with

$$
v_x = 2\,\text{m/s} \qquad \omega_i = \frac{v_x}{R_g} = \frac{2}{0.35} = 5.714\,\text{rad/s} \tag{2.434}
$$

$$
\delta = 0 \tag{2.435}
$$

At time $t = 0$, we apply an increasing torque on the rear wheel.

$$
T_1 = T_2 = 0 \qquad T_3 = T_4 = 50t\,\text{N m} \tag{2.436}
$$

The rear longitudinal slip ratio will increase and reaches its saturation; then the rear wheels start spinning. Figure 2.48 depicts the longitudinal slip ratios. The saturation limit of the tires is at $s_s = 0.1$. The slip ratio of the rear tires s_3 and s_4 increases linearly from zero up to the point at $s_s = 0.1$. At that point, tire starts spinning on the ground and the applied torque increases the wheels' angular velocity. Therefore, the slip ratio s_r increases rapidly. There is no torque on the front wheels and therefore the front wheels' slip ratio s_1 and s_2 remains unsaturated although it will not be zero. Figure 2.49 depicts the forward velocity of the vehicle v_x.

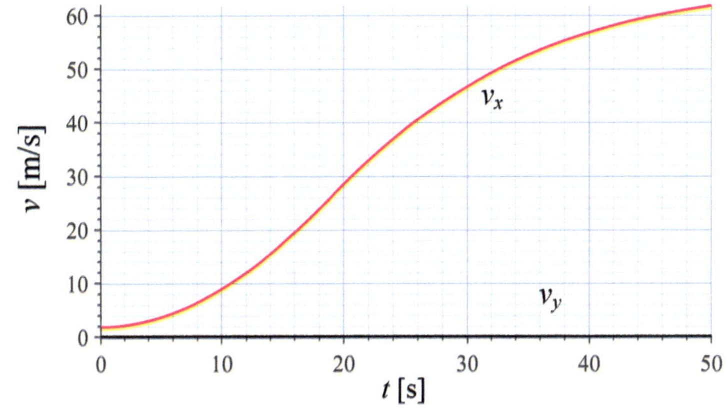

Fig. 2.49 The forward velocity of the vehicle v_x of a four-wheel planar vehicle

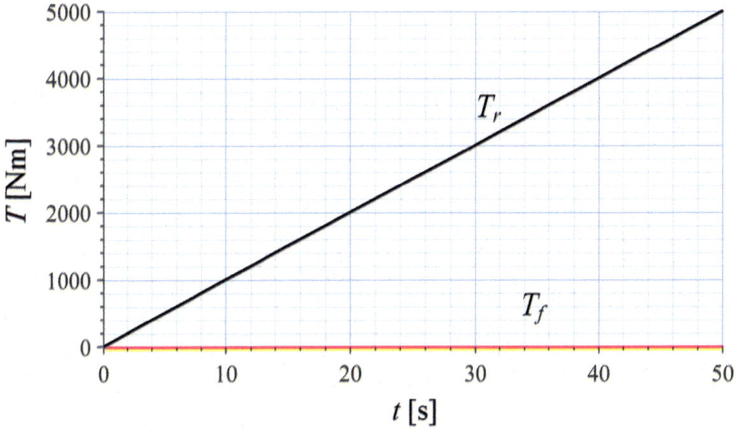

Fig. 2.50 The torques on the wheels of a four-wheel planar vehicle

The torques on the wheels are shown in Fig. 2.50. Because of variable acceleration of the vehicle, the vertical load on the wheels is not constant. Figure 2.51 illustrates the vertical loads, and the traction forces are shown in Fig. 2.52.

Example 76 Increasing front torque, straight driving.

Consider a vehicle with the data given in (2.429) and assume the vehicle is moving slowly straight with

$$v_x = 2\,\mathrm{m/s} \qquad \omega_i = \frac{v_x}{R_g} = \frac{2}{0.35} = 5.714\,\mathrm{rad/s} \tag{2.437}$$

$$\delta = 0 \tag{2.438}$$

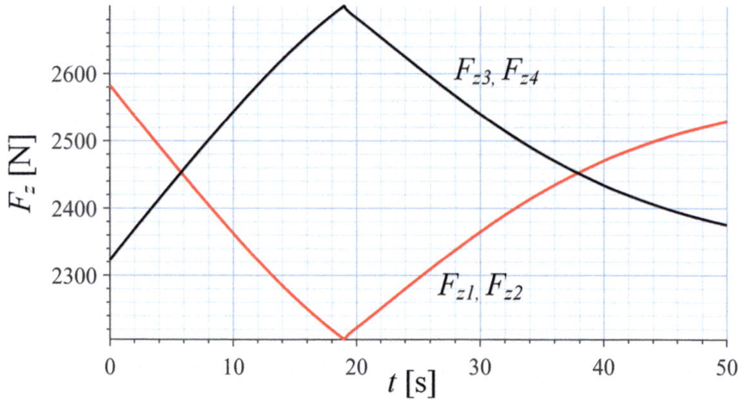

Fig. 2.51 The vertical load on the wheels of a four-wheel planar vehicle

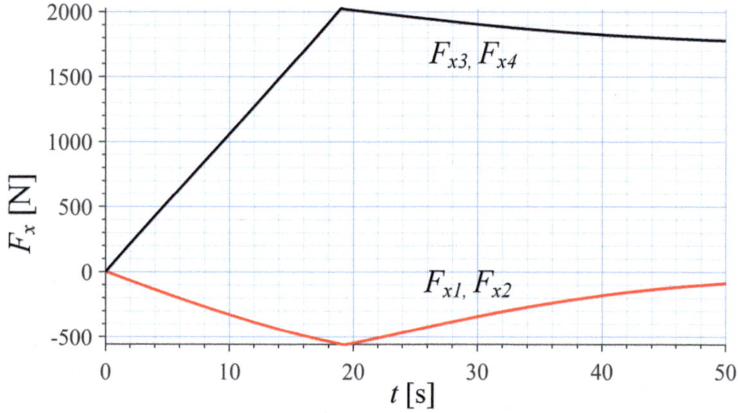

Fig. 2.52 The traction forces on the wheels of a four-wheel planar vehicle

At time $t = 0$, we apply an increasing torque on the front wheel

$$T_1 = T_2 = 50t \text{ N m} \qquad T_3 = T_4 = 0 \tag{2.439}$$

to discover the differences of planar four-wheel vehicle response to front or rear wheel drive. Figure 2.53 depicts the longitudinal slip ratios s_i. Figure 2.54 illustrates the angular velocities of the wheels ω_i. Figure 2.55 depicts the velocity components of the vehicle v_x and v_y. Figure 2.56 illustrates variation of the vertical loads F_{z_i}. Figure 2.57 depicts the longitudinal forces F_{x_i} on tires.

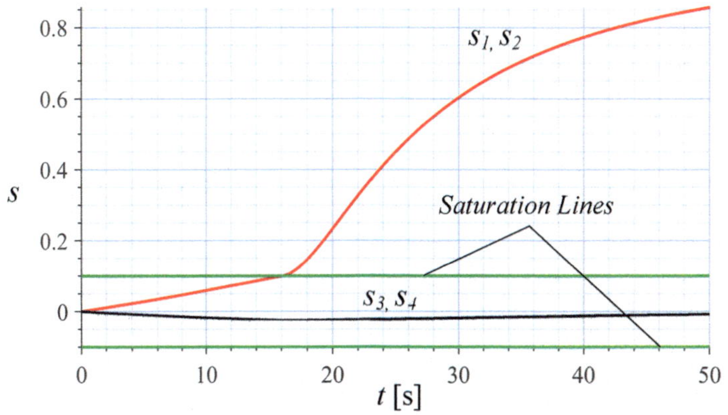

Fig. 2.53 The longitudinal slip ratios s_i of the tires of a four-wheel planar vehicle

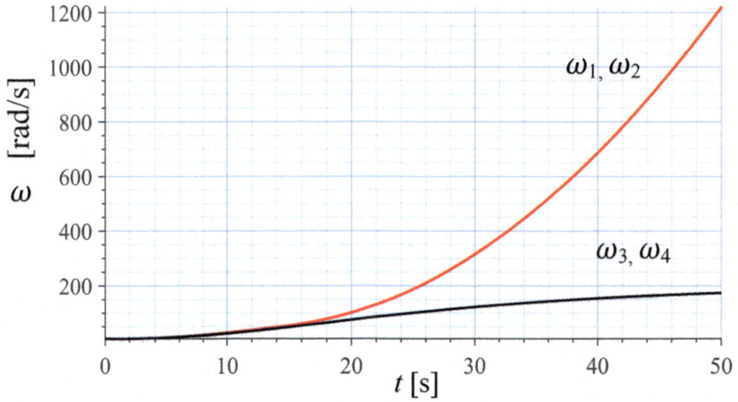

Fig. 2.54 The angular velocities of the wheels ω_i of a four-wheel planar vehicle

Example 77 Increasing steer angle and front torque, slip saturation.

Consider a vehicle with the following information

$$\begin{gathered}
m = 1000\,\text{kg} \qquad Iz = 2000\,\text{kg m}^2 \qquad I_1 = I_2 = I_3 = I_4 = 30\,\text{kg m}^2 \\
C_{\alpha_1} = C_{\alpha_2} = C_{\alpha_3} = C_{\alpha_4} = 8.5 \qquad C_{s_1} = C_{s_2} = C_{s_3} = C_{s_4} = 7.5 \\
\alpha_s = 5\,\text{deg} \qquad s_s = 0.1 \qquad C_{\alpha s} = C_{s\alpha} = 0.5 \\
a_1 = 1.35\,\text{m} \qquad a_2 = 1.5\,\text{m} \qquad l = 2.85\,\text{m} \qquad h = 0.9\,\text{m} \\
b_{1f} = b_{1r} = b_{2f} = b_{2r} = 0.9\,\text{m} \qquad w_f = w_r = 1.8\,\text{m} \qquad R_g = 0.35\,\text{m}
\end{gathered} \tag{2.440}$$

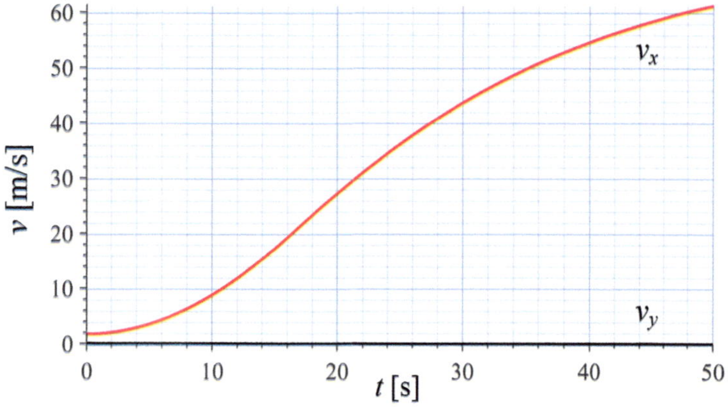

Fig. 2.55 The forward velocity components v_x and v_y of a four-wheel planar vehicle

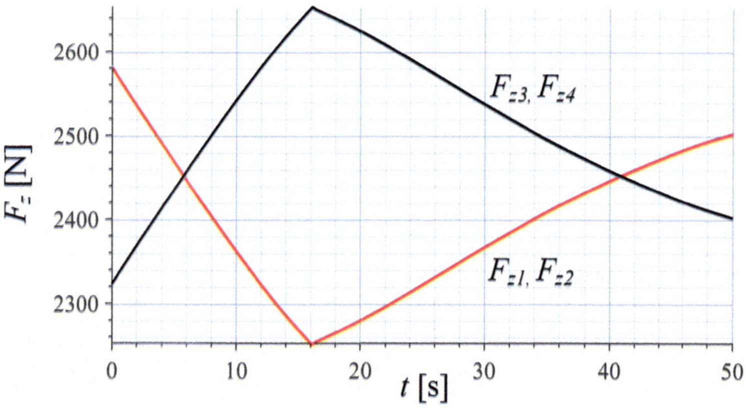

Fig. 2.56 The vertical load F_{z_i} on the wheels of a four-wheel planar vehicle

and assume the vehicle is moving very slowly straight.

$$v_x = 2\,\mathrm{m/s} \qquad \omega_i = \frac{v_x}{R_g} = \frac{2}{0.35} = 5.714\,\mathrm{rad/s} \tag{2.441}$$

At time $t = 0$, a linearly increasing torque will be applied on the front wheels up to $T_1 = T_2 = 750\,\mathrm{N\,m}$ and keep constant after that. The steer angle is also linearly increasing at a very low rate up to $\delta = 0.5\,\mathrm{deg}$ and remains constant after that.

$$T_1 = T_2 = \begin{cases} 50t\,\mathrm{N\,m} & 0 < t < 15\,\mathrm{s} \\ 750\,\mathrm{N\,m} & 15\,\mathrm{s} < t \end{cases} \qquad T_3 = T_4 = 0 \tag{2.442}$$

$$\delta = \begin{cases} 0.05t\,\mathrm{deg} = \frac{0.05\pi}{180}t\,\mathrm{rad} & 0 < t < 10\,\mathrm{s} \\ 0.5\,\mathrm{deg} = \frac{0.5\pi}{180}\,\mathrm{rad} & 10\,\mathrm{s} < t \end{cases} \tag{2.443}$$

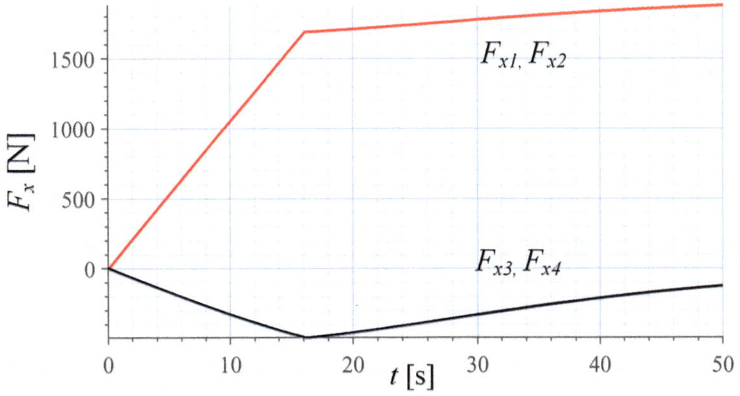

Fig. 2.57 The traction forces F_{x_i} on tires of a four-wheel planar vehicle

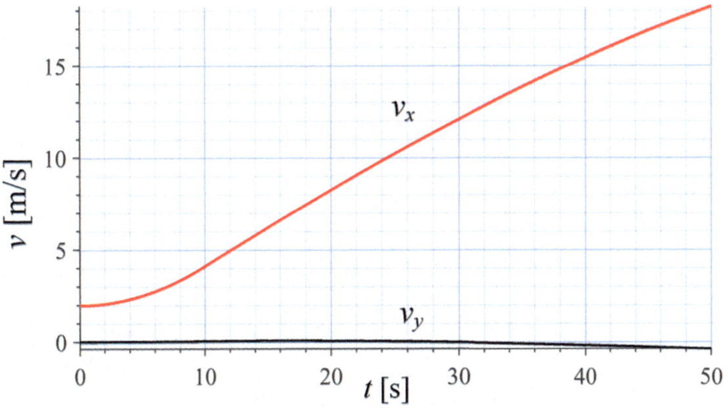

Fig. 2.58 The forward velocity components of a four-wheel planar vehicle v_x and v_y

The front torques increase and go beyond the limit of front wheels capability in producing traction force. As a result, the front tire number 1 slip ratio $s_1 > s_0$ and hence its traction force F_{x_1} will be saturated.

Figure 2.58 depicts the velocity components of the vehicle v_x and v_y, measured in body coordinate B-frame. Figure 2.59 illustrates the angular velocities of the wheels ω_i. Figure 2.60 illustrates the sideslip angles of the wheels α_i. Figure 2.61 depicts the longitudinal slip ratios s_i for the tires. The acceleration components a_X and a_Y of the vehicle are plotted in Fig. 2.62. Figure 2.63 depicts the forces F_x and F_y on the vehicle at its mass center. Figure 2.64 illustrates variation of the vertical loads F_{z_i}. Figure 2.65 illustrates the longitudinal forces F_{x_i} on tires. Figure 2.66 shows the lateral forces F_{y_i} on tires. Figure 2.67 illustrates the path of the vehicle.

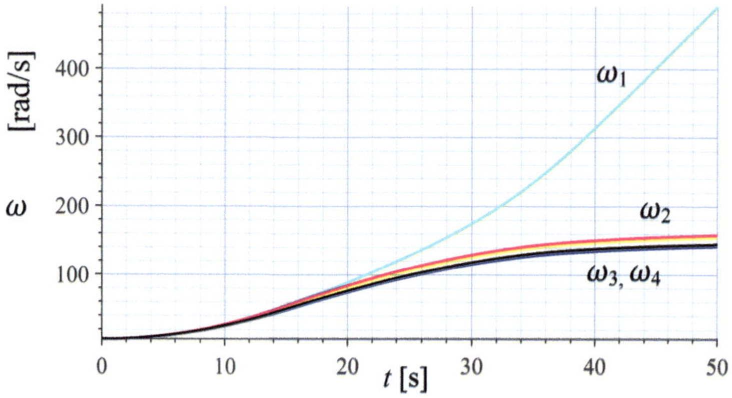

Fig. 2.59 The angular velocities of a four-wheel planar vehicle wheels ω_i

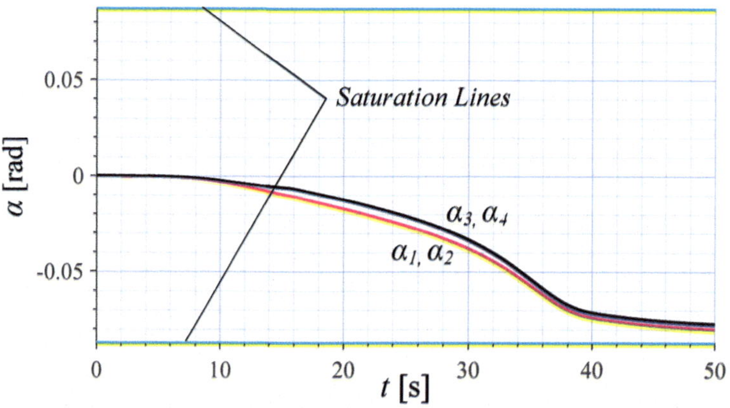

Fig. 2.60 The sideslip angles of a four-wheel planar vehicle wheels α_i

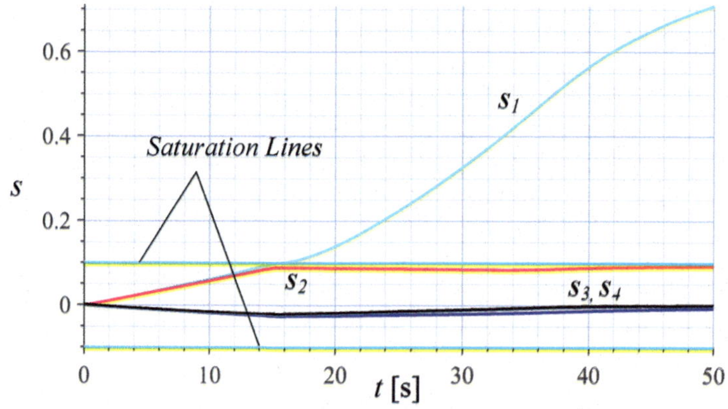

Fig. 2.61 The longitudinal slip ratios s_i of a four-wheel planar vehicle tires

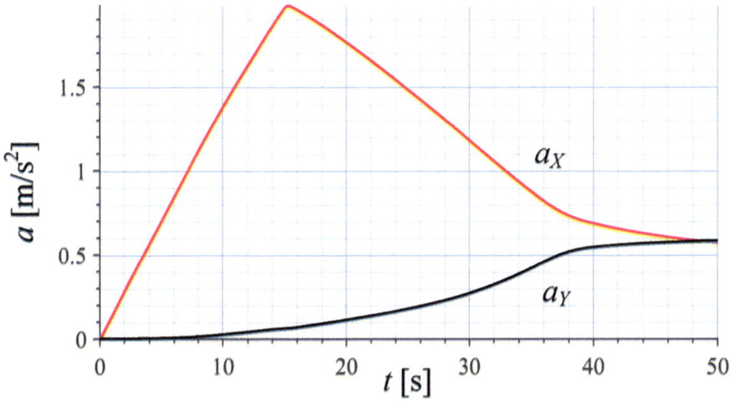

Fig. 2.62 The acceleration components a_X and a_Y/g of a four-wheel planar vehicle

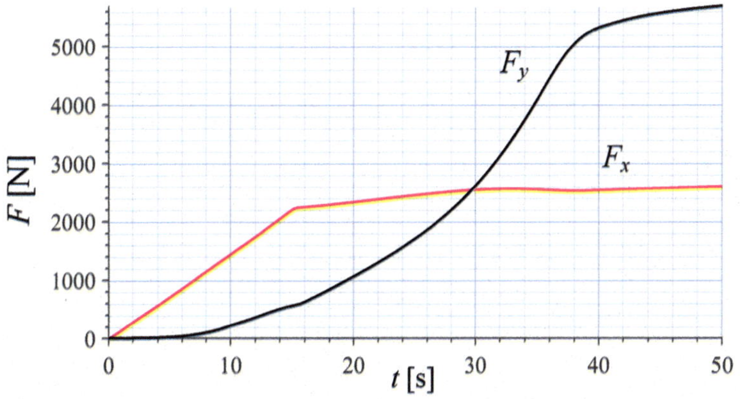

Fig. 2.63 The forces F_x and F_y on the vehicle at its mass center

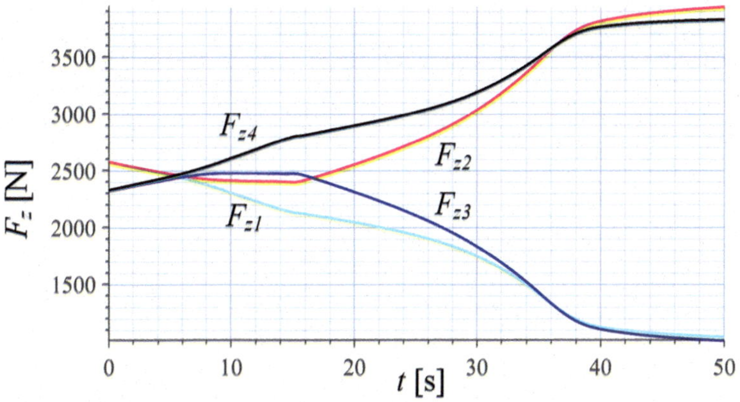

Fig. 2.64 The vertical load F_{z_i} of a four-wheel planar vehicle wheels

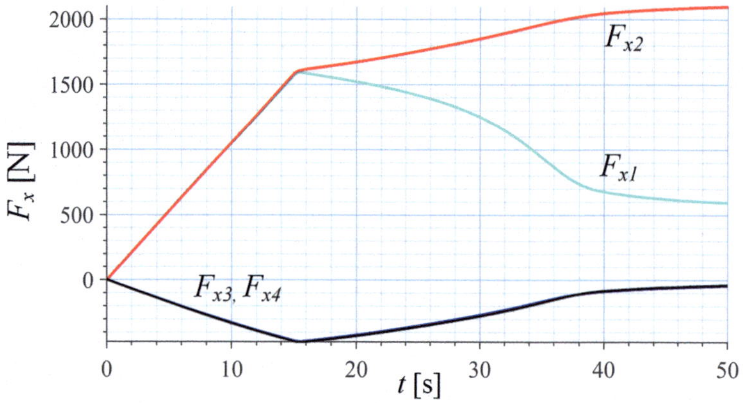

Fig. 2.65 The traction forces F_{xi} of a four-wheel planar vehicle tires

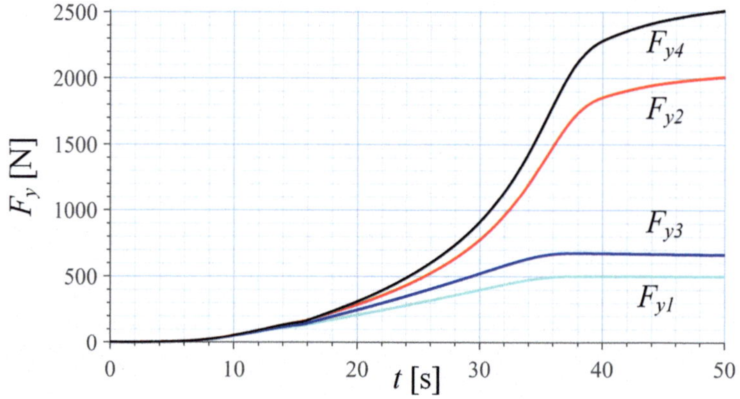

Fig. 2.66 The lateral forces F_{y_i} of a four-wheel planar vehicle tires

Example 78 Increasing steer angle and rear torque.

Consider a vehicle with the following parameters.

$$
\begin{gathered}
m = 1000\,\text{kg} \qquad Iz = 2000\,\text{kg m}^2 \qquad I_1 = I_2 = I_3 = I_4 = 30\,\text{kg m}^2 \\
C_{\alpha_1} = C_{\alpha_2} = C_{\alpha_3} = C_{\alpha_4} = 8.5 \qquad C_{s_1} = C_{s_2} = C_{s_3} = C_{s_4} = 7.5 \\
\alpha_s = 5\,\text{deg} \qquad s_s = 0.1 \qquad C_{\alpha s} = C_{s\alpha} = 0.5 \\
a_1 = 1.35\,\text{m} \qquad a_2 = 1.5\,\text{m} \qquad l = 2.85\,\text{m} \qquad h = 0.9\,\text{m} \\
b_{1f} = b_{1r} = b_{2f} = b_{2r} = 0.9\,\text{m} \qquad w_f = w_r = 1.8\,\text{m} \qquad R_g = 0.35\,\text{m}
\end{gathered}
\tag{2.444}
$$

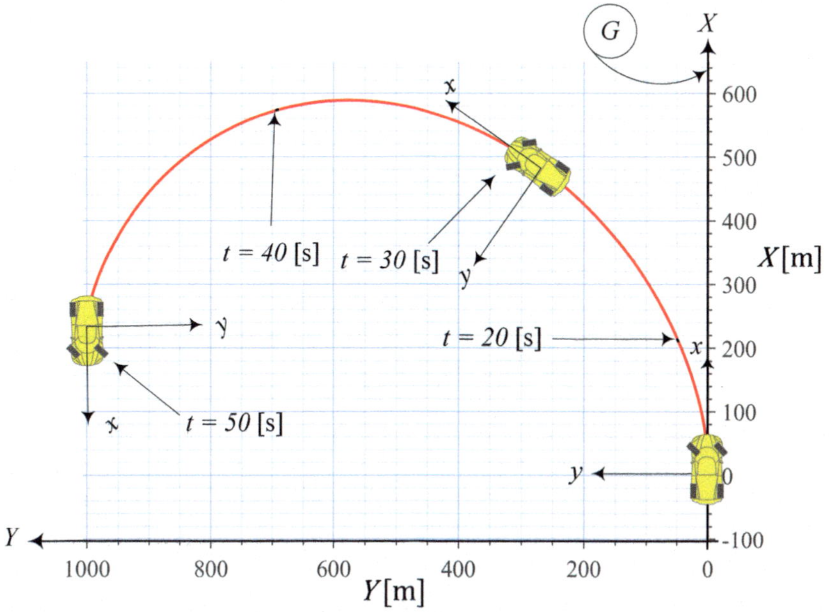

Fig. 2.67 Path of motion of a four-wheel planar vehicle for $0 < t < 50\,\mathrm{s}$ by increasing steer angle and applying front torque and having slip ratio saturation

The vehicle is moving slowly straight with

$$v_x = 2\,\mathrm{m/s} \qquad \omega_i = \frac{v_x}{R_g} = \frac{2}{0.35} = 5.714\,\mathrm{rad/s} \tag{2.445}$$

$$C_A = 0.8 \tag{2.446}$$

At time $t = 0$, we apply an increasing torque on the rear wheel as well as an increasing steer angle for limited times as expressed below.

$$T_2 = T_4 = \begin{cases} 15t\,\mathrm{N\,m} & 0 < t < 10\,\mathrm{s} \\ 150\,\mathrm{N\,m} & 10\,\mathrm{s} < t \end{cases} \qquad T_1 = T_3 = 0 \tag{2.447}$$

$$\delta = \begin{cases} 0.1t\,\mathrm{deg} = \frac{0.1\pi}{180}t\,\mathrm{rad} & 0 < t < 15\,\mathrm{s} \\ 1.5\,\mathrm{deg} = \frac{1.5\pi}{180}\,\mathrm{rad} & 15\,\mathrm{s} < t \end{cases} \tag{2.448}$$

Figure 2.68 depicts the velocity components of the vehicle v_x and v_y. Figure 2.69 illustrates the angular velocities of the front and rear wheels ω_i. Figure 2.70 illustrates the sideslip angles of the front and rear wheels α_i. Figure 2.71 depicts the longitudinal slip ratios s_i of tires. The acceleration components a_x and a_y of the vehicle are plotted in Fig. 2.72. The angular accelerations of the wheels $\dot{\omega}_i$ are shown in Fig. 2.73. Figure 2.74 depicts the resultant forces F_x and F_y on the

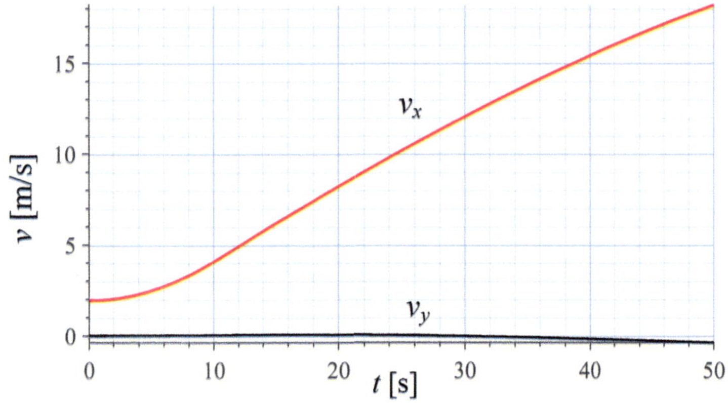

Fig. 2.68 The velocity components of a four-wheel planar vehicle v_x and v_y

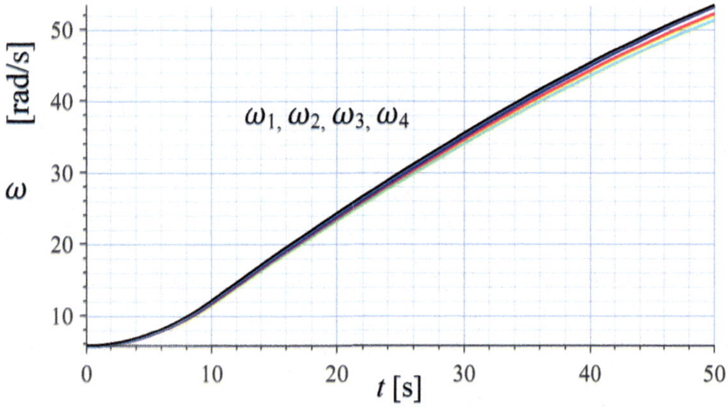

Fig. 2.69 The angular velocities ω_i of a four-wheel planar vehicle

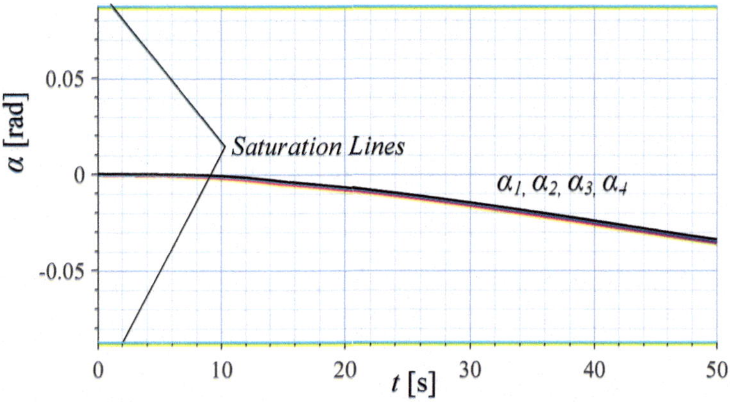

Fig. 2.70 The sideslip angles α_i of a four-wheel planar vehicle

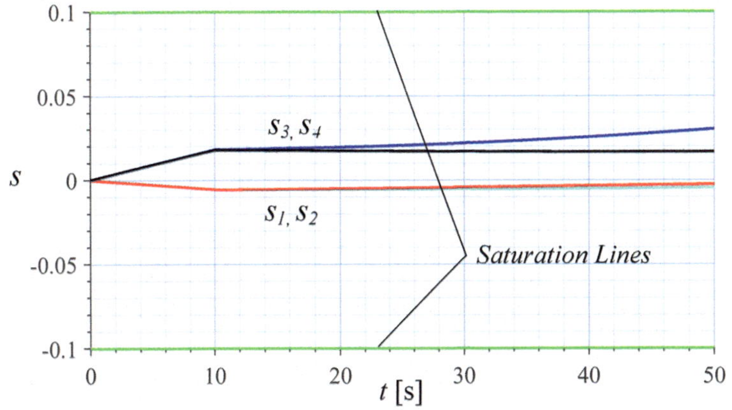

Fig. 2.71 The longitudinal slip ratios s_i of a four-wheel planar vehicle

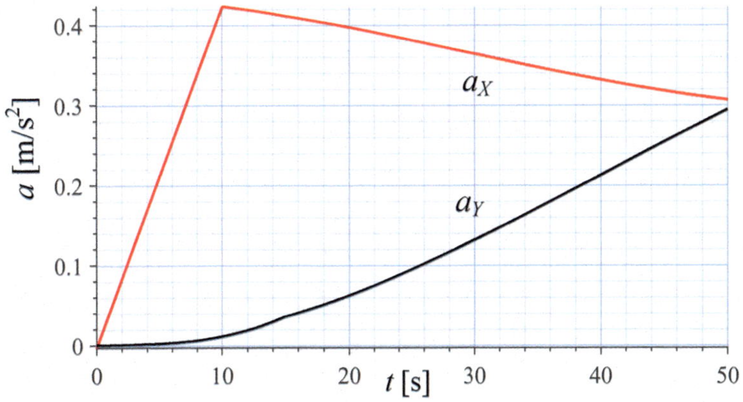

Fig. 2.72 The acceleration components a_X and a_Y/g of a four-wheel planar vehicle

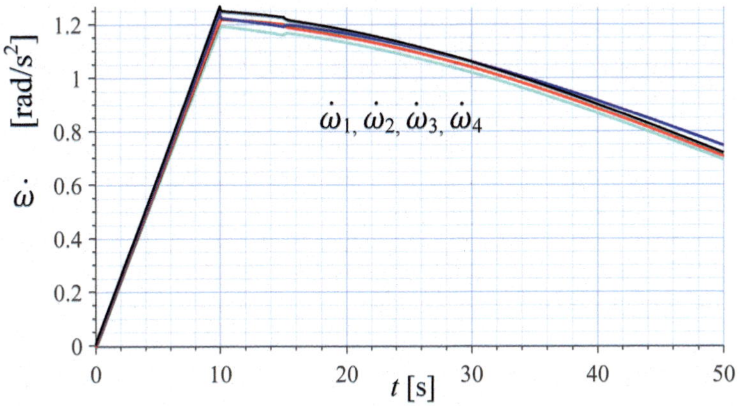

Fig. 2.73 The angular accelerations of the wheels $\dot{\omega}_i$ of a four-wheel planar vehicle

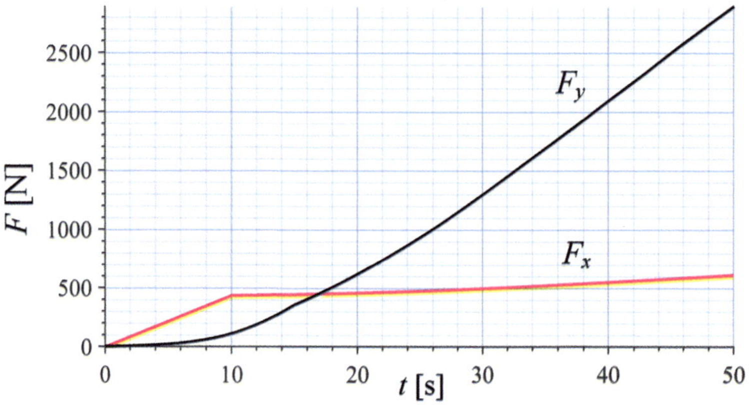

Fig. 2.74 The forces F_x and F_y of a four-wheel planar vehicle at its mass center

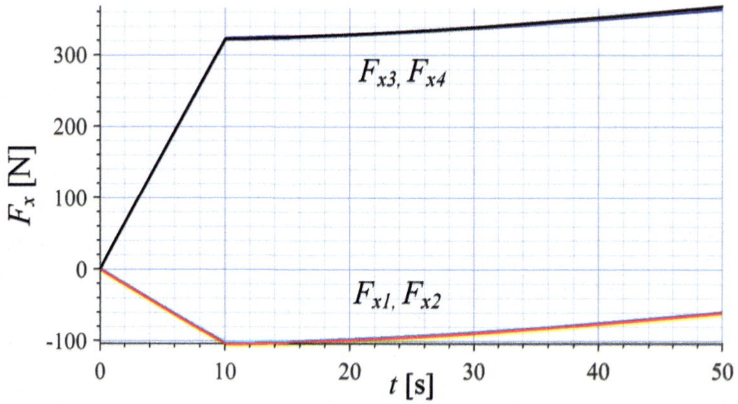

Fig. 2.75 The longitudinal forces F_{x_i} on a four-wheel planar vehicle at its mass center

vehicle at its mass center. Figure 2.75 illustrates the longitudinal forces F_{x_i} and Fig. 2.76 illustrates the lateral forces F_{y_i} on tires. Figure 2.77 illustrates variation of the vertical loads F_{zf} and F_{zr}. Figure 2.78 illustrates the path of the vehicle.

Example 79 No friction on left side and applying rear torque.

To show that the four-wheel planar vehicle mathematical model is capable to analyze special cases, we assume that left side of the vehicle is moving on a no friction pavement while an increasing torque is applied on the rear wheels. The friction coefficients of the tires number 1 and 3 would be zero.

Let us consider a vehicle with the following parameters.

$$m = 1000\,\text{kg} \qquad Iz = 2000\,\text{kg}\,\text{m}^2 \qquad I_1 = I_2 = I_3 = I_4 = 30\,\text{kg}\,\text{m}^2$$

$$C_{\alpha_1} = C_{\alpha_3} = 0 \qquad C_{\alpha_2} = C_{\alpha_4} = 8.5 \qquad C_{\alpha s} = C_{s\alpha} = 0.5$$

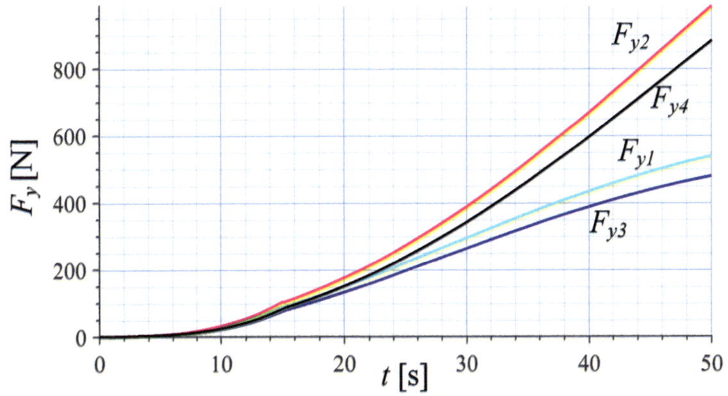

Fig. 2.76 The lateral forces F_{y_i} on tires of a four-wheel planar vehicle

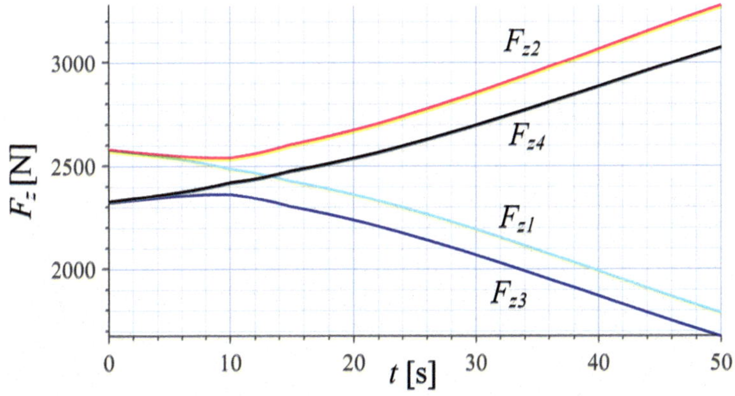

Fig. 2.77 The vertical load F_{z_i} on front and rear wheels of a four-wheel planar vehicle

$$
\begin{aligned}
&C_{s_1} = C_{s_3} = 0 \qquad C_{s_2} = C_{s_4} = 7.5 \qquad \alpha_s = 5\,\text{deg} \qquad s_s = 0.1 \\
&a_1 = 1.35\,\text{m} \qquad a_2 = 1.5\,\text{m} \\
&b_{1f} = b_{1r} = b_{2f} = b_{2r} = 0.9\,\text{m} \qquad l = 2.85\,\text{m} \qquad h = 0.9\,\text{m} \\
&w_f = 1.8\,\text{m} \qquad w_r = 1.8\,\text{m} \qquad R_g = 0.35\,\text{m}
\end{aligned}
\tag{2.449}
$$

The vehicle is moving slowly straight with

$$
v_x = 2\,\text{m/s} \qquad \omega_i = \frac{v_x}{R_g} = \frac{2}{0.35} = 5.714\,\text{rad/s} \tag{2.450}
$$

$$
C_A = 0 \qquad \delta = 0 \tag{2.451}
$$

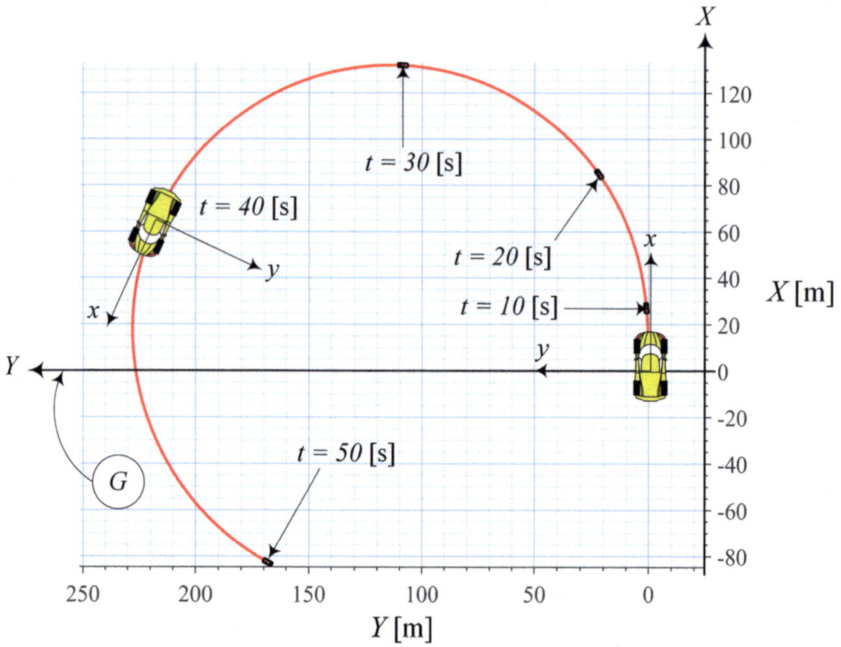

Fig. 2.78 Path of motion of a four-wheel planar vehicle for $0 < t < 50$ s by increasing steer angle up to 1.5 deg and applying rear torque up to 300 N m

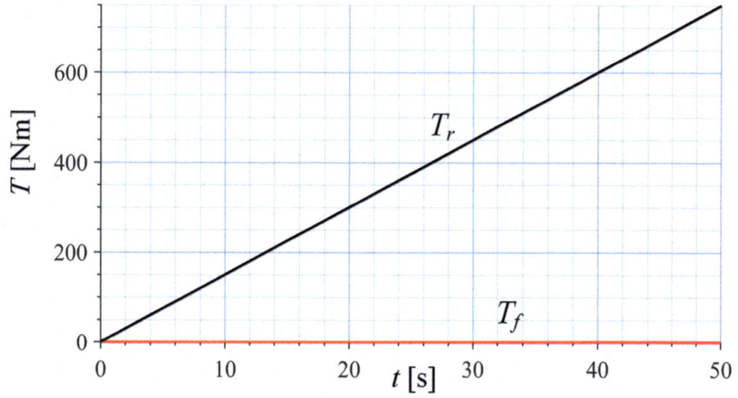

Fig. 2.79 Torque history on the wheels of a four-wheel vehicle with no friction on left side

At time $t = 0$, we apply an increasing torque on the rear wheels as expressed below.

$$T_1 = T_3 \qquad T_4 = T_2 = 7.5t \text{ N m} \tag{2.452}$$

Figure 2.79 depicts the applied torque history on the wheels of the vehicle, and Fig. 2.80 illustrates the traction forces F_{x_i} under each wheel. Figure 2.81 depicts the velocity components of the vehicle v_x and v_y. Figure 2.82 illustrates the angular

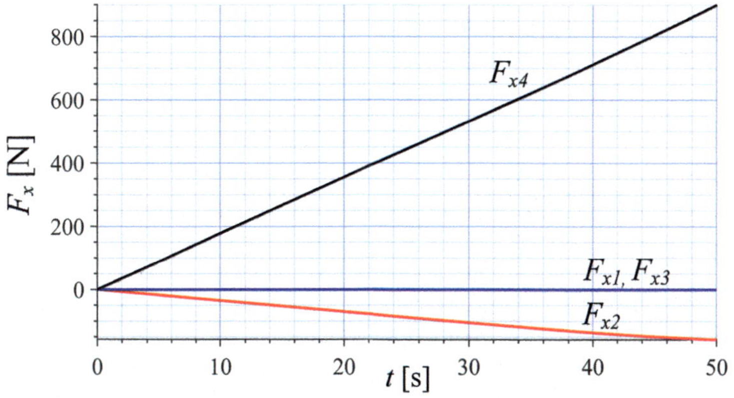

Fig. 2.80 F_{x_i} of a four-wheel vehicle with no friction on left side

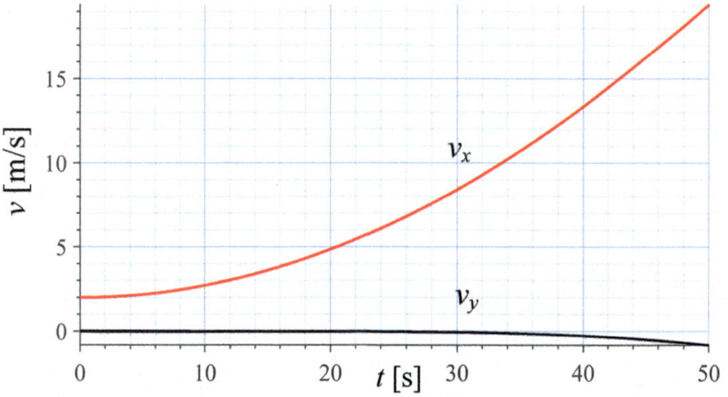

Fig. 2.81 v_x and v_y of a four-wheel vehicle with no friction on left side

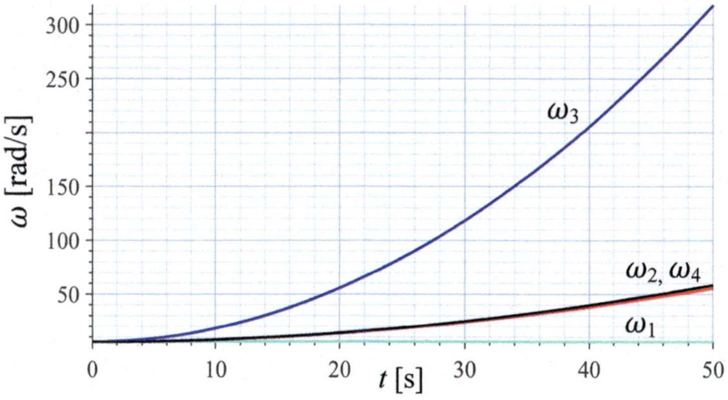

Fig. 2.82 ω_i of a four-wheel vehicle with no friction on left side

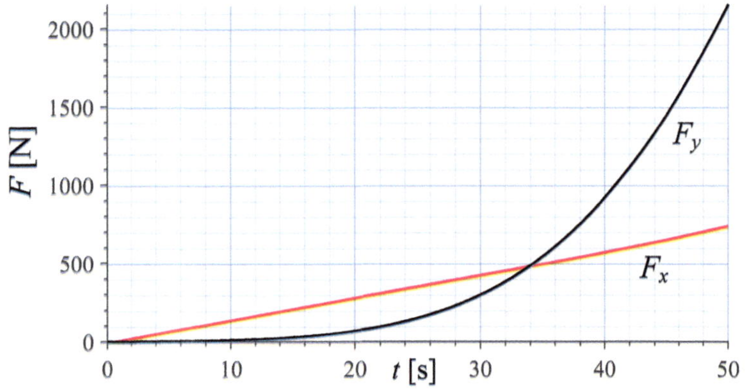

Fig. 2.83 F_x and F_y on a four-wheel vehicle with no friction on left side

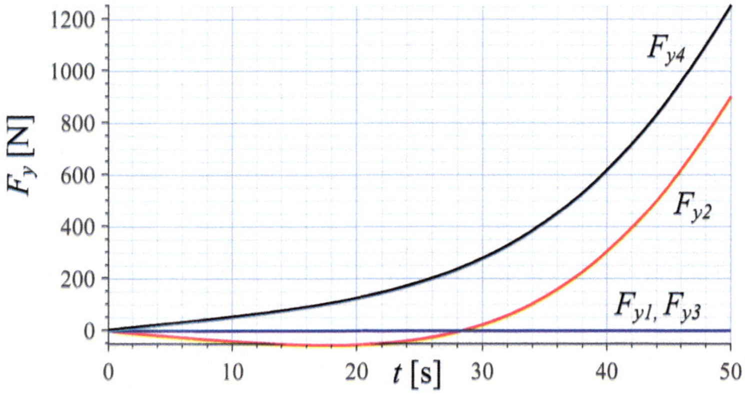

Fig. 2.84 F_{y_i} of a four-wheel vehicle with no friction on left side

velocities of the front and rear wheels ω_i. Figure 2.83 depicts the resultant forces F_x and F_y on the vehicle at its mass center. Figure 2.84 illustrates the lateral forces F_{y_i} on tires. Figure 2.85 illustrates variation of the vertical loads F_{z_i}. Figure 2.86 illustrates the path of motion of the vehicle comparing a front-wheel-drive (FWD) and rear-wheel-drive (RWD) cases.

Example 80 No friction on left side and braking rear torque.

In this example we assume a vehicle is moving at a high speed and that left side of the vehicle is moving on a no friction pavement. We apply an increasing negative rear torque. The friction coefficients of the tires number 1 and 3 would be zero. Let us consider the vehicle with the following parameters.

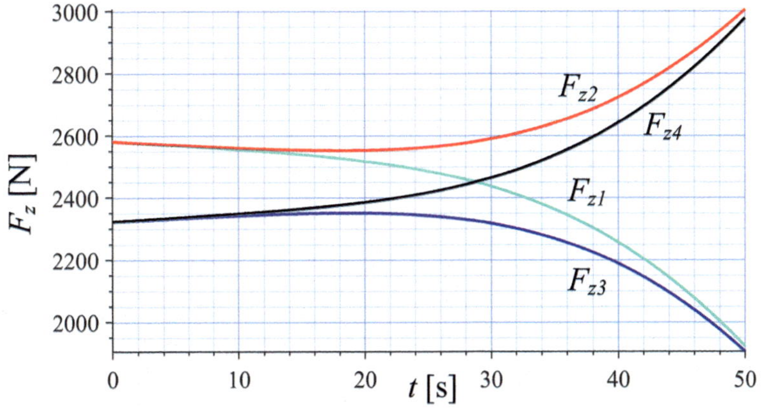

Fig. 2.85 F_{z_i} of a four-wheel vehicle with no friction on left side

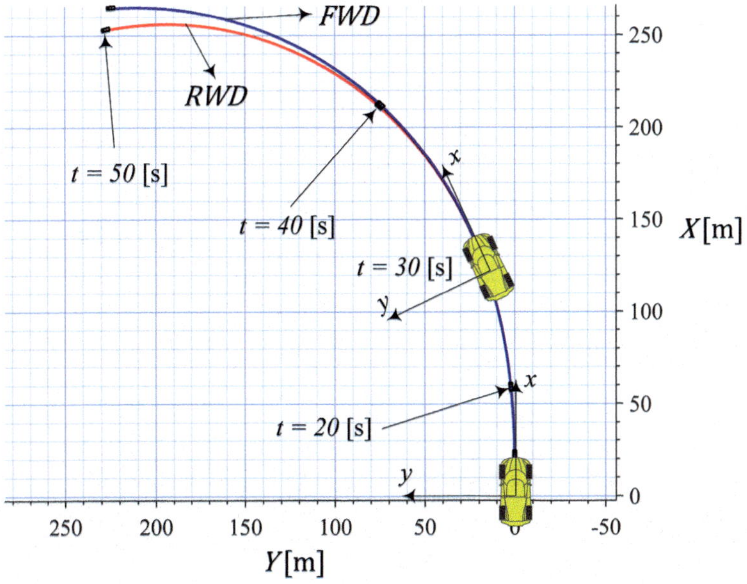

Fig. 2.86 Path of motion of a four-wheel vehicle with no friction on left side

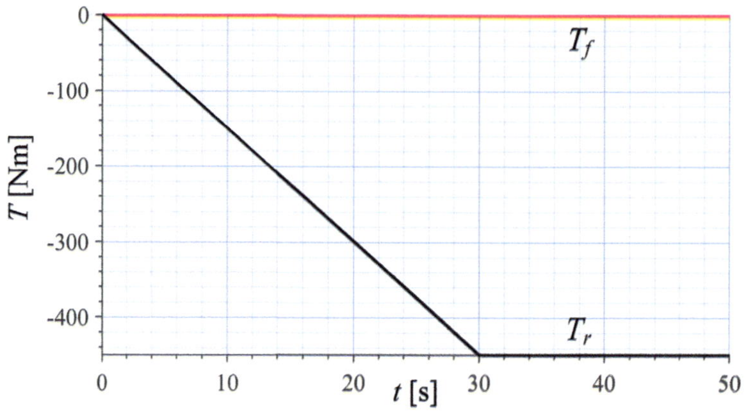

Fig. 2.87 Torque history on the wheels of a four-wheel vehicle with no friction on left side and a braking rear torque

$$\begin{array}{llll} m = 1000\,\text{kg} & Iz = 2000\,\text{kg m}^2 & I_1 = I_2 = I_3 = I_4 = 30\,\text{kg m}^2 & \\ C_{\alpha_1} = C_{\alpha_3} = 0 & C_{\alpha_2} = C_{\alpha_4} = 8.5 & C_{\alpha s} = C_{s\alpha} = 0.5 & \\ C_{s_1} = C_{s_3} = 0 & C_{s_2} = C_{s_4} = 7.5 & & \\ \alpha_s = 5\,\text{deg} & s_s = 0.1 & a_1 = 1.35\,\text{m} & a_2 = 1.5\,\text{m} \\ l = 2.85\,\text{m} & h = 0.9\,\text{m} & b_{1f} = b_{1r} = b_{2f} = b_{2r} = 0.9\,\text{m} & \\ w_f = w_r = 1.8\,\text{m} & R_g = 0.35\,\text{m} & & \end{array} \tag{2.453}$$

The vehicle is moving fast straight.

$$v_x = 20\,\text{m/s} \qquad \omega_i = \frac{v_x}{R_g} = \frac{20}{0.35} = 57.14\,\text{rad/s} \tag{2.454}$$

$$C_A = 0 \qquad \delta = 0 \tag{2.455}$$

At time $t = 0$, we apply an increasing negative torque on the rear wheels as below.

$$T_1 = T_2 \qquad T_4 = T_3 = -7.5t\,\text{N m} \tag{2.456}$$

Figure 2.87 depicts the applied torque history on the wheels of the vehicle. Figure 2.88 depicts the velocity components of the vehicle v_x and v_y. Figure 2.89 illustrates the angular velocities of the front and rear wheels ω_i. Figure 2.90 depicts the resultant forces F_x and F_y on the vehicle at its mass center. Figure 2.91 illustrates the longitudinal forces F_{x_i} and Fig. 2.92 illustrates the lateral forces F_{y_i} on tires. Figure 2.93 illustrates variation of the vertical loads F_{z_i}. Figure 2.94 illustrates the path of motion of the vehicle comparing a front-wheel-brake (FWB) and rear-wheel-brake (RWB) cases.

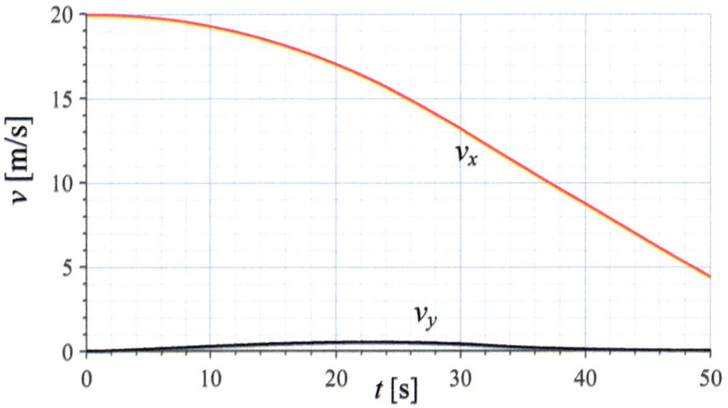

Fig. 2.88 Velocities v_x and v_y of the four-wheel vehicle with no friction on left side and a braking rear torque

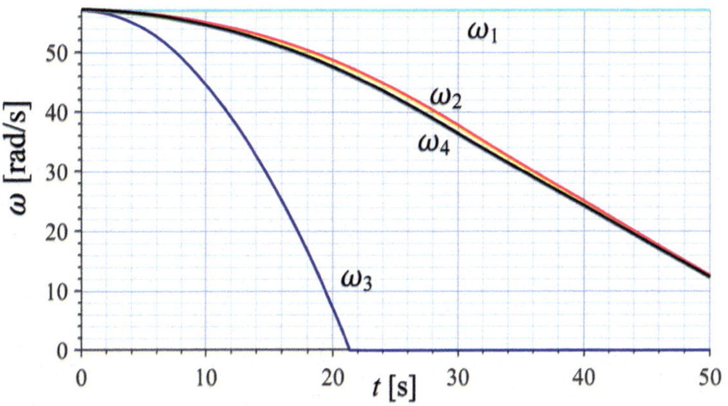

Fig. 2.89 Wheel angular velocities ω_i of the four-wheel vehicle with no friction on left side and a braking rear torque

Example 81 Only one-wheel torque.

Let us examine a vehicle that due to some reasons, only one of the wheel actuators is working. Let us consider the vehicle with the following parameters.

$$m = 1000\,\text{kg} \qquad Iz = 2000\,\text{kg}\,\text{m}^2 \qquad I_1 = I_2 = I_3 = I_4 = 30\,\text{kg}\,\text{m}^2$$
$$C_{\alpha_1} = C_{\alpha_2} = C_{\alpha_3} = C_{\alpha_4} = 8.5 \qquad C_{s_1} = C_{s_2} = C_{s_3} = C_{s_4} = 7.5$$
$$\alpha_s = 5\,\text{deg} \qquad s_s = 0.1 \qquad C_{\alpha s} = C_{s\alpha} = 0.5 \tag{2.457}$$
$$a_1 = 1.35\,\text{m} \qquad a_2 = 1.5\,\text{m} \qquad l = 2.85\,\text{m} \qquad h = 0.9\,\text{m}$$
$$b_{1f} = b_{1r} = b_{2f} = b_{2r} = 0.9\,\text{m} \qquad w_f = w_r = 1.8\,\text{m} \qquad R_g = 0.35\,\text{m}$$

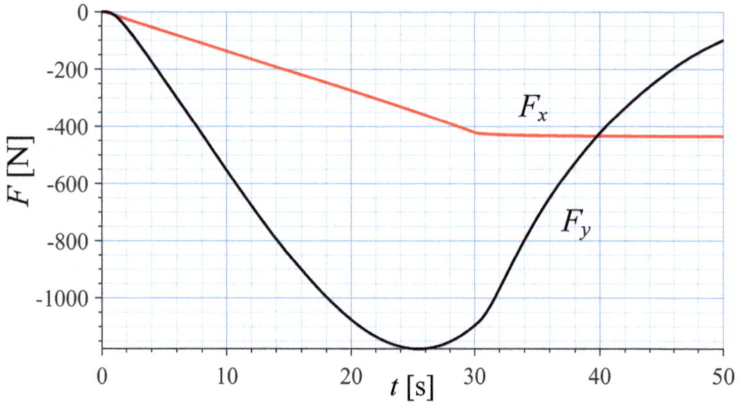

Fig. 2.90 Applied forces F_x and F_y on the four-wheel vehicle with no friction on left side and a braking rear torque

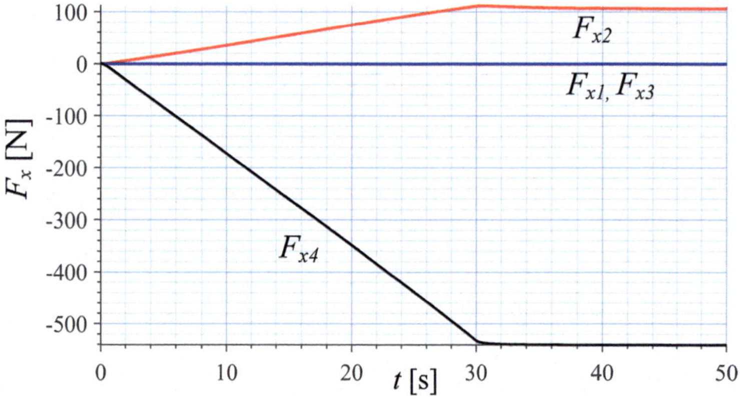

Fig. 2.91 Wheel forces F_{x_i} on the four-wheel vehicle with no friction on left side and a braking rear torque

The vehicle is moving slowly straight.

$$v_x = 2\,\mathrm{m/s} \qquad \omega_i = \frac{v_x}{R_g} = \frac{2}{0.35} = 5.714\,\mathrm{rad/s} \tag{2.458}$$

$$C_A = 0.8 \qquad \delta = 0 \tag{2.459}$$

At time $t = 0$, we apply an increasing torque on the rear wheels as below.

$$T_1 = T_2 = T_3 \qquad\qquad T_4 = 7.5t\,\mathrm{N\,m} \tag{2.460}$$

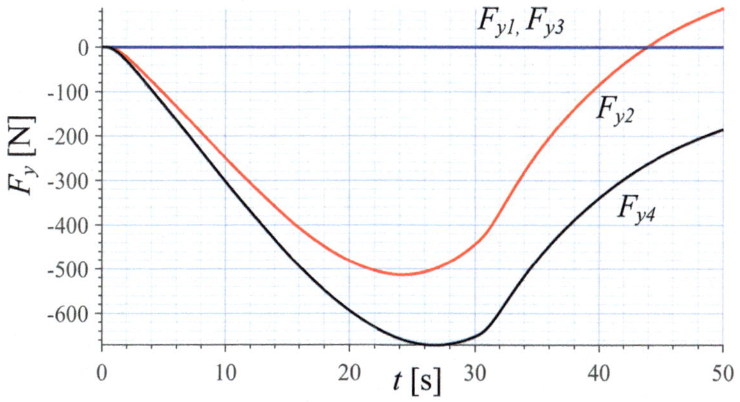

Fig. 2.92 Wheel forces F_{y_i} on the four-wheel vehicle with no friction on left side and a braking rear torque

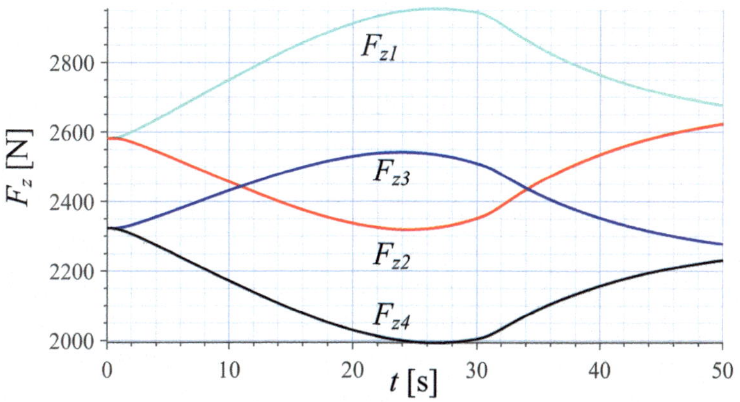

Fig. 2.93 Wheel forces F_{z_i} on the four-wheel vehicle with no friction on left side and a braking rear torque

Figure 2.95 depicts the applied torque history on the wheels of the vehicle, and Fig. 2.96 illustrates the traction forces F_{x_i} under each wheel. Figure 2.97 depicts the velocity components of the vehicle v_x and v_y. Figure 2.98 illustrates the angular velocities of the front and rear wheels ω_i. The global acceleration components a_X and a_Y of the vehicle are plotted in Fig. 2.99. Figure 2.100 depicts the resultant forces F_x and F_y on the vehicle at its mass center. Figure 2.101 illustrates the lateral forces F_{y_i} on tires. Figure 2.102 illustrates variation of the vertical loads F_{z_i}. Figure 2.103 illustrates the path of the vehicle.

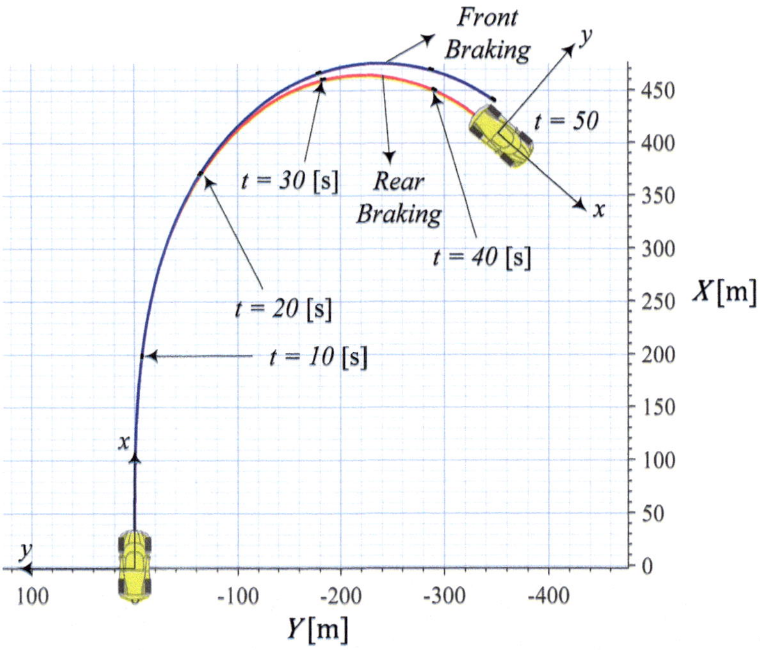

Fig. 2.94 Path of motion of the four-wheel vehicle with no friction on left side and a braking rear torque

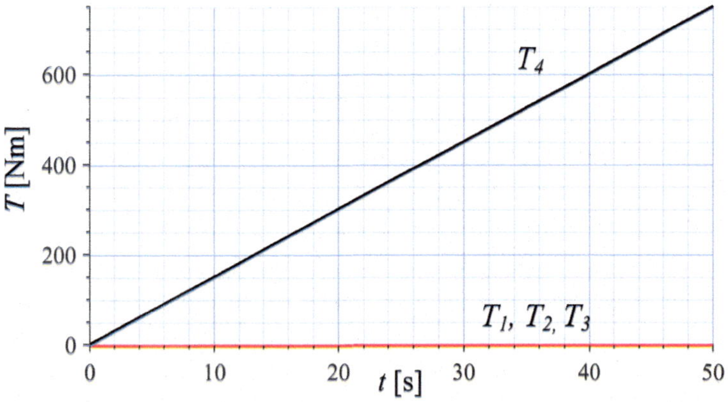

Fig. 2.95 Applied torques T_i on the wheels of a four-wheel planar vehicle with only $T_4 \neq 0$

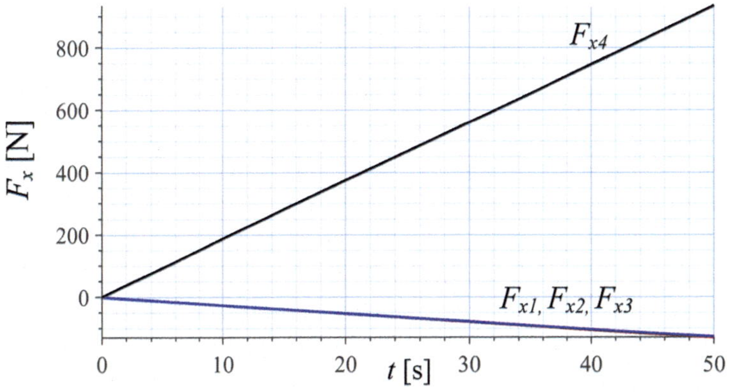

Fig. 2.96 Traction forces F_{x_i} on the wheels of a four-wheel planar vehicle with only $T_4 \neq 0$

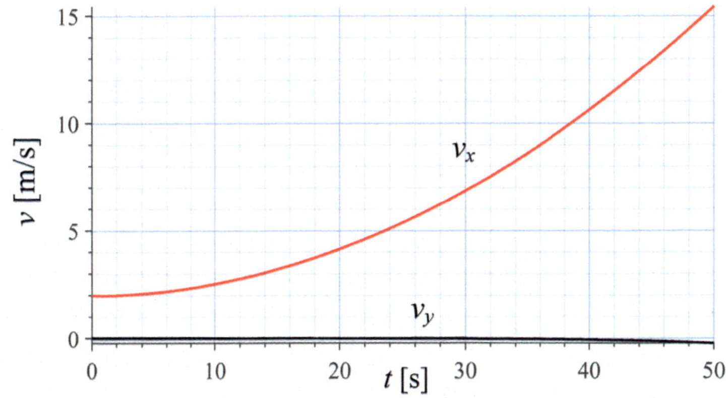

Fig. 2.97 The velocity components of a four-wheel planar vehicle v_x and v_y with only $T_4 \neq 0$

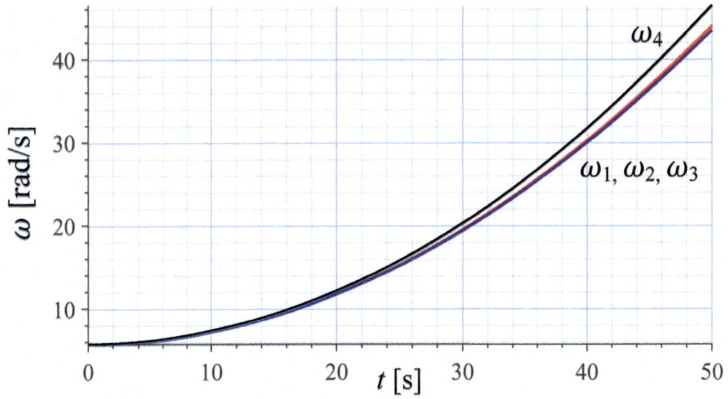

Fig. 2.98 The angular velocities of a four-wheel planar vehicle ω_i with only $T_4 \neq 0$

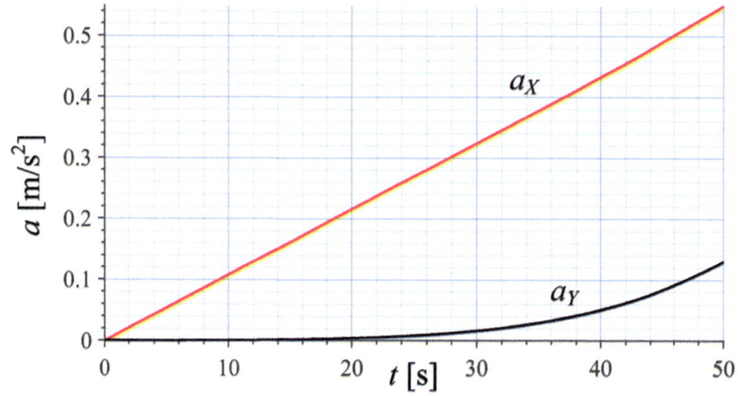

Fig. 2.99 The global acceleration components of a four-wheel planar vehicle a_X and a_Y/g with only $T_4 \neq 0$

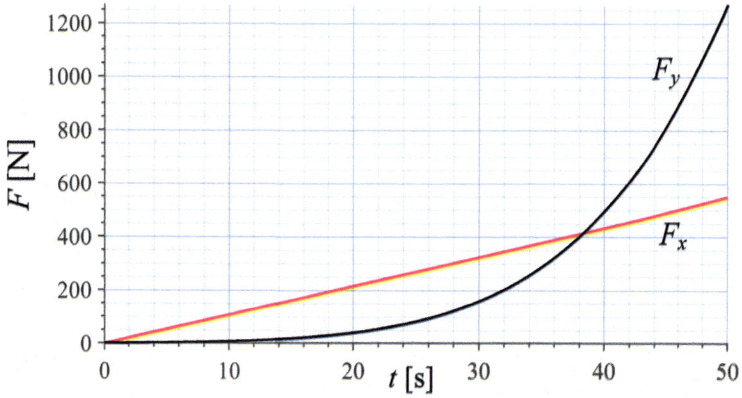

Fig. 2.100 The force components of a four-wheel planar vehicle F_x and F_y with only $T_4 \neq 0$

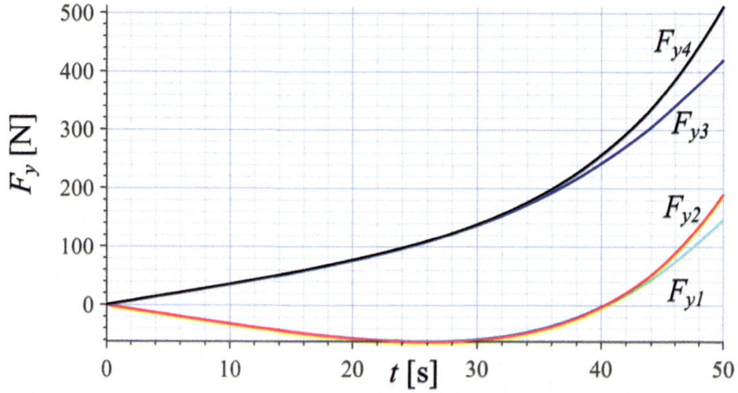

Fig. 2.101 The forces F_{y_i} of a four-wheel planar vehicle with only $T_4 \neq 0$

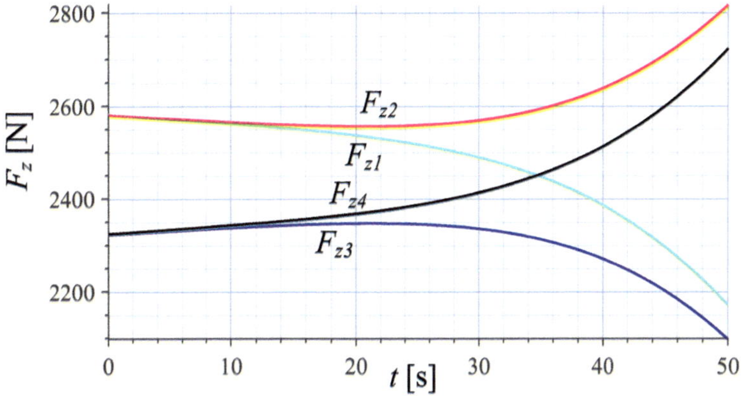

Fig. 2.102 The forces F_{z_i} of a four-wheel planar vehicle with only $T_4 \neq 0$

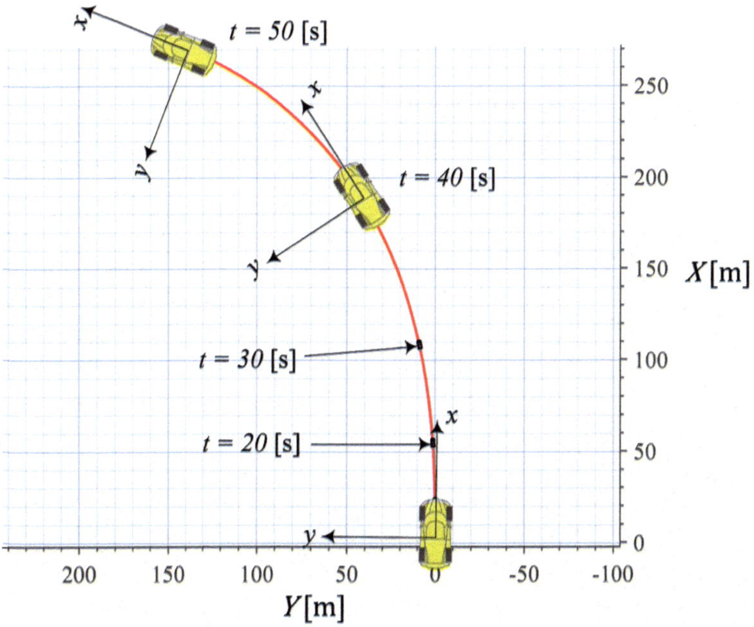

Fig. 2.103 Path of motion of a four-wheel planar vehicle with only $T_4 \neq 0$

2.7 Chapter Summary

In this chapter we study the planar model of vehicles to examine maneuvering by steering as well as the wheel torque control. The wheel torque and steer angle are the inputs and the longitudinal velocity, lateral velocity, and yaw rate are the main output variables of the planar vehicle dynamics model. The planar vehicle dynamic model is the simplest applied modeling in which we assume the vehicle remains parallel to the ground and has no roll, no pitch, and no bounce motions. The planar motion of vehicles has three degrees of freedom: translation in the x and y directions, and a rotation about the z-axis. The longitudinal velocity v_x along the x-axis, the lateral velocity v_y along the y-axis, and the yaw rate $r = \dot{\psi}$ about the z-axis are the outputs of the dynamic equations of motion.

By ignoring the roll motion as well as the lateral load transfer between left and right wheels, we define a simplified two-wheel model for the vehicle.

The four-wheel planar vehicle model is an extension to the two-wheel planar vehicle model to include the lateral weight transfer. The four-wheel planar model provides us with better simulation of drifting vehicles. This model is capable to simulate drift of vehicles as well as simulation of different tire-wheel interaction for all four tires of a vehicle.

2.8 Key Symbols

$a \equiv \ddot{x}$	Acceleration
a_i	Longitudinal distance of the axle number i from y-axis
A_f	Frontal area of vehicle
b_i	Lateral distance of tire number i from longitudinal x-axis
$B(Cxyz)$	Vehicle coordinate frame
C	Mass center
C_A	Aerodynamic coefficient
C_α	Tire sideslip coefficient
$C_{\alpha f}$	Front sideslip coefficient
$C_{\alpha i}$	Sideslip coefficient of tire number i
$C_{\alpha s}$	Tire lateral force drop factor
$C_{\alpha r}$	Rear sideslip coefficient
C_γ	Camber coefficient, camber stiffness
C_s	Tire slip ratio coefficient
C_{sf}	Front slip ratio coefficient
C_{sr}	Rear slip ratio coefficient
$C_{s\alpha}$	Tire longitudinal force drop factor
C_D	Drag coefficient
C_{sf}	Front slip ratio coefficient
C_{si}	Slip ratio coefficient of tire number i

C_{sr}	Rear slip ratio coefficient
C_β	Coefficient between F_y and β at steady-state
C_δ	Coefficient between F_y and δ at steady-state
C_κ	Coefficient between F_y and κ at steady-state
$\mathbf{d}$	Location vector
D_β	Coefficient between M_z and β at steady-state
D_δ	Coefficient between M_z and δ at steady-state
D_κ	Coefficient between M_z and κ at steady-state
F_A	Aerodynamic force
F_x	Longitudinal force, forward force, traction force
$\mathbf{F}_i$	Force vector of tire number i
F_{xi}	Longitudinal force of tire number i
F_y	Lateral force of vehicle
F_{yi}	Lateral force of tire number i
F_{yf}	Front lateral force
F_{yr}	Rear lateral force
F_{zi}	Normal force, vertical force of tire number i
$\mathbf{F}, \mathbf{M}$	Force system
$\mathbf{g}, g$	Gravitational acceleration
$G(OXYZ)$	Global coordinate frame
h	Height of mass center from the ground
H	Heaviside function
I	Mass moment
I_i, I_{wi}	Wheel number i mass moment
K	Stability factor
l	Wheel base
$\mathbf{L}$	Moment of momentum
m	Mass
M_x	Roll moment, bank moment, tilting torque
M_y	Pitch moment
M_z	Yaw moment, aligning moment
o, O	Origin of a coordinate frame
$p = \dot{\varphi}$	Roll rate
$\mathbf{p}$	Momentum
$q = \dot{\theta}$	Pitch rate
q_i	Generalized coordinate
$r = \dot{\psi}$	Yaw rate
$\mathbf{r}$	Position vector
R, R_g	Tire radius
${}^G R_B$	Rotation matrix to go from B frame to G frame
s	Longitudinal slip

s_s	Saturation value of longitudinal slip
S	Saturation function
$S_\kappa = \kappa/\delta$	Curvature response
$S_\beta = \beta/\delta$	Sideslip response
t	Time
T	Tire coordinate frame
T_i	Wheel torque
$v \equiv \dot{x},\ \mathbf{v}$	Velocity
$\mathbf{v}_{wind}$	Wind velocity
v_{x_i}	Longitudinal velocity of wheel number i
v_{y_i}	Lateral velocity of wheel number i
w	Wheelbase
w_f	Front wheelbase
w_r	Rear wheelbase
$x, y, z,\ \mathbf{x}$	Displacement
X, Y, Z	Global displacement
α	Sideslip angle
α_s	Sideslip angle saturation
β	Global sideslip angle
β	Vehicle sideslip angle, attitude angle
β_f	Sideslip angle of front wheel
β_i	Sideslip angle of wheel number i
β_r	Sideslip angle of rear wheel
$\beta + \psi$	Cruise angle
δ	Steer angle
δ_f	Front steer angle
δ_i	Front steer angle of wheel number i
δ_r	Rear steer angle
θ	Pitch angle
$\dot{\theta} = q$	Pitch rate
$\kappa = 1/\rho$	Curvature
λ	Eigenvalue
ρ	Radius of rotation, air density
φ	Roll angle
$\dot{\varphi} = p$	Roll rate
ψ	Yaw angle
$\dot{\psi} = r$	Yaw rate
ψ	Heading angle
$\omega, \boldsymbol{\omega}$	Angular velocity
ω_i	Angular velocity of wheel number i
$\dot{\omega}, \dot{\boldsymbol{\omega}}$	Angular acceleration
$\dot{\omega}_i$	Angular acceleration of wheel number i

Exercises

1. Graphical view of tire force and vehicle dynamics.
   Consider a vehicle with the following data.

$$C_{\alpha f} = 8.5 \qquad C_{\alpha r} = 8.5 \qquad \alpha_s = 5\,\text{deg} \qquad s_s = 0.1$$
$$C_{sf} = 7.5 \qquad C_{sr} = 7.5 \qquad C_{\alpha s} = 0.5 \qquad C_{s\alpha} = 0.5 \tag{2.461}$$

and the tire force equations:

$$\frac{F_x}{F_z} = C_s s \sqrt{1 - C_{s\alpha}\left(\frac{\alpha}{\alpha_s}\right)^2} \qquad |\alpha| < \alpha_s \qquad |s| < s_s \tag{2.462}$$

$$\frac{F_y}{F_z} = -C_\alpha \alpha \sqrt{1 - C_{\alpha s}\left(\frac{s}{s_s}\right)^2} \qquad |\alpha| < \alpha_s \qquad |s| < s_s \tag{2.463}$$

(a) Plot F_x/F_z as functions of α and s for $-2\alpha_s < \alpha < 2\alpha_s$ and $-2s_s < s < 2s_s$. Replot F_x/F_z for $C_{\alpha s} = 0$ and $C_{s\alpha} = 0$ and compare the plots.
(b) Plot F_z/F_z as functions of α and s for $-2\alpha_s < \alpha < 2\alpha_s$ and $-2s_s < s < 2s_s$. Replot F_y/F_z for $C_{\alpha s} = 0$ and $C_{s\alpha} = 0$ and compare the plots.
(c) If there are any sharp edges in any of the above plots, explain what would you expect to see in the plot of F_x, F_y, F_z, T_i.
(d) Chose a set of inputs of δ and T_i to show your predictions.

2. Wind force in B-frame.
   Transform the result of the wind force ${}^G\mathbf{F}_A$ of Eq. (2.48) into B-frame and find the result of Eq. (2.49).

(a) Determine the force system that applies on the two-wheel model of the car.

$$F_y = C_r\, r + C_\beta\, \beta + C_\delta\, \delta \tag{2.464}$$
$$M_z = D_r\, r + D_\beta\, \beta + D_\delta\, \delta \tag{2.465}$$

(b) Write the equations of motion of the car as

$$\begin{aligned} F_x &= m\, \dot{v}_x - mr\, v_y \\ F_y &= m\, \dot{v}_y + mr\, v_x \\ M_z &= \dot{r}\, I_z \end{aligned} \tag{2.466}$$

(c) Derive the force system coefficient that the velocity is measured in km/h instead of m/s.

3. Velocity drop with aerodynamic resistance.
   (a) Analytically determine if the forward velocity decreasing in Fig. 2.10 is linear or is a curve?
   (b) Analytically determine if the forward velocity decreasing in Fig. 2.47 is linear or is a curve?
   (c) Are the curves in Figs. 2.10 and 2.47 exactly the same?
4. Nonlinear tire behavior.
   Let us assume that the sideslip coefficient of the tires of a vehicle is nonlinear such that its lateral force generation capacity drops at high α according to:

$$C_\alpha = C_1 - C_2\alpha^2 \tag{2.467}$$

   (a) Develop the force system for the bicycle vehicle model.
   (b) Derive the equations of motion for a front wheel steerable planar bicycle vehicle model.
5. Nonlinear equation for $F_y = F_y(\alpha)$.
   Accept the nonlinear functions of

$$F_{y_f} = -C_{1f}\,\alpha_f + C_{2f}\,\alpha_f^3 \qquad F_{y_r} = -C_{1r}\,\alpha_r + C_{2r}\,\alpha_r^3 \tag{2.468}$$

   and develop the force system for the planar bicycle vehicle model.
6. $3D$ global wind force.
   Assume the wind has a global velocity ${}^G\mathbf{v}_{wind}$ and a vehicle is moving with a velocity ${}^G\mathbf{v}$.

$$ {}^G\mathbf{v}_{wind} = \begin{bmatrix} v_{wx} \\ v_{wy} \\ v_{wz} \end{bmatrix} \qquad {}^G\mathbf{v} = \begin{bmatrix} v_x \\ v_y \\ 0 \end{bmatrix} \tag{2.469}$$

   The wind is blowing at angle η with respect to the ground XY-plane and angle θ with respect to the global XZ-plane. The vehicle x-axis is at angle ψ with respect to X-axis.
   (a) Determine the wind velocity expression in G-frame in terms of v_w..
   (b) Determine the wind force on the vehicle in G-frame in terms of v and v_w.

$${}^G\mathbf{F}_A = C_A\left({}^G\mathbf{v} - {}^G\mathbf{v}_w\right)^2 \tag{2.470}$$

   (c) Determine the wind force on the vehicle in B-frame in terms of v and v_w.
   (d) Determine the changes on F_{z_i} for two-wheel planar vehicle model.
   (e) Determine the changes on F_{z_i} for four-wheel planar vehicle model.
7. ★Initial camber set.
   To increase directional stability of vehicles, specially the race cars, we design their suspensions such that there exists a set inward camber angle γ_0. The

inward camber means at $\delta = 0$, we have $\gamma_1 = \gamma_0$, $\gamma_2 = -\gamma_0$, $\gamma_3 = \gamma_0$, $\gamma_4 = -\gamma_0$, assuming front and rear suspensions initial cambers are equal. Derive the total planar force system on the vehicle in the body coordinate frame B expressed in Eqs. (2.59)–(2.61) with this suspension design.

8. Front and rear steer bicycle plane vehicle model.
   Re-derive the bicycle plane vehicle force components equations in (2.91)–(2.93) for a front δ_f and rear δ_r steerable bicycle vehicle.
9. Front and rear steer two-wheel plane vehicle model.
   The equations of motion of the two-wheel plane vehicle model are given in (2.203)–(2.207).

   (a) Re-derive the equations for a front δ_f and rear δ_r steerable vehicle.
   (b) Re-derive the equations for a front $\delta_r = \delta_f/10$. In this parallel steering design, we will only have one input steer angle.
   (c) Re-derive the equations for a front $\delta_r = -\delta_f/10$. In this opposite steering design, we will only have one input steer angle.

10. Aerodynamic effect.
   Consider a vehicle is moving in a windy environment with a wind velocity of:

$$^B\mathbf{v}_{wind} = -v_{wx}\,\hat{\imath} \tag{2.471}$$

   Assume the vehicle initially is moving at

$$v_x = 20\,\mathrm{m/s} \qquad \delta = 0 \qquad T_f = T_r = 0 \tag{2.472}$$

$$\omega_f = \omega_r = \frac{v_x}{R_g} = \frac{20}{0.35} = 57.143\,\mathrm{rad/s} \tag{2.473}$$

   The other parameters of the vehicle are given in Eq. (2.247) and the wind force on vehicle is calculated by

$$F_x = C_A\left(v_x - (-v_{wx}\sin\omega t)\right)^2 \tag{2.474}$$

   (a) Assume $C_D = 0.8$ and determine the forward velocity of the vehicle. Plot the forward velocity v_x to the time at which $v_x = 10\,\mathrm{m/s}$.
   (b) ★Indicating the time at which v_x decreases to $v_x = 10\,\mathrm{m/s}$ by t_h. Plot t_h versus C_D.

11. Alternative aerodynamic effect.
   Consider a vehicle is moving in a stormy environment with an alternative wind velocity.

$$^B\mathbf{v}_{wind} = -v_{wx}\sin\omega t\,\hat{\imath} \tag{2.475}$$

   Assume the vehicle initially is moving at

$$v_x = 20\,\mathrm{m/s} \qquad \delta = 0 \qquad T_f = T_r = 0 \tag{2.476}$$

$$\omega_f = \omega_r = \frac{v_x}{R_g} = \frac{20}{0.35} = 57.143\,\mathrm{rad/s} \tag{2.477}$$

The other parameters of the vehicle are given in Eq. (2.247) and the wind force on vehicle is calculated by

$$F_x = C_A \left(v_x - (-v_{wx}\sin\omega t)\right)^2 \tag{2.478}$$

$$C_D = 0.8 \tag{2.479}$$

(a) Assume $\omega = 0.01/\,\mathrm{s}$ and determine the forward velocity of the vehicle. Plot the forward velocity v_x to the time at which $v_x = 10\,\mathrm{m/s}$.
(b) ★Indicating the time at which v_x decreases to $v_x = 10\,\mathrm{m/s}$ by t_h, and plot t_h versus ω.

12. ★Front locked wheel.
Consider a vehicle with the data given in (2.251). Assume the vehicle is moving slowly straight with

$$v_x = 2\,\mathrm{m/s} \qquad \omega_r = \frac{v_x}{R_g} = \frac{2}{0.35} = 5.714\,\mathrm{rad/s} \tag{2.480}$$

At time $t = 0$, we apply an increasing torque on the rear wheel

$$T_f = 0 \qquad T_r = 100t\,\mathrm{N\,m} \tag{2.481}$$

while the steer angle is kept at zero $\delta = 0$ and the front wheel is locked at $\omega_f = 0$.
Solve the equations of motion numerically and plot T_f, T_r, s_f, s_r, F_f, F_r, v_x, a_x, ω_r, for $0 < t < 50\,\mathrm{s}$.

13. ★Rear locked wheel.
Consider a vehicle with the data given in (2.251). Assume the vehicle is moving slowly straight with

$$v_x = 2\,\mathrm{m/s} \qquad \omega_f = \frac{v_x}{R_g} = \frac{2}{0.35} = 5.714\,\mathrm{rad/s} \tag{2.482}$$

At time $t = 0$, we apply an increasing torque on the front wheel

$$T_r = 0 \qquad T_f = 100t\,\mathrm{N\,m} \tag{2.483}$$

while the steer angle is kept at zero $\delta = 0$ and the rear wheel is locked at $\omega_r = 0$.
Solve the equations of motion numerically and plot T_f, T_r, F_f, F_r, s_f, s_r, v_x, a_x, ω_f, for $0 < t < 50\,\mathrm{s}$.

14. Increasing steer angle and front torque.
Use the date in Example 70 and repeat the example for,

(a) $C_{\alpha s} = 0.2$, $C_{s\alpha} = 0.8$
(b) $C_{\alpha s} = 0.8$, $C_{s\alpha} = 0.2$

15. Increasing steer angle and rear torque.
Use the date in Example 72 and repeat the example for,

(a) $C_{\alpha s} = 0.2$, $C_{s\alpha} = 0.8$
(b) $C_{\alpha s} = 0.8$, $C_{s\alpha} = 0.2$

16. Steady-state stability condition.
Plot S_κ and S_β versus v_x for the vehicle of Example 70 if

(a) $C_{\alpha f} = 5C_{\alpha r}$, $C_{\alpha r} = 8.5$.
(b) $C_{\alpha f} = C_{\alpha r}/5$, $C_{\alpha r} = 8.5$.

17. Four-wheel plane vehicle, increasing rear torque, straight driving.
This exercise simulates changing one tire with a new or different tire.

(a) Repeat Example 75 assuming $C_{\alpha_1} = 10$, $C_{\alpha_2} = 8.5$, $C_{\alpha_3} = 8.5$, $C_{\alpha_4} = 8.5$.
(b) ★Determine the required steer angle $\delta = \delta(v_x)$ such that the vehicle moves straight.

18. Four-wheel plane vehicle, increasing steer angle and front torque.
Repeat Example 75 assuming $C_{\alpha_1} = 10$, $C_{\alpha_2} = 8.5$, $C_{\alpha_3} = 8.5$, $C_{\alpha_4} = 8.5$.

19. ★Increasing steer angle and opposite rear torque.
Use the date in Example 78 and assume opposite left and right rear torques.

$$T_3 = \begin{cases} -30t \text{ N m} & 0 < t < 10 \text{ s} \\ -300 \text{ N m} & 10 \text{ s} < t \end{cases} \qquad T_1 = 0 \tag{2.484}$$

$$T_4 = \begin{cases} 30t \text{ N m} & 0 < t < 10 \text{ s} \\ 300 \text{ N m} & 10 \text{ s} < t \end{cases} \qquad T_2 = 0 \tag{2.485}$$

(a) Does the vehicle move?
(b) Solve the equations of motion numerically and plot the same graphs of the example to compare.

20. No friction on one tire and applying front torque.
In Example 80 assume the wheel number 1 has no friction.

$$\begin{array}{llll} C_{\alpha_1} = 0 & C_{\alpha_2} = 8.5 & C_{\alpha_3} = 8.5 & C_{\alpha_4} = 8.5 \\ C_{s_1} = 0 & C_{s_2} = 7.5 & C_{s_3} = 7.5 & C_{s_4} = 7.5 \end{array} \tag{2.486}$$

(a) Repeat the example and plot the same graphs to compare.
(b) ★Determine the required steer angle $\delta = \delta(v_x)$ such that the vehicle moves straight.

Chapter 3
Vehicle Roll Dynamics

We study the roll vehicle dynamic model in this chapter. The roll dynamic model of vehicles has forward, lateral, yaw, and roll motions. The model of a rollable rigid vehicle is more exact and more realistic compared to the vehicle planar model. Using this model, we are able to analyze the roll behavior of a vehicle while maneuvering.

3.1 Equations of Motion and Degrees of Freedom

The *Newton–Euler equations of motion* of the roll vehicle dynamics in the body coordinate frame B are:

$$F_x = m\,\dot{v}_x - mr\,v_y \tag{3.1}$$

$$F_y = m\,\dot{v}_y + mr\,v_x \tag{3.2}$$

$$M_z = I_z\,\dot{\omega}_z = I_z\,\dot{r} \tag{3.3}$$

$$M_x = I_x\,\dot{\omega}_x = I_x\,\dot{p} \tag{3.4}$$

$$T_i = I_{w_i}\,\dot{\omega}_{w_i} + R_w\,F_{x_i} \tag{3.5}$$

The roll vehicle dynamic model is well expressed by four kinematic variables: the forward motion x, the lateral motion y, the roll angle φ, and the yaw angle ψ, plus four equations for the dynamics of each wheel. In the roll model of vehicle dynamics, we do not consider vertical movement z and pitch motion θ (Jazar 2017, 2011).

Figure 3.1 illustrates a roll vehicle model with a body coordinate frame $B(xyz)$ at the mass center C, its force system, and its kinematic variables. The body coordinate frame is assumed to be the principal coordinate frame of the vehicle. The x-axis

R. N. Jazar, *Advanced Vehicle Dynamics*,
https://doi.org/10.1007/978-3-030-13062-6_3

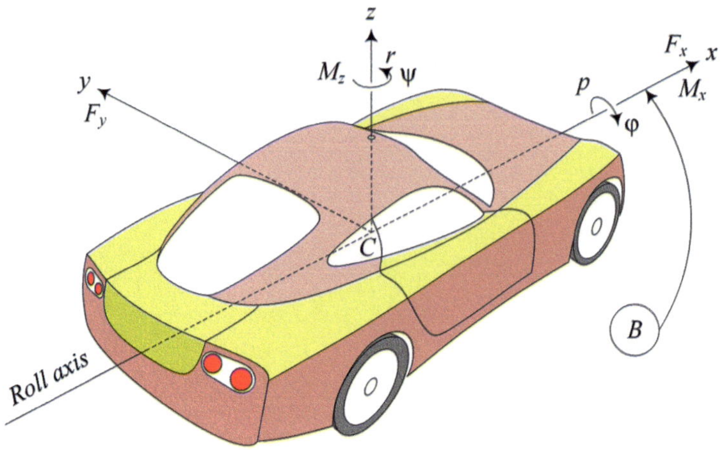

Fig. 3.1 The roll vehicle dynamic model has four degrees of freedom indicated by: x, y, φ, ψ

is a longitudinal axis passing through C and directed forward. The y-axis goes laterally to the left from the driver's viewpoint. The z-axis makes the coordinate system a right-hand triad. The z-axis is perpendicular to the ground, opposite to the gravitational acceleration **g** on a flat horizontal road (Cossalter 2002; Dai et al. 2016, 2017a,b).

Angular orientation of the vehicle is expressed by three angles: *roll* φ, *pitch* θ, and *yaw* ψ, and the vehicle angular velocities are expressed by their rates: *roll rate* p, *pitch rate* q, and *yaw rate* r.

$$p = \dot{\varphi} = \omega_x \tag{3.6}$$

$$q = \dot{\theta} = \omega_y \tag{3.7}$$

$$r = \dot{\psi} = \omega_z \tag{3.8}$$

The vehicle force system (**F**, **M**) is the resultant of external forces and moments that the vehicle receives from the ground and environment. The forces in the body coordinate frame are expressed by:

$${}^B\mathbf{F} = F_x\hat{\imath} + F_y\hat{\jmath} + F_z\hat{k} \tag{3.9}$$

$${}^B\mathbf{M} = M_x\hat{\imath} + M_y\hat{\jmath} + M_z\hat{k} \tag{3.10}$$

In roll vehicle dynamic model we assume:

$$\theta = 0 \tag{3.11}$$

$$q = \dot{\theta} = \omega_y = 0 \tag{3.12}$$

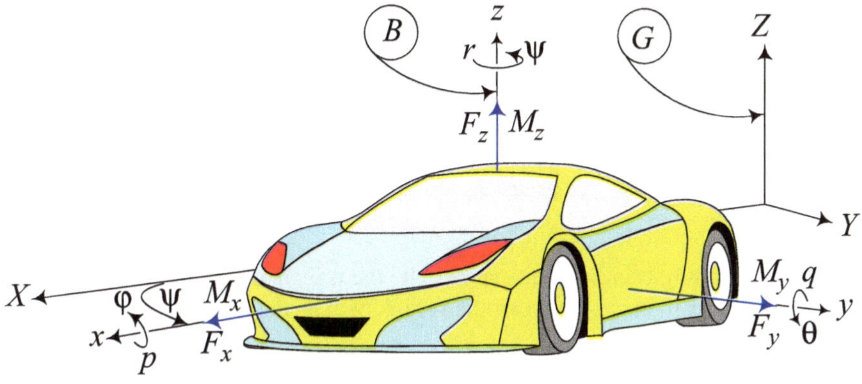

Fig. 3.2 The vehicle roll dynamic model and its degrees of freedom and force system

Proof Consider the vehicle in Fig. 3.2. A global coordinate frame G is fixed on the ground, and a local coordinate frame B is attached to the vehicle at its mass center C. The orientation of the frame B can be expressed by the heading angle ψ between the X and x axes measured from X, and the roll angle φ about the x-axis, between the Z and z axes measured from Z. The global position vector of the mass center is denoted by ${}^{G}\mathbf{d}$.

The force system in the body coordinate frame is:

$$\begin{aligned} {}^{B}\mathbf{F} &= {}^{B}R_G\,{}^{G}\mathbf{F} = {}^{B}R_G\left(m\,{}^{G}\mathbf{a}_B\right) = m\,{}^{B}_{G}\mathbf{a}_B \\ &= m\,{}^{B}\dot{\mathbf{v}}_B + m\,{}^{B}_{G}\boldsymbol{\omega}_B \times {}^{B}\mathbf{v}_B \end{aligned} \tag{3.13}$$

$$ {}^{B}\mathbf{M} = \frac{{}^{G}d}{dt}\,{}^{B}\mathbf{L} = {}^{B}_{G}\dot{\mathbf{L}}_B = {}^{B}\dot{\mathbf{L}} + {}^{B}_{G}\boldsymbol{\omega}_B \times {}^{B}\mathbf{L} \tag{3.14}$$

$$ = {}^{B}I\,{}^{B}_{G}\dot{\boldsymbol{\omega}}_B + {}^{B}_{G}\boldsymbol{\omega}_B \times \left({}^{B}I\,{}^{B}_{G}\boldsymbol{\omega}_B\right) \tag{3.15}$$

Ignoring the vertical movement at the mass center, the velocity ${}^{B}\mathbf{v}$ and acceleration ${}^{B}\dot{\mathbf{v}}$ of the vehicle, expressed in the body coordinate frame are:

$$ {}^{B}\mathbf{v} = \begin{bmatrix} v_x & v_y & 0 \end{bmatrix}^T \tag{3.16}$$

$$ {}^{B}\dot{\mathbf{v}} = \begin{bmatrix} \dot{v}_x & \dot{v}_y & 0 \end{bmatrix}^T \tag{3.17}$$

where v_x is the forward component and v_y is the lateral component of the vehicle velocity vector ${}^{B}\mathbf{v}$. Ignoring pitch motion of the vehicle, the angular velocity and its rate are:

$$
{}_G^B\boldsymbol{\omega}_B = \begin{bmatrix} \omega_x \\ 0 \\ \omega_z \end{bmatrix} = \begin{bmatrix} \dot{\varphi} \\ 0 \\ \dot{\psi} \end{bmatrix} = \begin{bmatrix} p \\ 0 \\ r \end{bmatrix} \tag{3.18}
$$

$$
{}_G^B\dot{\boldsymbol{\omega}}_B = \begin{bmatrix} \dot{\omega}_x \\ 0 \\ \dot{\omega}_z \end{bmatrix} = \begin{bmatrix} \dot{p} \\ 0 \\ \dot{r} \end{bmatrix} \tag{3.19}
$$

Assuming the body coordinate frame B to be the principal coordinate frame of the vehicle, the mass moment matrix of the vehicle will be diagonal:

$$
{}^BI = \begin{bmatrix} I_x & 0 & 0 \\ 0 & I_y & 0 \\ 0 & 0 & I_z \end{bmatrix} = \begin{bmatrix} I_1 & 0 & 0 \\ 0 & I_2 & 0 \\ 0 & 0 & I_3 \end{bmatrix} \tag{3.20}
$$

Substituting the above vectors and matrices in the Eqs. (3.13) and (3.15) provides us with the following equations:

$$
{}^B\mathbf{F} = m\ {}^B\dot{\mathbf{v}}_B + m\ {}_G^B\boldsymbol{\omega}_B \times {}^B\mathbf{v}_B \tag{3.21}
$$

$$
\begin{bmatrix} F_x \\ F_y \\ F_z \end{bmatrix} = m \begin{bmatrix} \dot{v}_x \\ \dot{v}_y \\ 0 \end{bmatrix} + m \begin{bmatrix} p \\ 0 \\ r \end{bmatrix} \times \begin{bmatrix} v_x \\ v_y \\ 0 \end{bmatrix}
$$

$$
= \begin{bmatrix} m\dot{v}_x - mrv_y \\ m\dot{v}_y + mrv_x \\ mpv_y \end{bmatrix} \tag{3.22}
$$

$$
{}^B\mathbf{M} = {}^BI\ {}_G^B\dot{\boldsymbol{\omega}}_B + {}_G^B\boldsymbol{\omega}_B \times \left({}^BI\ {}_G^B\boldsymbol{\omega}_B \right) \tag{3.23}
$$

$$
\begin{bmatrix} M_x \\ M_y \\ M_z \end{bmatrix} = \begin{bmatrix} I_1 & 0 & 0 \\ 0 & I_2 & 0 \\ 0 & 0 & I_3 \end{bmatrix} \begin{bmatrix} \dot{p} \\ 0 \\ \dot{r} \end{bmatrix}
$$

$$
+ \begin{bmatrix} p \\ 0 \\ r \end{bmatrix} \times \left(\begin{bmatrix} I_1 & 0 & 0 \\ 0 & I_2 & 0 \\ 0 & 0 & I_3 \end{bmatrix} \begin{bmatrix} p \\ 0 \\ r \end{bmatrix} \right)
$$

$$
= \begin{bmatrix} I_1\dot{p} \\ (I_1 - I_3)\,pr \\ I_3\dot{r} \end{bmatrix} \tag{3.24}
$$

The first two equations of (3.22) are the translational equations of motion in the x and y directions.

$$\begin{bmatrix} F_x \\ F_y \end{bmatrix} = \begin{bmatrix} m\dot{v}_x - mrv_y \\ m\dot{v}_y + mrv_x \end{bmatrix} \tag{3.25}$$

and the first and third Euler equations (3.24) are the rotational equations of motion about the x and z axes.

$$\begin{bmatrix} M_x \\ M_z \end{bmatrix} = \begin{bmatrix} I_1\dot{\omega}_x \\ I_3\dot{\omega}_z \end{bmatrix} \tag{3.26}$$

The third Newton's equation

$$mpv_y = F_z \tag{3.27}$$

provides the compatibility condition to keep the vehicle on the road and calculates the change in F_z because of motion. The second Euler equation

$$(I_1 - I_3)\,pr = M_y \tag{3.28}$$

is another compatibility condition that provides the required pitch moment to keep the vehicle on the road.

The last equation (3.5) is the result of analysis of individual wheels of the vehicle. The free-body-diagram of the wheel number i is shown in Fig. 2.3. The traction force F_x and a roll resistance force F_R are applied on the tire in the x-direction. Employing the in-wheel torque T_i and a resultant traction force F_{x_i}, the moment equation of motion of the wheel number i would be:

$$T_i = I_{w_i}\,\dot{\omega}_{w_i} + R_{w_i}\,F_{x_i} \tag{3.29}$$

where I_w is the mass moment of the wheel about its spin axis, ω_w is the angular velocity of the wheel about its spin axis, and R_w is the equivalent tire radius. For simplicity, we may replace R_w with tire geometric radius R_g considering them to be close to each other. ■

Example 82 Five degrees of freedom vehicle motion.

Assume a vehicle that has forward, lateral translations along the x-axis, y-axis as well as rotations about the x-axis, y-axis and the z-axis. Such a vehicle will have five degrees of freedom, and its mathematical model is considered full vehicle dynamic model for handling study. To develop the equations of motion of such a vehicle, we need to define the kinematic characteristics as follows:

$$^B\mathbf{v} = \begin{bmatrix} v_x & v_y & 0 \end{bmatrix}^T \tag{3.30}$$

$$^B\dot{\mathbf{v}} = \begin{bmatrix} \dot{v}_x & \dot{v}_y & 0 \end{bmatrix}^T \tag{3.31}$$

$$ {}^B_G\boldsymbol{\omega}_B = \begin{bmatrix} \omega_x & \omega_y & \omega_z \end{bmatrix}^T \tag{3.32}$$

$$ {}^B_G\dot{\boldsymbol{\omega}}_B = \begin{bmatrix} \dot{\omega}_x & \dot{\omega}_y & \dot{\omega}_z \end{bmatrix}^T \tag{3.33}$$

The acceleration vector of the vehicle in the body coordinate is (Jazar 2010a):

$$\begin{aligned} {}^B\mathbf{F} &= {}^B R_G \, {}^G\mathbf{F} = {}^B R_G \left(m \, {}^G\mathbf{a}_B \right) = m \, {}^B_G\mathbf{a}_B \\ &= m \, {}^B\dot{\mathbf{v}}_B + m \, {}^B_G\boldsymbol{\omega}_B \times {}^B\mathbf{v}_B \end{aligned} \tag{3.34}$$

$$ {}^B\mathbf{M} = \frac{{}^G d}{dt} \, {}^B\mathbf{L} = {}^B_G\dot{\mathbf{L}}_B = {}^B\dot{\mathbf{L}} + {}^B_G\boldsymbol{\omega}_B \times {}^B\mathbf{L} \tag{3.35}$$

$$ = {}^B I \, {}^B_G\dot{\boldsymbol{\omega}}_B + {}^B_G\boldsymbol{\omega}_B \times \left({}^B I \, {}^B_G\boldsymbol{\omega}_B \right) \tag{3.36}$$

$$ {}^B\mathbf{a} = {}^B\dot{\mathbf{v}}_B + {}^B_G\boldsymbol{\omega}_B \times {}^B\mathbf{v}_B = \begin{bmatrix} \dot{v}_x + \omega_y v_z - \omega_z v_y \\ \dot{v}_y + \omega_z v_x - \omega_x v_z \\ \dot{v}_z + \omega_x v_y - \omega_y v_x \end{bmatrix} \tag{3.37}$$

and therefore, the Newton's equations of motion for the vehicle are:

$$\begin{aligned} {}^B\mathbf{F} &= m \, {}^B\mathbf{a} = m \, {}^B\dot{\mathbf{v}}_B + m \, {}^B_G\boldsymbol{\omega}_B \times {}^B\mathbf{v}_B \\ \begin{bmatrix} F_x \\ F_y \\ F_z \end{bmatrix} &= m \begin{bmatrix} \dot{v}_x + \omega_y v_z - \omega_z v_y \\ \dot{v}_y + \omega_z v_x - \omega_x v_z \\ \dot{v}_z + \omega_x v_y - \omega_y v_x \end{bmatrix} \end{aligned} \tag{3.38}$$

Similarly, the Euler's equations of motion will be:

$$\begin{aligned} {}^B\mathbf{M} &= {}^B I \, {}^B_G\dot{\boldsymbol{\omega}}_B + {}^B_G\boldsymbol{\omega}_B \times \left({}^B I \, {}^B_G\boldsymbol{\omega}_B \right) \\ \begin{bmatrix} M_x \\ M_y \\ M_z \end{bmatrix} &= \begin{bmatrix} \dot{\omega}_x I_1 - \omega_y \omega_z I_2 + \omega_y \omega_z I_3 \\ \dot{\omega}_y I_2 + \omega_x \omega_z I_1 - \omega_x \omega_z I_3 \\ \dot{\omega}_z I_3 - \omega_x \omega_y I_1 + \omega_x \omega_y I_2 \end{bmatrix} \end{aligned} \tag{3.39}$$

in which we assumed the mass moment matrix to be principal.

$$ {}^B I = \begin{bmatrix} I_1 & 0 & 0 \\ 0 & I_2 & 0 \\ 0 & 0 & I_3 \end{bmatrix} \tag{3.40}$$

Equations (3.38) and (3.39) are the force system expression of a vehicle with five degrees of freedom.

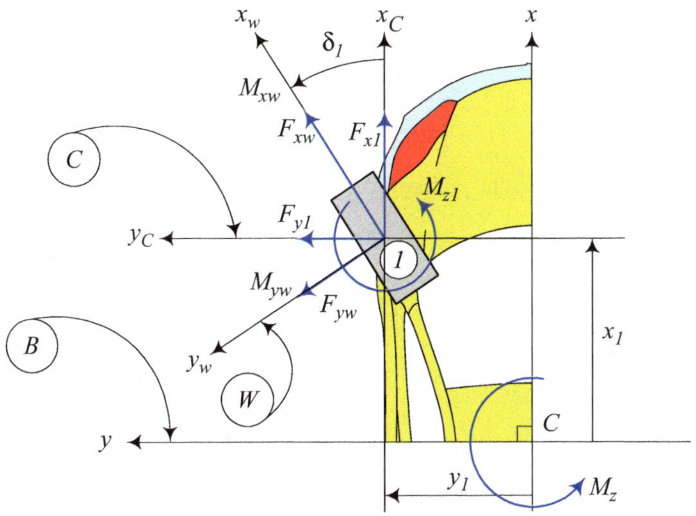

Fig. 3.3 The force system at the tireprint of tire number 1, and their resultant force system at C

3.2 Tire Force System

Figure 3.3 depicts the wheel number 1 of a vehicle and its force system $\left(F_{x_1}, F_{y_1}, M_{z_1}\right)$. The force components are generated at the center of tireprint in the T-frame $\left(F_{x_{T_1}}, F_{y_{T_1}}, M_{z_{T_1}}\right)$ and will be transformed to the C-frame (x_1, y_1, z_1) at the tire center. Then the forces in C-frames of all wheels will be transformed to the B-frame (x, y, z) at the mass center of the vehicle to determine the resultant applied forces on the vehicle (Jazar 2017; Popp and Schiehlen 2010). Let the force system at the tireprint in the tire coordinate frame T to be:

$$^T\mathbf{F}_w = \begin{bmatrix} F_{x_{T_i}} & F_{y_{T_i}} & F_{z_{T_i}} \end{bmatrix}^T \tag{3.41}$$

$$^T\mathbf{M}_w = \begin{bmatrix} M_{x_{T_i}} & M_{y_{T_i}} & M_{z_{T_i}} \end{bmatrix}^T \tag{3.42}$$

then the force system at the center of the wheel in W-frame will be:

$$^W\mathbf{F}_w = {}^W R_T \, {}^T\mathbf{F}_w = \begin{bmatrix} F_{xw_i} \\ F_{yw_i} \\ F_{zw_i} \end{bmatrix} = \begin{bmatrix} F_{x_{T_i}} \\ F_{y_{T_i}} \cos\gamma_i + F_{z_{T_i}} \sin\gamma_i \\ F_{z_{T_i}} \cos\gamma_i - F_{y_{T_i}} \sin\gamma_i \end{bmatrix} \tag{3.43}$$

$$
{}^{W}\mathbf{M}_{w} = \begin{bmatrix} M_{xW_i} \\ M_{yW_i} \\ M_{zW_i} \end{bmatrix} = \begin{bmatrix} M_{xT_i} + RF_{yT_i}\cos\gamma + RF_{zT_i}\sin\gamma \\ M_{yT_i}\cos\gamma - RF_{xT_i} + M_{zT_i}\sin\gamma \\ M_{zT_i}\cos\gamma - M_{yT_i}\sin\gamma \end{bmatrix} \tag{3.44}
$$

where R is the tire radius.

The force system at the center of the wheel number i in the wheel-body coordinate frame C is:

$$
{}^{C}\mathbf{F}_{w} = \begin{bmatrix} F_{x_i} \\ F_{y_i} \\ F_{z_i} \end{bmatrix} = \begin{bmatrix} F_{xW_i}\cos\delta_1 - F_{yW_i}\sin\delta_1 \\ F_{yW_i}\cos\delta_1 + F_{xW_i}\sin\delta_1 \\ F_{zW_i} \end{bmatrix} \tag{3.45}
$$

$$
{}^{C}\mathbf{M}_{w} = \begin{bmatrix} M_{x_i} \\ M_{y_i} \\ M_{z_i} \end{bmatrix} = \begin{bmatrix} M_{xW_i}\cos\delta_1 - M_{yW_i}\sin\delta_1 \\ M_{yW_i}\cos\delta_1 + M_{xW_i}\sin\delta_1 \\ M_{zW_i} \end{bmatrix} \tag{3.46}
$$

In this analysis, we ignored the components of the tire moment at the tireprint $M_{y_{w_i}}$ compared to other moments.

The total important force system on the rigid vehicle in the body coordinate frame to analyze the roll model of rigid vehicle is

$$
{}^{B}F_{x} = \sum_i F_{x_i} = \sum_i F_{xW_i}\cos\delta_i - \sum_i F_{yW_i}\sin\delta_i \tag{3.47}
$$

$$
{}^{B}F_{y} = \sum_i F_{y_i} = \sum_i F_{yW_i}\cos\delta_i + \sum_i F_{xW_i}\sin\delta_i \tag{3.48}
$$

$$
{}^{B}M_{x} = \sum_i M_{x_i} + \sum_i y_i F_{z_i} - \sum_i z_i F_{y_i} \tag{3.49}
$$

$$
{}^{B}M_{z} = \sum_i M_{z_i} + \sum_i x_i F_{y_i} - \sum_i y_i F_{x_i} \tag{3.50}
$$

Proof A tire coordinate frame T is set at the center of the tireprint at the intersection of tire-plane and the ground. The z_T-axis is always perpendicular to the ground and upward. The tire T-frame follows the steer angle rotation about the z_T-axis but it does not follow the spin and camber rotations of the tire. A W-frame is also attached to the center of the wheel that follows every motion of the wheel except the spin. The C-frame is set at the center of the neutral wheel and is parallel to the body coordinate frame B. The W-frame and C-frame will be coincident when the wheel is at the rest position. The C-frame is motionless with respect to the B-frame and does not follow any motion of the wheel. The vehicle body coordinate frame $B\,(x, y, z)$ is attached to the vehicle at its mass center. The wheel force system that are generated in T-frame must be transformed to the C-frame and then B-frame to develop the vehicle equations of motion.

Let us assume that the force system in the T-frame at the tireprint of the wheel number i to be:

$$ {}^{T}\mathbf{F}_{w} = \begin{bmatrix} F_{xT_i} & F_{yT_i} & F_{zT_i} \end{bmatrix}^{T} \tag{3.51} $$

$$ {}^{T}\mathbf{M}_{w} = \begin{bmatrix} M_{xT_i} & M_{yT_i} & M_{zT_i} \end{bmatrix}^{T} \tag{3.52} $$

The rotation matrix between T-frame and W-frame is:

$$ {}^{W}R_{T} = \begin{bmatrix} 1 & 0 & 0 \\ 0 & \cos\gamma_i & \sin\gamma_i \\ 0 & -\sin\gamma_i & \cos\gamma_i \end{bmatrix} \tag{3.53} $$

and the rotation matrix between the W-frame and the C-frame is:

$$ {}^{C}R_{W} = \begin{bmatrix} \cos\delta_i & -\sin\delta_i & 0 \\ \sin\delta_i & \cos\delta_i & 0 \\ 0 & 0 & 1 \end{bmatrix} \tag{3.54} $$

Therefore, the force system in the W-frame is:

$$ {}^{W}\mathbf{F}_{w} = {}^{W}R_{T}\,{}^{T}\mathbf{F}_{w} = \begin{bmatrix} F_{xT_i} \\ F_{yT_i}\cos\gamma_i + F_{zT_i}\sin\gamma_i \\ F_{zT_i}\cos\gamma_i - F_{yT_i}\sin\gamma_i \end{bmatrix} = \begin{bmatrix} F_{xW_i} \\ F_{yW_i} \\ F_{zW_i} \end{bmatrix} \tag{3.55} $$

$$ \begin{aligned} {}^{W}\mathbf{M}_{w} &= {}^{W}R_{T}\,{}^{T}\mathbf{M}_{w} + {}^{W}\mathbf{R} \times {}^{W}_{T}\mathbf{F}_{w} \\ &= {}^{W}R_{T}\,{}^{T}\mathbf{M}_{w} + (-R)\,{}^{W}\hat{k} \times {}^{W}R_{T}\,{}^{T}\mathbf{F}_{w} \\ &= {}^{W}R_{T} \begin{bmatrix} M_{xT_i} \\ M_{yT_i} \\ M_{zT_i} \end{bmatrix} + \begin{bmatrix} 0 \\ 0 \\ -R \end{bmatrix} \times {}^{W}R_{T} \begin{bmatrix} F_{xT_i} \\ F_{yT_i} \\ F_{zT_i} \end{bmatrix} \\ &= \begin{bmatrix} M_{xT_i} + RF_{yT_i}\cos\gamma + RF_{zT_i}\sin\gamma \\ M_{yT_i}\cos\gamma - RF_{xT_i} + M_{zT_i}\sin\gamma \\ M_{zT_i}\cos\gamma - M_{yT_i}\sin\gamma \end{bmatrix} = \begin{bmatrix} M_{xW_i} \\ M_{yW_i} \\ M_{zW_i} \end{bmatrix} \end{aligned} \tag{3.56} $$

where ${}^{W}\mathbf{R}$ is the position vector of the T-frame in W-frame which is equal to radius of the wheel.

$$ {}^{W}\mathbf{R} = -R\,{}^{W}\hat{k} = \begin{bmatrix} 0\ 0\ -R \end{bmatrix}^{T} \tag{3.57} $$

The force system in the C-frame is then calculated by transformation.

$$
{}^{C}\mathbf{F}_{w} = {}^{C}R_{W}\,{}^{W}\mathbf{F}_{w} \tag{3.58}
$$

$$
\begin{aligned}
\begin{bmatrix} F_{x_i} \\ F_{y_i} \\ F_{z_i} \end{bmatrix} &= \begin{bmatrix} \cos\delta_i & -\sin\delta_i & 0 \\ \sin\delta_i & \cos\delta_i & 0 \\ 0 & 0 & 1 \end{bmatrix} \begin{bmatrix} F_{xW_i} \\ F_{yW_i} \\ F_{zW_i} \end{bmatrix} \\
&= \begin{bmatrix} F_{xW_i}\cos\delta_i - F_{yW_i}\sin\delta_i \\ F_{yW_i}\cos\delta_i + F_{xW_i}\sin\delta_i \\ F_{zW_i} \end{bmatrix}
\end{aligned} \tag{3.59}
$$

$$
\begin{aligned}
{}^{C}\mathbf{M}_{w} &= {}^{C}R_{W}\,{}^{W}\mathbf{M}_{w} \\
\begin{bmatrix} M_{x_i} \\ M_{y_i} \\ M_{z_i} \end{bmatrix} &= \begin{bmatrix} \cos\delta_i & -\sin\delta_i & 0 \\ \sin\delta_i & \cos\delta_i & 0 \\ 0 & 0 & 1 \end{bmatrix} \begin{bmatrix} M_{xW_i} \\ M_{yW_i} \\ M_{zW_i} \end{bmatrix} \\
&= \begin{bmatrix} M_{xW_i}\cos\delta_i - M_{yW_i}\sin\delta_i \\ M_{yW_i}\cos\delta_i + M_{xW_i}\sin\delta_i \\ M_{zW_i} \end{bmatrix}
\end{aligned} \tag{3.60}
$$

We transform the force system of each tire to the body coordinate frame B, located at the vehicle mass center to calculate the total applied force system on the vehicle.

$$
{}^{B}\mathbf{F} = \sum_i {}^{C}\mathbf{F}_{w} = \sum_i F_{x_i}\,\hat{\imath} + \sum_i F_{y_i}\,\hat{\jmath} + \sum_i F_{z_i}\,\hat{k} \tag{3.61}
$$

$$
\begin{aligned}
{}^{B}\mathbf{M} &= \sum_i {}^{C}\mathbf{M}_{w} + \sum_i {}^{B}\mathbf{r}_i \times {}^{B}\mathbf{F}_{w_i} \\
&= \sum_i M_{x_i}\,\hat{\imath} + \sum_i M_{y_i}\,\hat{\jmath} + \sum_i M_{z_i}\,\hat{k} + \sum_i \begin{bmatrix} y_i F_{z_i} - z_i F_{y_i} \\ z_i F_{x_i} - x_i F_{z_i} \\ x_i F_{y_i} - y_i F_{x_i} \end{bmatrix}
\end{aligned} \tag{3.62}
$$

where ${}^{B}\mathbf{r}_i$ is the position vector of the wheel number i in B-frame.

$$
{}^{B}\mathbf{r}_i = \begin{bmatrix} x_i & y_i & z_i \end{bmatrix}^T = \begin{bmatrix} a_i & b_i & z_i \end{bmatrix}^T \tag{3.63}
$$

$$
z_i \simeq 0 \tag{3.64}
$$

Assuming

$$ {}^{T}\mathbf{F}_{w} \simeq \begin{bmatrix} F_{xT_i} & F_{yT_i} & 0 \end{bmatrix}^{T} \tag{3.65}$$

$$ {}^{T}\mathbf{M}_{w} \simeq \begin{bmatrix} M_{xT_i} & 0 & M_{zT_i} \end{bmatrix}^{T} \tag{3.66}$$

and expanding Eqs. (3.61) and (3.62) provide us with the total force system on the roll vehicle dynamic model.

$$ {}^{B}F_{x} = \sum_{i} F_{xw_i} \cos\delta_i - \sum_{i} F_{yw_i} \sin\delta_i \tag{3.67}$$

$$ {}^{B}F_{y} = \sum_{i} F_{yw_i} \cos\delta_i + \sum_{i} F_{xw_i} \sin\delta_i \tag{3.68}$$

$$ {}^{B}M_{x} = \sum_{i} M_{x_i} + \sum_{i} y_i F_{z_i} - \sum_{i} z_i F_{y_i} \tag{3.69}$$

$$ {}^{B}M_{z} = \sum_{i} M_{z_i} + \sum_{i} x_i F_{y_i} - \sum_{i} y_i F_{x_i} \tag{3.70}$$

■

Example 83 Tire force system in W-frame.

If the force system at the tireprint is

$$ {}^{T}\mathbf{F} = F_{xT_i}\, {}^{T}\hat{\imath} + F_{yT_i}\, {}^{T}\hat{\jmath} + F_{zT_i}\, {}^{T}\hat{k} \tag{3.71}$$

$$ {}^{T}\mathbf{M} = M_{xT_i}\, {}^{T}\hat{\imath} + M_{zT_i}\, {}^{T}\hat{k} \tag{3.72}$$

then the force system in the W-frame at the center of the wheel would be

$$ {}^{W}\mathbf{F}_{w} = {}^{W}R_{T}\, {}^{T}\mathbf{F}_{w} = {}^{T}R_{W}^{T}\, {}^{T}\mathbf{F}_{w} \tag{3.73}$$

$$ \begin{bmatrix} F_{xW_i} \\ F_{yW_i} \\ F_{zW_i} \end{bmatrix} = \begin{bmatrix} 1 & 0 & 0 \\ 0 & \cos\gamma & -\sin\gamma \\ 0 & \sin\gamma & \cos\gamma \end{bmatrix}^{T} \begin{bmatrix} F_{xT_i} \\ F_{yT_i} \\ F_{zT_i} \end{bmatrix} = \begin{bmatrix} F_{xT_i} \\ F_{yT_i}\cos\gamma + F_{zT_i}\sin\gamma \\ F_{zT_i}\cos\gamma - F_{yT_i}\sin\gamma \end{bmatrix} \tag{3.74}$$

$$ {}^{W}\mathbf{M} = {}^{T}R_{W}^{T} \left({}^{T}\mathbf{M} - {}^{T}\mathbf{r}_{o} \times {}^{T}\mathbf{F} \right) \tag{3.75}$$

$$\begin{bmatrix} M_{xW_i} \\ M_{yW_i} \\ M_{zW_i} \end{bmatrix} = \begin{bmatrix} M_{xT_i} + RF_{yT_i} \cos\gamma_i + RF_{zT_i} \sin\gamma_i \\ M_{zT_i} \sin\gamma_i - RF_{xT_i} \\ M_{zT_i} \cos\gamma_i \end{bmatrix} \tag{3.76}$$

where

$$^T\mathbf{r}_o = \begin{bmatrix} 0 \\ -R\sin\gamma_i \\ R\cos\gamma_i \end{bmatrix} \tag{3.77}$$

The wheel force system at zero camber, $\gamma = 0$, reduces to

$$^W\mathbf{F}_w = {}^W R_T \, {}^T\mathbf{F}_w = \begin{bmatrix} F_{xT_i} \\ F_{yT_i} \\ F_{zT_i} \end{bmatrix} \tag{3.78}$$

$$^W\mathbf{M} = {}^T R_W^T \left({}^T\mathbf{M} - {}^T\mathbf{r}_o \times {}^T\mathbf{F} \right) = \begin{bmatrix} M_{xT_i} + RF_{yT_i} \\ -RF_{xT_i} \\ M_{zT_i} \end{bmatrix} \tag{3.79}$$

Example 84 Tire force system in C-frame.

Considering force $^T\mathbf{F}$ under the tire number i in the T-frame is:

$$^T\mathbf{F}_w = \begin{bmatrix} F_{xT_i} & F_{yT_i} & F_{zT_i} \end{bmatrix}^T \tag{3.80}$$

then the force in W-frame would be (3.74)

$$^W\mathbf{F}_w = {}^W R_T \, {}^T\mathbf{F}_w = \begin{bmatrix} F_{xT_i} \\ F_{yT_i} \cos\gamma_i + F_{zT_i} \sin\gamma_i \\ F_{zT_i} \cos\gamma_i - F_{yT_i} \sin\gamma_i \end{bmatrix} \tag{3.81}$$

and the force in C-frame would be

$$^C\mathbf{F}_w = {}^C R_W \, {}^W\mathbf{F}_w \tag{3.82}$$

$$\begin{bmatrix} F_{x_i} \\ F_{y_i} \\ F_{z_i} \end{bmatrix} = \begin{bmatrix} F_{xT_i} \cos\delta_i - \left(F_{yT_i} \cos\gamma_i + F_{zT_i} \sin\gamma_i \right) \sin\delta_i \\ F_{xT_i} \sin\delta_i + \left(F_{yT_i} \cos\gamma_i + F_{zT_i} \sin\gamma_i \right) \cos\delta_i \\ F_{zT_i} \cos\gamma_i - F_{yT_i} \sin\gamma_i \end{bmatrix}$$

The moment $^T\mathbf{M}$ under the tire number i in the T-frame is:

$$^T\mathbf{M}_w = \begin{bmatrix} M_{xT_i} & M_{yT_i} & M_{zT_i} \end{bmatrix}^T \tag{3.83}$$

Therefore, the moment in W-frame is

$$\begin{aligned} {}^W\mathbf{M} &= {}^W R_T \, {}^T\mathbf{M} + {}^W\mathbf{R} \times {}^W_T\mathbf{F} \\ &= {}^W R_T \, {}^T\mathbf{M} - R \, {}^W\hat{k} \times {}^W R_T \, {}^T\mathbf{F} \\ &= \begin{bmatrix} M_{x_{T_i}} + RF_{y_{T_i}} \cos\gamma + RF_{z_{T_i}} \sin\gamma \\ M_{y_{T_i}} \cos\gamma - RF_{x_{T_i}} + M_{z_{T_i}} \sin\gamma \\ M_{z_{T_i}} \cos\gamma - M_{y_{T_i}} \sin\gamma \end{bmatrix} \end{aligned} \tag{3.84}$$

and the moment in C-frame would be

$${}^C\mathbf{M} = {}^C R_W \, {}^W\mathbf{M} \tag{3.85}$$

$$= \begin{bmatrix} -\left(M_{y_{T_i}} \cos\gamma - RF_{x_{T_i}} + M_{z_{T_i}} \sin\gamma\right) \sin\delta_1 \\ \left(M_{x_{T_i}} + RF_{y_{T_i}} \cos\gamma + RF_{z_{T_i}} \sin\gamma\right) \sin\delta_1 \\ +\left(M_{y_{T_i}} \cos\gamma - RF_{x_{T_i}} + M_{z_{T_i}} \sin\gamma\right) \cos\delta_1 \\ M_{z_{T_i}} \cos\gamma - M_{y_{T_i}} \sin\gamma \end{bmatrix}$$

3.3 Bicycle Roll Vehicle Force Components

Figure 3.4 illustrates the force system of each wheel in the wheel-body coordinate frame C, as well as the force system on the roll vehicle model in the body coordinate frame B. The wheel forces are acting at the wheel center of a front-wheel-steering vehicle. When we ignore the pitch motions of the vehicle, the angle between the z and Z axes is the roll angle φ. The z and Z axes are parallel at $\varphi = 0$.

Ignoring the lateral load transfer between left and right wheels, we may define a simplified *two-wheel model* for the roll vehicle model as is shown in Fig. 3.5. The two-wheel roll vehicle model is also called the *bicycle roll model*. However, the two-wheel roll vehicle model does not act similar to the roll behavior of traditional bicycles (Ellis 1994; Jazar 2017).

The force system applied on a bicycle roll model of vehicle at its mass center C is:

$$\begin{aligned} F_x = {}& F_{x_f} \cos\left(\delta + \delta_{\varphi_f}\right) + F_{x_r} \cos\delta_{\varphi_r} \\ & - F_{y_f} \sin\left(\delta + \delta_{\varphi_f}\right) - F_{y_r} \sin\delta_{\varphi_r} \end{aligned} \tag{3.86}$$

$$\begin{aligned} F_y = {}& F_{y_f} \cos\left(\delta + \delta_{\varphi_f}\right) + F_{y_r} \cos\delta_{\varphi_r} \\ & + F_{x_f} \sin\left(\delta + \delta_{\varphi_f}\right) + F_{x_r} \sin\delta_{\varphi_r} \end{aligned} \tag{3.87}$$

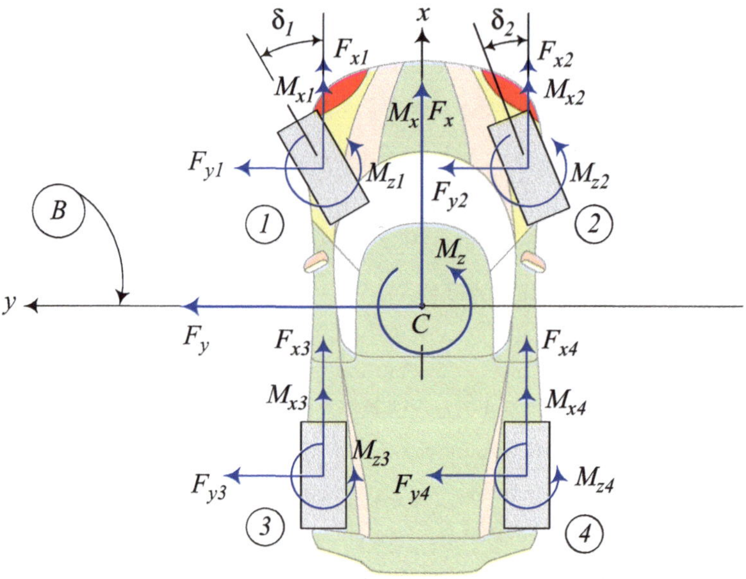

Fig. 3.4 The acting forces at the wheel center of a front-wheel-steering four-wheel roll model vehicle

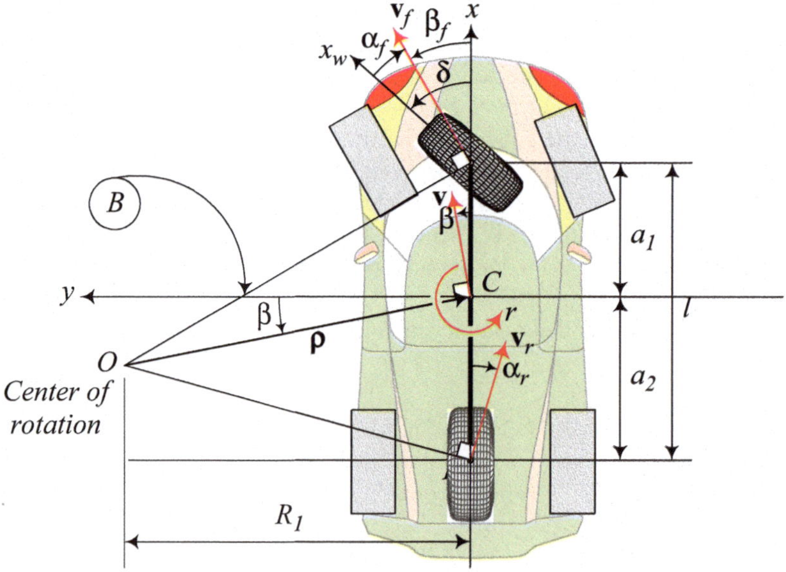

Fig. 3.5 A two-wheel model for a roll vehicle model

$$M_x = C_{T_f} F_{y_f} + C_{T_r} F_{y_r} - k_\varphi \varphi - c_\varphi \dot{\varphi} \tag{3.88}$$

$$M_z = a_1 \left(F_{x_f} \sin\left(\delta + \delta_{\varphi_f}\right) + F_{y_f} \cos\varphi \cos\left(\delta + \delta_{\varphi_f}\right)\right)$$
$$-a_2 \left(F_{x_r} \sin\delta_{\varphi_r} + F_{y_r} \cos\delta_{\varphi_r} \cos\varphi\right) \tag{3.89}$$

$$T_f = I_f \dot{\omega}_f + R_w F_{x_f} \tag{3.90}$$

$$T_r = I_r \dot{\omega}_r + R_w F_{x_r} \tag{3.91}$$

where

$$F_{x_f} = F_{z_f} C_{sf} S\left(s_f - s_s\right) \sqrt{1 - C_{s\alpha} \left(\frac{S\left(\alpha_f - \alpha_s\right)}{\alpha_s}\right)^2} \tag{3.92}$$

$$F_{x_r} = F_{z_r} C_{sr} S\left(s_r - s_s\right) \sqrt{1 - C_{s\alpha} \left(\frac{S\left(\alpha_r - \alpha_s\right)}{\alpha_s}\right)^2} \tag{3.93}$$

$$F_{y_f} = -F_{z_f} C_{\alpha f} S\left(\alpha_f - \alpha_s\right) \sqrt{1 - C_{\alpha s} \left(\frac{S\left(s_f - s_s\right)}{s_s}\right)^2} - C_{\varphi_f} \varphi \tag{3.94}$$

$$F_{y_r} = -F_{z_r} C_{\alpha r} S\left(\alpha_r - \alpha_s\right) \sqrt{1 - C_{\alpha s} \left(\frac{S\left(s_r - s_s\right)}{s_s}\right)^2} - C_{\varphi_r} \varphi \tag{3.95}$$

and S is the saturation function (1.60).

$$S(x - x_0) = \begin{cases} x_0 & x_0 < x \\ x & -x_0 < x < x_0 \\ -x_0 & x < -x_0 \end{cases} \tag{3.96}$$

The tire force characteristics are:

$$C_{\alpha f} = C_{\alpha_1} = C_{\alpha_2} \tag{3.97}$$

$$C_{\alpha r} = C_{\alpha_3} = C_{\alpha_4} \tag{3.98}$$

$$C_{sf} = C_{s_1} = C_{s_2} \tag{3.99}$$

$$C_{sr} = C_{s_3} = C_{s_4} \tag{3.100}$$

$$\alpha_f = \arctan\left(\frac{v_y}{v_x} + \frac{a_1}{v_x} r - C_{\beta_f} \frac{p}{v_x}\right) - \delta - C_{\delta\varphi_f} \varphi \tag{3.101}$$

$$\alpha_r = \arctan\left(\frac{v_y}{v_x} - \frac{a_2}{v_x}r - C_{\beta_r}\frac{p}{v_x}\right) - C_{\delta\varphi_r}\varphi \tag{3.102}$$

$$\beta = \arctan\frac{v_y}{v_x} \tag{3.103}$$

$$\beta_i = \arctan\frac{v_y + x_i\, r - C_{\beta_i} p}{v_x} \tag{3.104}$$

$$\alpha_i = \beta_i - \delta_a = \beta_i - \delta_i - \delta_{\varphi_i} = \beta_i - \delta_i - C_{\delta\varphi_i}\varphi \tag{3.105}$$

$$s_f = \frac{R_g\,\omega_f - v_{xTf}}{R_g\,\omega_f\,H(R_g\,\omega_f - v_{xTf}) + v_{xTf}\,H(v_{xTf} - R_g\,\omega_f)} \tag{3.106}$$

$$s_r = \frac{R_g\,\omega_r - v_x}{R_g\,\omega_r\,H(R_g\,\omega_r - v_x) + v_x\,H(v_x - R_g\,\omega_r)} \tag{3.107}$$

$$v_{xTf} = v_x\cos\left(\delta + \delta_{\varphi_f}\right) + \left(v_y + a_1 r\right)\sin\left(\delta + \delta_{\varphi_f}\right) \tag{3.108}$$

$$\dot{\omega}_f = \frac{T_f - R_{wf}F_{x_f}}{I_f} \tag{3.109}$$

$$\dot{\omega}_r = \frac{T_r - R_{wr}F_{x_r}}{I_r} \tag{3.110}$$

C_{β_i} is the *wheel slip coefficient* to determine the effect of vehicle roll on sideslip angles.

The vertical load on front and rear wheels is:

$$F_{z_f} = F_{z_1} + F_{z_2} = mg\frac{a_2}{l} - m\left(\dot{v}_x - r\,v_y\right)\frac{h}{l} \tag{3.111}$$

$$F_{z_r} = F_{z_3} + F_{z_4} = mg\frac{a_1}{l} + m\left(\dot{v}_x - r\,v_y\right)\frac{h}{l} \tag{3.112}$$

$$l = a_1 + a_2 \tag{3.113}$$

The applied torques on front and rear wheels are T_f and T_r. The front and rear wheels' mass moments about their spin axes are indicated by I_f and I_r. The forces $\left(F_{x_f}, F_{y_f}\right)$ and $\left(F_{x_r}, F_{y_r}\right)$ are the planar forces at the tireprint of the front and rear wheels and we assume them to be at the wheel center.

If the steer angle of the steering mechanism is denoted by δ, then the actual steer angle δ_a is:

$$\delta_a = \delta + \delta_\varphi \tag{3.114}$$

where δ_φ is the *roll-steering* angle.

$$\delta_\varphi = C_{\delta\varphi}\varphi \tag{3.115}$$

The roll-steering angle δ_φ is proportional to the roll angle φ and the coefficient $C_{\delta\varphi}$ is called the *roll-steering* coefficient. The roll steering happens because of the suspension mechanisms that generates some steer angle when deflected. The *tire sideslip angle* α_i of each tire of a rollable vehicle is then equal to:

$$\alpha_i = \beta_i - \delta_a = \beta_i - \delta_i - \delta_{\varphi_i} = \beta_i - \delta_i - C_{\delta\varphi_i}\varphi \tag{3.116}$$

where the *wheel sideslip angle* β_i is the angle between the wheel velocity vector $\mathbf{v}_i$ and the vehicle body x-axis.

Proof Employing tire force equations (3.47) and (3.48) and rewriting them for a bicycle vehicle provide us with Eqs. (3.86) and (3.87). To derive the force equations (3.86)–(3.89) on the bicycle roll vehicle model, we begin by defining the cot-average δ of the outer δ_o and inner δ_i steer angles or δ_2 and δ_1 as the only steer angle of the vehicle.

$$\cot\delta = \frac{\cot\delta_o + \cot\delta_i}{2} = \frac{\cot\delta_2 + \cot\delta_1}{2} \tag{3.117}$$

We also define sideslip coefficients $C_{\alpha f}$ and $C_{\alpha r}$ and slip ratio coefficients C_{sf} and C_{sr} for the front and rear tires as the average between the left and right tires.

$$C_{\alpha f} = \left(C_{\alpha_1} + C_{\alpha_2}\right)/2 \tag{3.118}$$

$$C_{\alpha r} = \left(C_{\alpha_3} + C_{\alpha_4}\right)/2 \tag{3.119}$$

$$C_{sf} = \left(C_{s_1} + C_{s_2}\right)/2 \tag{3.120}$$

$$C_{sr} = \left(C_{s_3} + C_{s_4}\right)/2 \tag{3.121}$$

Assuming the left and right tires to be identical, we have:

$$C_{\alpha f} = C_{\alpha_1} = C_{\alpha 2} \tag{3.122}$$

$$C_{\alpha r} = C_{\alpha_3} = C_{\alpha 4} \tag{3.123}$$

$$C_{sf} = C_{s_1} = C_{s_2} \tag{3.124}$$

$$C_{sr} = C_{s_3} = C_{s_4} \tag{3.125}$$

We may also assume similar force components to be equal on the left and right tires and adding them up to make the longitudinal and lateral forces $\left(F_{x_f}, F_{y_f}\right)$ and $\left(F_{x_r}, F_{y_r}\right)$ on front and rear tires.

$$F_{x_f} = F_{x_1} + F_{x_2} \qquad F_{x_1} = F_{x_2} \tag{3.126}$$

$$F_{x_r} = F_{x_3} + F_{x_4} \qquad F_{x_3} = F_{x_4} \tag{3.127}$$

$$F_{y_f} = F_{y_1} + F_{y_2} \qquad F_{y_1} = F_{y_2} \tag{3.128}$$

$$F_{y_r} = F_{y_3} + F_{y_4} \qquad F_{y_3} = F_{y_4} \tag{3.129}$$

$$F_{z_f} = F_{z_1} + F_{z_2} \qquad F_{z_1} = F_{z_2} \tag{3.130}$$

$$F_{z_r} = F_{z_3} + F_{z_4} \qquad F_{z_3} = F_{z_4} \tag{3.131}$$

The longitudinal and lateral forces on front and rear wheels will be expanded to the following equations by employing the elliptical combined tire force models (1.202) and (1.203), and proportional-saturation tire force behaviors (1.67) and (1.152).

$$\begin{aligned} F_{x_f} &= F_{x_1} + F_{x_2} \\ &= F_{z_f} C_{sf}\, S\left(s_f - s_s\right) \sqrt{1 - C_{s\alpha} \left(\frac{S\left(\alpha_f - \alpha_s\right)}{\alpha_s}\right)^2} \end{aligned} \tag{3.132}$$

$$\begin{aligned} F_{x_r} &= F_{x_3} + F_{x_4} \\ &= F_{z_r} C_{sr}\, S\left(s_r - s_s\right) \sqrt{1 - C_{s\alpha} \left(\frac{S\left(\alpha_r - \alpha_s\right)}{\alpha_s}\right)^2} \end{aligned} \tag{3.133}$$

$$\begin{aligned} F_{y_f} &= F_{y_1} + F_{y_2} \\ &= -F_{z_f} C_{\alpha f}\, S\left(\alpha_f - \alpha_s\right) \sqrt{1 - C_{\alpha s} \left(\frac{S\left(s_f - s_s\right)}{s_s}\right)^2} - C_{\varphi_f} \varphi \end{aligned} \tag{3.134}$$

$$\begin{aligned} F_{y_r} &= F_{y_3} + F_{y_4} \\ &= -F_{z_r} C_{\alpha r}\, S\left(\alpha_r - \alpha_s\right) \sqrt{1 - C_{\alpha s} \left(\frac{S\left(s_r - s_s\right)}{s_s}\right)^2} - C_{\varphi_r} \varphi \end{aligned} \tag{3.135}$$

Assuming

$$F_{z_1} = F_{z_2} \qquad F_{z_3} = F_{z_4} \tag{3.136}$$

and using the vertical weight transfer equations in body frame (1.462)–(1.465), the vertical load on front and rear tires are:

$$F_{z_f} = F_{z_1} + F_{z_2} = m\left(\frac{a_2}{l} g - \frac{h}{l}\left(\dot{v}_x - r\, v_y\right)\right) \tag{3.137}$$

$$F_{z_r} = F_{z_3} + F_{z_4} = m\left(\frac{a_1}{l} g + \frac{h}{l}\left(\dot{v}_x - r\, v_y\right)\right) \tag{3.138}$$

$$l = a_1 + a_2 \tag{3.139}$$

A rolled vehicle introduces some new reactions in the tires of the vehicle that must be considered in development of the dynamic equations of motion. The most important reactions are:

1. Roll-thrust $F_{y\varphi}$.
   An extra lateral force appears because of the vehicle roll. Tire roll-thrust is assumed to be proportional to the vehicle roll angle φ.

$$F_{y\varphi} = -C_{\varphi}\varphi \tag{3.140}$$

$$C_{\varphi} = \lim_{\varphi \longrightarrow 0} \frac{d\left(-F_y\right)}{d\varphi} \tag{3.141}$$

2. Roll-steer angle δ_{φ}.
   An extra steer angle appears because of the wheel roll. The roll steer is a result of suspension mechanisms that provide some steer angle when the vehicle rolls and the mechanism deflects. The wheel roll steering is assumed to be proportional to the vehicle roll angle φ.

$$\delta_{\varphi} = C_{\delta\varphi}\varphi \tag{3.142}$$

$$C_{\delta\varphi} = \lim_{\varphi \longrightarrow 0} \frac{d\delta}{d\varphi} \tag{3.143}$$

   Therefore, the actual steer angle δ_a of such a tire will be:

$$\delta_a = \delta + \delta_{\varphi} \tag{3.144}$$

   Depending on the design of suspension and steering mechanism, the coefficient $C_{\delta\varphi}$ may be positive or negative.

   Assume the wheel number i of a vehicle is located at:

$${}^{B}\mathbf{r}_i = \left[\, x_i \; y_i \; z_i \,\right]^T \tag{3.145}$$

The velocity of the wheel number i is

$${}^{B}\mathbf{v}_i = {}^{B}\mathbf{v} + {}^{B}_{G}\boldsymbol{\omega}_B \times {}^{B}\mathbf{r}_i \tag{3.146}$$

where ${}^{B}\mathbf{v}$ is the velocity vector of the vehicle at its mass center C, and ${}^{B}_{G}\boldsymbol{\omega}_B$ is the angular velocity of the vehicle expressed in the B-frame.

$${}^{B}\boldsymbol{\omega} = \dot{\varphi}\hat{\imath} + \dot{\psi}\hat{k} = p\hat{\imath} + r\hat{k} \tag{3.147}$$

Equation (3.146) provides us with the following velocity vector for the wheel number i expressed in the vehicle coordinate frame at B.

$$\begin{bmatrix} v_{x_i} \\ v_{y_i} \\ v_{z_i} \end{bmatrix} = \begin{bmatrix} v_x \\ v_y \\ 0 \end{bmatrix} + \begin{bmatrix} \dot{\varphi} \\ 0 \\ \dot{\psi} \end{bmatrix} \times \begin{bmatrix} x_i \\ y_i \\ z_i \end{bmatrix} = \begin{bmatrix} v_x - \dot{\psi} y_i \\ v_y - \dot{\varphi} z_i + \dot{\psi} x_i \\ \dot{\varphi} y_i \end{bmatrix} \tag{3.148}$$

Consider a bicycle model for the rollable vehicle to have

$$^B\mathbf{r}_f = \left[a_1 \; 0 \; z_f\right]^T \tag{3.149}$$

$$^B\mathbf{r}_r = \left[-a_2 \; 0 \; z_r\right]^T \tag{3.150}$$

therefore we have

$$\begin{bmatrix} v_{x_f} \\ v_{y_f} \\ v_{z_f} \end{bmatrix} = \begin{bmatrix} v_x \\ v_y - z_f p + a_1 r \\ 0 \end{bmatrix} \tag{3.151}$$

$$\begin{bmatrix} v_{x_r} \\ v_{y_r} \\ v_{z_r} \end{bmatrix} = \begin{bmatrix} v_x \\ v_y - z_r p - a_2 r \\ 0 \end{bmatrix} \tag{3.152}$$

The wheel slip angle β_i for the wheel i is defined as the angle between the wheel velocity vector $\mathbf{v}_i$ and the vehicle x-axis.

$$\beta_i = \arctan\left(\frac{v_{y_i}}{v_{x_i}}\right) = \arctan\frac{v_y - \dot{\varphi} z_i + \dot{\psi} x_i}{v_x} \tag{3.153}$$

If the tire number i has a steer angle δ_i, then its sideslip angle α_i would be:

$$\alpha_i = \beta_i - \delta_a = \arctan\left(\frac{v_y - \dot{\varphi} z_i + \dot{\psi} x_i}{v_x}\right) - \left(\delta_i + C_{\delta\varphi_i}\varphi\right) \tag{3.154}$$

The wheel slip angle β_i for the front and rear wheels of a two-wheel vehicle, β_f and β_r, is

$$\beta_f = \arctan\left(\frac{v_{y_f}}{v_{x_f}}\right) = \arctan\frac{v_y - z_f p + a_1 r}{v_x} \tag{3.155}$$

$$\beta_r = \arctan\left(\frac{v_{y_r}}{v_{x_r}}\right) = \arctan\frac{v_y - z_r p - a_2 r}{v_x} \tag{3.156}$$

We will use the *vehicle slip angle* β to relate β_i to the vehicle's dynamic variables.

$$\beta = \arctan\left(\frac{v_y}{v_x}\right) \tag{3.157}$$

Although we may assume that the center of the wheels is on the xy-plane and the z_i coordinate of the wheels of vehicle to zero, they are not constant for a rollable vehicle. To show the effect of variation of z_i, we substitute it by coefficient C_{β_i} called the *tire roll rate coefficient*, and define coefficients C_{β_f} and C_{β_r} to express the change in v_{y_i} because of roll rate p.

$$v_{y_i} = v_y + r x_i - C_{\beta_i} p \tag{3.158}$$

The coefficient C_{β_i} should be determined by experiment for any given vehicle.

$$C_{\beta_i} = \lim_{p \longrightarrow 0} \frac{d v_{y_i}}{dp} \tag{3.159}$$

Therefore,

$$\beta_f = \arctan\left(\frac{v_y + a_1 r - C_{\beta_f} p}{v_x}\right) \tag{3.160}$$

$$\beta_r = \arctan\left(\frac{v_y - a_2 r - C_{\beta_r} p}{v_x}\right) \tag{3.161}$$

and

$$\alpha_f = \beta_f - \delta_f = \arctan\left(\frac{v_y + a_1 r - C_{\beta_f} p}{v_x}\right) - \left(\delta + C_{\delta\varphi_f}\varphi\right) \tag{3.162}$$

$$\alpha_r = \beta_r - \delta_r = \arctan\left(\frac{v_y - a_2 r - C_{\beta_r} p}{v_x}\right) - C_{\delta\varphi_r}\varphi \tag{3.163}$$

Assuming small angles for slip angles β_f, β, and β_r, then the tire sideslip angles for the front and rear wheels, α_f and α_r, may be approximated as

$$\begin{aligned}\alpha_f &= \frac{1}{v_x}\left(v_y + a_1 r - z_f p\right) - \delta - \delta_{\varphi_f} \\ &= \beta + a_1 \frac{r}{v_x} - C_{\beta_f} \frac{p}{v_x} - \delta - C_{\delta\varphi_f}\varphi\end{aligned} \tag{3.164}$$

$$\begin{aligned}\alpha_r &= \frac{1}{v_x}\left(v_y - a_2 r - z_r p\right) - \delta_{\varphi_r} \\ &= \beta - a_2 \frac{r}{v_x} - C_{\beta_r} \frac{p}{v_x} - C_{\delta\varphi_r}\varphi\end{aligned} \tag{3.165}$$

Also we substitute s from (1.57),

$$s = \frac{R_g\,\omega_w - v_x}{R_g\,\omega_w\,H(R_g\,\omega_w - v_x) + v_x\,H(v_x - R_g\,\omega_w)} \tag{3.166}$$

however, the slip ratio of the front tire needs to be adapted as the steer angle will change velocity of the tire in the x_T-direction as is shown in Fig. 1.13. The velocity of the front wheel center in the x-direction of its local C-frame is

$$v_{xT_f} = v_x \cos\left(\delta + \delta_{\varphi_f}\right) + \left(v_y + a_1 r\right)\sin\left(\delta + \delta_{\varphi_f}\right) \tag{3.167}$$

Therefore, the longitudinal slip ratios of the front and rear tires are:

$$s_f = \frac{R_g\,\omega_f - v_{xT_f}}{R_g\,\omega_f\,H(R_g\,\omega_f - v_{xT_f}) + v_{xT_f}\,H(v_{xT_f} - R_g\,\omega_f)} \tag{3.168}$$

$$s_r = \frac{R_g\,\omega_r - v_x}{R_g\,\omega_r\,H(R_g\,\omega_r - v_x) + v_x\,H(v_x - R_g\,\omega_r)} \tag{3.169}$$

Backward substitution of these equations completes Eqs. (3.86) and (3.87).

To derive Eqs. (3.88) and (3.89), we calculate the resultant roll and yaw moments M_x, M_z because of tire forces.

$$\begin{aligned}
{}^B\mathbf{M} &= \sum_{i=1}^{2} \mathbf{r}_i \times {}^B\mathbf{F}_i = \sum_{i=1}^{2} \mathbf{r}_i \times {}^C R_T\,{}^T\mathbf{F}_i \\
&= \sum_{i=1}^{2} {}^B\mathbf{M}_i + \begin{bmatrix} a_1 \\ 0 \\ 0 \end{bmatrix} \times {}^C R_{T_f} \begin{bmatrix} F_{x_f} \\ F_{y_f} \\ 0 \end{bmatrix} + \begin{bmatrix} -a_2 \\ 0 \\ 0 \end{bmatrix} \times {}^C R_{T_r} \begin{bmatrix} F_{x_r} \\ F_{y_r} \\ 0 \end{bmatrix} \\
&= \begin{bmatrix} M_{x_f} + M_{x_r} \\ M_{y_f} + M_{y_r} + a_1 F_{y_f}\sin\varphi - a_2 F_{y_r}\sin\varphi \\ a_1\left(F_{x_f}\sin\left(\delta + \delta_{\varphi_f}\right) + F_{y_f}\cos\varphi\cos\left(\delta + \delta_{\varphi_f}\right)\right) \\ -a_2\left(F_{x_r}\sin\delta_{\varphi_r} + F_{y_r}\cos\delta_{\varphi_r}\cos\varphi\right) \end{bmatrix}
\end{aligned} \tag{3.170}$$

where

$$\begin{aligned}
{}^C R_T = {}^C R_W\,{}^W R_T &= \begin{bmatrix} \cos\delta_a & -\sin\delta_a & 0 \\ \sin\delta_a & \cos\delta_a & 0 \\ 0 & 0 & 1 \end{bmatrix} \begin{bmatrix} 1 & 0 & 0 \\ 0 & \cos\varphi & \sin\varphi \\ 0 & -\sin\varphi & \cos\varphi \end{bmatrix} \\
&= \begin{bmatrix} \cos\delta_a & -\cos\varphi\sin\delta_a & -\sin\delta_a\sin\varphi \\ \sin\delta_a & \cos\delta_a\cos\varphi & \cos\delta_a\sin\varphi \\ 0 & -\sin\varphi & \cos\varphi \end{bmatrix}
\end{aligned} \tag{3.171}$$

and therefore,

$$
{}^{C}R_{T_f} = \begin{bmatrix} \cos\left(\delta + \delta_{\varphi_f}\right) & -\cos\varphi\sin\left(\delta + \delta_{\varphi_f}\right) & -\sin\left(\delta + \delta_{\varphi_f}\right)\sin\varphi \\ \sin\left(\delta + \delta_{\varphi_f}\right) & \cos\left(\delta + \delta_{\varphi_f}\right)\cos\varphi & \cos\left(\delta + \delta_{\varphi_f}\right)\sin\varphi \\ 0 & -\sin\varphi & \cos\varphi \end{bmatrix} \tag{3.172}
$$

$$
{}^{C}R_{T_r} = \begin{bmatrix} \cos\delta_{\varphi_r} & -\cos\varphi\sin\delta_{\varphi_r} & -\sin\delta_{\varphi_r}\sin\varphi \\ \sin\delta_{\varphi_r} & \cos\delta_{\varphi_r}\cos\varphi & \cos\delta_{\varphi_r}\sin\varphi \\ 0 & -\sin\varphi & \cos\varphi \end{bmatrix} \tag{3.173}
$$

The M_{x_f} and M_{x_r} are the moments appeared at the tireprint because of lateral shift of the resultant vertical force F_{z_i}, and also because of the lateral forces under each tire. Let us assume that the slip and camber moments are proportional to the wheels' lateral forces and write them as

$$
M_{x_f} = C_{T_f} F_{y_f} \tag{3.174}
$$

$$
M_{x_r} = C_{T_r} F_{y_r} \tag{3.175}
$$

where C_{T_f} and C_{T_r} are the overall *torque coefficient* of the front and rear wheels, respectively.

$$
C_{T_f} = \lim_{F_{y_f} \longrightarrow 0} \frac{dM_x}{dF_{y_f}} \qquad C_{T_r} = \lim_{F_{y_r} \longrightarrow 0} \frac{dM_x}{dF_{y_r}} \tag{3.176}
$$

The contribution in M_x would also be the effect of the left and right suspension stiffness and damping characteristics due to roll angle of the vehicle body. To keep the effect of roll moment for the bicycle model, we introduce the suspension deflection roll moments $-wc_f\,\dot{\varphi} - wk_f\,\varphi$, assuming $w = w_f = w_r$. This part of roll moment is due to change in normal force of the left and right wheels as a result of force change in springs and dampers. These unbalanced forces generate a roll stiffness moment that is proportional to the vehicle's roll angle and roll rate,

$$
M_{x_k} = -k_\varphi \varphi \tag{3.177}
$$

$$
M_{x_c} = -c_\varphi \dot{\varphi} \tag{3.178}
$$

where k_φ and c_φ are the roll stiffness and roll damping of the vehicle.

$$
k_\varphi = wk = w\left(k_f + k_r\right) = \lim_{\varphi \longrightarrow 0} \frac{d\left(-M_x\right)}{d\varphi} \tag{3.179}
$$

$$
c_\varphi = wc = w\left(c_f + c_r\right) = \lim_{\dot{\varphi} \longrightarrow 0} \frac{d\left(-M_x\right)}{d\dot{\varphi}} \tag{3.180}
$$

w is the track of the vehicle and k and c are sum of the front and rear springs' stuffiness and shock absorbers damping. The coefficients k_φ and c_φ are called the *roll stiffness* and *roll damping*, respectively. Therefore, the applied roll moment on the vehicle can be summarized as

$$\begin{aligned} M_x &= M_{x_f} + M_{x_r} + M_{x_c} + M_{x_k} \\ &= C_{T_f} F_{y_f} + C_{T_r} F_{y_r} - w\left(c_f + c_r\right)\dot{\varphi} - w\left(k_f + k_r\right)\varphi \\ &\approx C_{T_f} F_{y_f} + C_{T_r} F_{y_r} - k_\varphi \varphi - c_\varphi \dot{\varphi} \end{aligned} \tag{3.181}$$

We ignore the generated M_{y_i} at the tireprint; then, we conclude the force equations (3.86) and (3.89).

In majority of maneuvers we may assume φ and δ to be very small such that the force equations may be approximated by the following equations.

$$F_x \approx F_{x_f} + F_{x_r} \tag{3.182}$$

$$F_y \approx F_{y_f} + F_{y_r} \tag{3.183}$$

$$M_x = C_{T_f} F_{y_f} + C_{T_r} F_{y_r} - k_\varphi \varphi - c_\varphi \dot{\varphi} \tag{3.184}$$

$$M_z \approx a_1 F_{y_f} - a_2 F_{y_r}$$

$$T_f = I_f \dot{\omega}_f + R_w F_{x_f} \tag{3.185}$$

$$T_r = I_r \dot{\omega}_r + R_w F_{x_r} \tag{3.186}$$

■

Example 85 The tireprint position vector in the B-frame.

In Eq. (3.170)

$$^B\mathbf{M} = \sum_{i=1}^{2} \mathbf{r}_i \times {}^C R_T \, {}^T\mathbf{F}_i$$

the $\mathbf{r}_i$ are supposed to be position vectors in the B-frame.

$$^B\mathbf{r}_1 = \begin{bmatrix} a_1 \\ 0 \\ 0 \end{bmatrix} \qquad ^B\mathbf{r}_2 = \begin{bmatrix} -a_2 \\ 0 \\ 0 \end{bmatrix} \tag{3.187}$$

because we moved ${}^T\mathbf{F}_i$ into the B-frame using transformation matrix ${}^C R_T$. However, we could have done the calculation in the T-frame and move the resultant moment into the B-frame. To do that calculation we needed to calculate the position vectors in the T-frame.

3.4 Two-Wheel Roll Vehicle Dynamics

We combine the equations of motion (2.1)–(2.4) with (3.86)–(3.91) for a two-wheel roll vehicle model, and express its motion by the following set of equations (Ellis 1994; Jazar 2017).

There are 6 first order differential equations of motion:

$$F_x - F_A = m\,\dot{v}_x - mr\,v_y \tag{3.188}$$

$$F_y = m\,\dot{v}_y + mr\,v_x \tag{3.189}$$

$$M_z = I_z\,\dot{r} \tag{3.190}$$

$$M_x = I_x\,\dot{p} \tag{3.191}$$

$$T_f = I_f\,\dot{\omega}_f + R_w\,F_{x_f} \tag{3.192}$$

$$T_r = I_r\,\dot{\omega}_r + R_w\,F_{x_r} \tag{3.193}$$

The left-hand sides of the force system are:

$$\begin{aligned} F_x = {} & F_{x_f}\cos\left(\delta + \delta_{\varphi_f}\right) + F_{x_r}\cos\delta_{\varphi_r} \\ & - F_{y_f}\sin\left(\delta + \delta_{\varphi_f}\right) - F_{y_r}\sin\delta_{\varphi_r} \end{aligned} \tag{3.194}$$

$$\begin{aligned} F_y = {} & F_{y_f}\cos\left(\delta + \delta_{\varphi_f}\right) + F_{y_r}\cos\delta_{\varphi_r} \\ & + F_{x_f}\sin\left(\delta + \delta_{\varphi_f}\right) + F_{x_r}\sin\delta_{\varphi_r} \end{aligned} \tag{3.195}$$

$$M_x = C_{T_f}\,F_{y_f} + C_{T_r}\,F_{y_r} - k_\varphi \varphi - c_\varphi \dot{\varphi} \tag{3.196}$$

$$\begin{aligned} M_z = {} & a_1\left(F_{x_f}\sin\left(\delta + \delta_{\varphi_f}\right) + F_{y_f}\cos\varphi\cos\left(\delta + \delta_{\varphi_f}\right)\right) \\ & - a_2\left(F_{x_r}\sin\delta_{\varphi_r} + F_{y_r}\cos\delta_{\varphi_r}\cos\varphi\right) \end{aligned} \tag{3.197}$$

The longitudinal and lateral forces on the front and rear wheels of the bicycle vehicle model are:

$$F_{x_f} = F_{z_f} C_{sf}\,S\left(s_f - s_s\right)\sqrt{1 - C_{s\alpha}\left(\frac{S\left(\alpha_f - \alpha_s\right)}{\alpha_s}\right)^2} \tag{3.198}$$

$$F_{x_r} = F_{z_r} C_{sr}\,S\left(s_r - s_s\right)\sqrt{1 - C_{s\alpha}\left(\frac{S\left(\alpha_r - \alpha_s\right)}{\alpha_s}\right)^2} \tag{3.199}$$

$$F_{y_f} = -F_{z_f} C_{\alpha f}\, S\left(\alpha_f - \alpha_s\right) \sqrt{1 - C_{\alpha s}\left(\frac{S\left(s_f - s_s\right)}{s_s}\right)^2} - C_{\varphi_f}\varphi \tag{3.200}$$

$$F_{y_r} = -F_{z_r} C_{\alpha r}\, S\left(\alpha_r - \alpha_s\right) \sqrt{1 - C_{\alpha s}\left(\frac{S\left(s_r - s_s\right)}{s_s}\right)^2} - C_{\varphi_r}\varphi \tag{3.201}$$

The longitudinal and lateral forces F_{x_i} and F_{y_i} are proportional to F_{z_f} and F_{z_r} which are:

$$F_{z_f} = F_{z_1} + F_{z_2} = mg\frac{a_2}{l} - m\left(\dot{v}_x - r\, v_y\right)\frac{h}{l} \tag{3.202}$$

$$F_{z_r} = F_{z_3} + F_{z_4} = mg\frac{a_1}{l} + m\left(\dot{v}_x - r\, v_y\right)\frac{h}{l} \tag{3.203}$$

$$l = a_1 + a_2 \tag{3.204}$$

The tire forces F_{x_i} and F_{y_i} are also functions of the sideslip angles α_i

$$\alpha_f = \beta_f - \delta_f = \arctan\left(\frac{v_y}{v_x} + \frac{a_1}{v_x}r - C_{\beta_f}\frac{p}{v_x}\right) - \delta - C_{\delta\varphi_f}\varphi \tag{3.205}$$

$$\alpha_r = \beta_r - \delta_r = \arctan\left(\frac{v_y}{v_x} - \frac{a_2}{v_x}r - C_{\beta_r}\frac{p}{v_x}\right) - C_{\delta\varphi_r}\varphi \tag{3.206}$$

where

$$\beta = \arctan\frac{v_y}{v_x} \tag{3.207}$$

$$\beta_f = \arctan\left(\frac{v_y + a_1 r - C_{\beta_f} p}{v_x}\right) \tag{3.208}$$

$$\beta_r = \arctan\left(\frac{v_y - a_2 r - C_{\beta_r} p}{v_x}\right) \tag{3.209}$$

The tire forces F_{x_i} and F_{y_i} are also functions of the slip ratios s_i.

$$s_f = \frac{R_g\,\omega_f - v_{xT_f}}{R_g\,\omega_f\, H(R_g\,\omega_f - v_{xT_f}) + v_{xT_f}\, H(v_{xT_f} - R_g\,\omega_f)} \tag{3.210}$$

$$s_r = \frac{R_g\,\omega_r - v_{xT_r}}{R_g\,\omega_r\, H(R_g\,\omega_r - v_{xT_r}) + v_{xT_r}\, H(v_{xT_r} - R_g\,\omega_r)} \tag{3.211}$$

$$v_{xT_f} = v_x \cos\left(\delta + \delta_{\varphi_f}\right) + \left(v_y + a_1 r - C_{\beta_f} p\right)\sin\left(\delta + \delta_{\varphi_f}\right) \tag{3.212}$$

$$v_{xT_r} = v_x \cos\delta_{\varphi_r} + \left(v_y - a_2 r - C_{\beta_r} p\right)\sin\delta_{\varphi_r} \tag{3.213}$$

Proof The Newton–Euler equations of motion for roll vehicle model in the local coordinate frame B, attached to the vehicle at its mass center C, are given in Eqs. (3.188)–(3.193). The right-hand side of the equations are B-expression of the vehicle translational and rotational accelerations, as well as the torque equations of the wheels of the vehicle.

The right-hand side of the equations are made of an equation chain starting with the B-expression of the applied forces on the vehicle (3.194)–(3.197) and end up with substituting the tire forces F_{x_f}, F_{x_r}, F_{y_f}, F_{y_r}, and followed by vertical forces F_{z_f}, F_{z_r}, and tire slips $\alpha_f, \alpha_r, s_f, s_r$. ■

Example 86 Two-wheel roll, increasing steer angle and front torque, slip saturation.
Consider a vehicle with the following data

$$\begin{aligned}
m &= 1000\,\text{kg} \qquad I_f = I_r = 30\,\text{kg m}^2 \qquad R_g = 0.35\,\text{m} \\
I_x &= 300\,\text{kg m}^2 \qquad I_z = 2000\,\text{kg m}^2 \qquad C_A = 0.8 \\
C_{\alpha f} &= C_{\alpha r} = 8.5 \qquad \alpha_s = 5\,\text{deg} \qquad C_{\alpha s} = C_{s\alpha} = 0.5 \\
C_{sf} &= C_{sr} = 7.5 \qquad s_s = 0.1 \qquad w_f = w_r = 1.8\,\text{m} \\
a_1 &= 1.35\,\text{m} \qquad a_2 = 1.5\,\text{m} \qquad h = 0.9\,\text{m} \\
k_\varphi &= 26{,}612\,\text{N m/rad} \qquad c_\varphi = 1700\,\text{N m s/rad} \\
C_{\beta_f} &= -0.4 \qquad C_{\beta_r} = -0.1 \qquad C_{T_f} = C_{T_r} = 0.4 \\
C_{\delta\varphi_f} &= C_{\delta\varphi_r} = 0.01 \qquad C_{\varphi_f} = 3200 \qquad C_{\varphi_r} = 0
\end{aligned} \tag{3.214}$$

and assume the vehicle is moving slowly straight.

$$v_x = 2\,\text{m/s} \qquad \omega_f = \omega_r = \frac{v_x}{R_g} = \frac{2}{0.35} = 5.714\,\text{rad/s} \tag{3.215}$$

At time $t = 0$, we apply a linearly increasing torque on the front wheel up to $T_f = 1500\,\text{N m}$ and keep constant after that. The steer angle is also linearly increasing at a very low rate up to $\delta = 0.5$ deg and remains constant after that.

$$T_f = \begin{cases} 100t\,\text{N m} & 0 < t < 15\,\text{s} \\ 1500\,\text{N m} & 15\,\text{s} \le t \end{cases} \qquad T_r = 0 \tag{3.216}$$

$$\delta = \begin{cases} 0.05t\,\text{deg} = \frac{0.05\pi}{180}t\,\text{rad} & 0 < t < 10\,\text{s} \\ 0.5\,\text{deg} = \frac{0.5\pi}{180}\,\text{rad} & 10\,\text{s} \le t \end{cases} \tag{3.217}$$

The front torque increases and goes beyond the limit of front wheel capability in producing traction force. Therefore, the front tire slip ratio s_f and hence its traction force F_{xf} will become saturated. This vehicle will have s_f to be saturated and all other slips unsaturated.

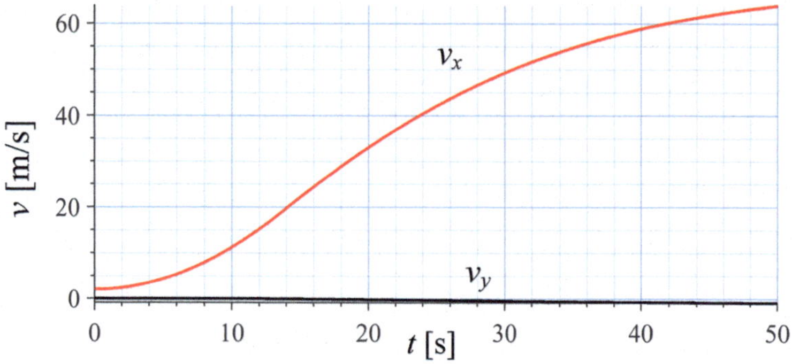

Fig. 3.6 The forward velocity components v_x and v_y of a bicycle roll vehicle model under front torque and steer angle

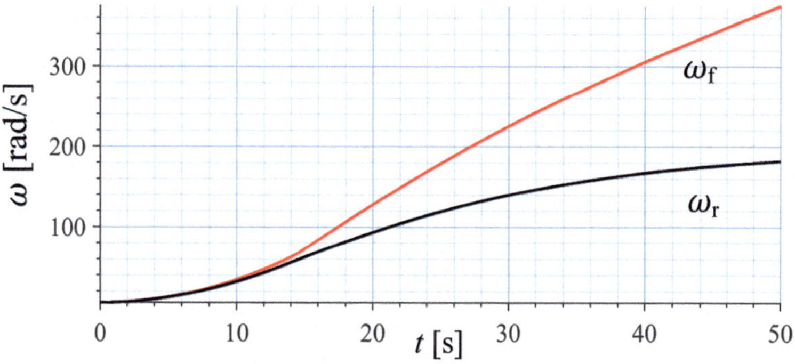

Fig. 3.7 The angular velocity components ω_f and ω_r of a bicycle roll vehicle model under front torque and steer angle

Figure 3.6 depicts the velocity components of the vehicle v_x and v_y, measured in body coordinate frame B. Figure 3.7 illustrates the angular velocities of the front and rear wheels ω_f and ω_r. Figure 3.8 illustrates the sideslip angles of the front and rear wheels α_f and α_r. Figure 3.9 depicts the longitudinal slip ratios s_f and s_r for the front and rear tires. The acceleration components a_X and a_Y of the vehicle are plotted in Fig. 3.10. Figure 3.11 depicts the forces F_x and F_y on the vehicle at its mass center. Figure 3.12 illustrates variation of the vertical loads F_{zf} and F_{zr}. Figure 3.13 illustrates the longitudinal forces F_{xf} and F_{xr} on front and rear tires. Figure 3.14 shows the lateral forces F_{yf} and F_{yr} on front and rear tires. The roll angle φ and roll rate p are shown in Figs. 3.15 and 3.16 respectively. Figure 3.17 depicts the angular accelerations $\dot{\omega}_f$ and $\dot{\omega}_r$ of the front and rear wheels. Figure 3.18 illustrates the path of the vehicle.

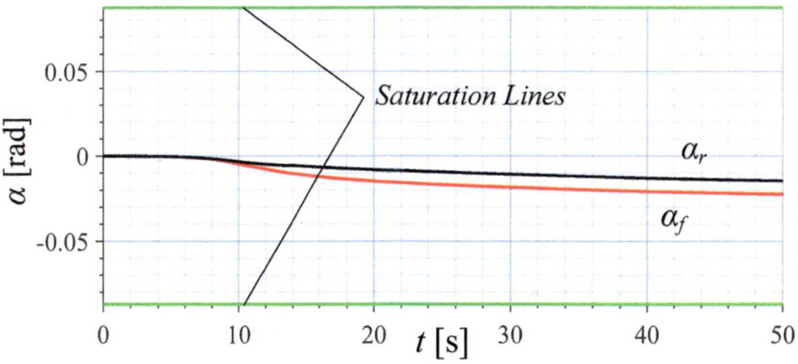

Fig. 3.8 The sideslip angles α_f and α_r of a bicycle roll vehicle model under front torque and steer angle

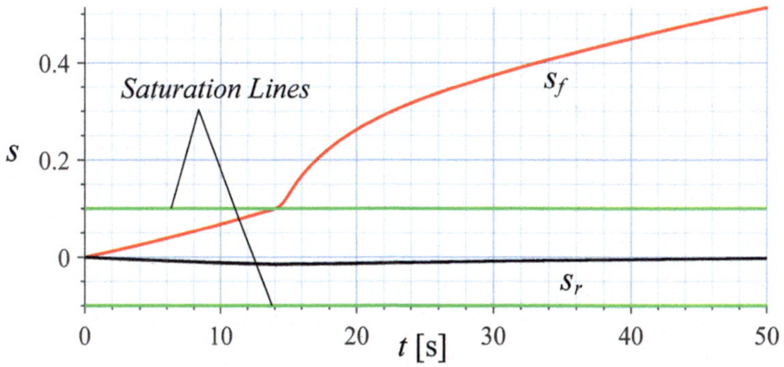

Fig. 3.9 The slip ratios s_f and s_r of a bicycle roll vehicle model under front torque and steer angle

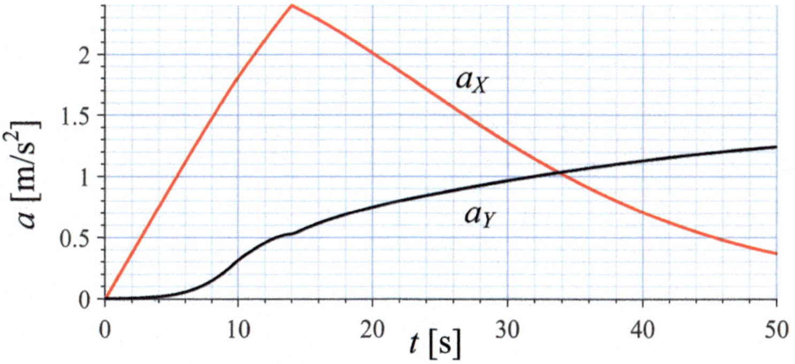

Fig. 3.10 The global acceleration components a_X and a_Y of a bicycle roll vehicle model under front torque and steer angle

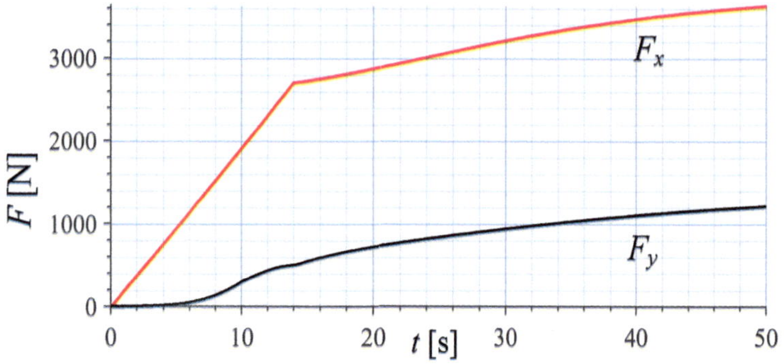

Fig. 3.11 The force components F_x and F_y of a bicycle roll vehicle model under front torque and steer angle

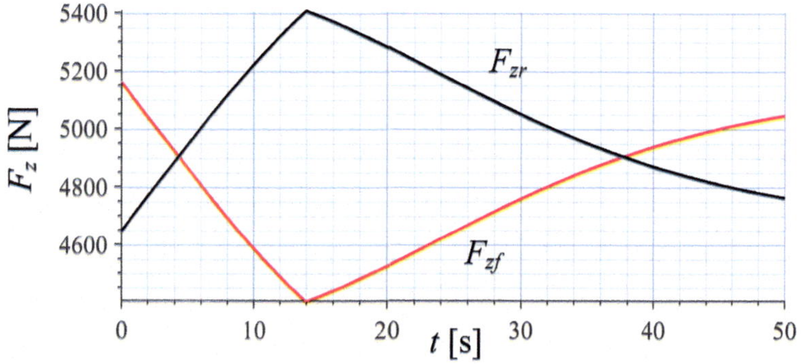

Fig. 3.12 The vertical forces F_{z_f} and F_{z_r} of a bicycle roll vehicle model under front torque and steer angle

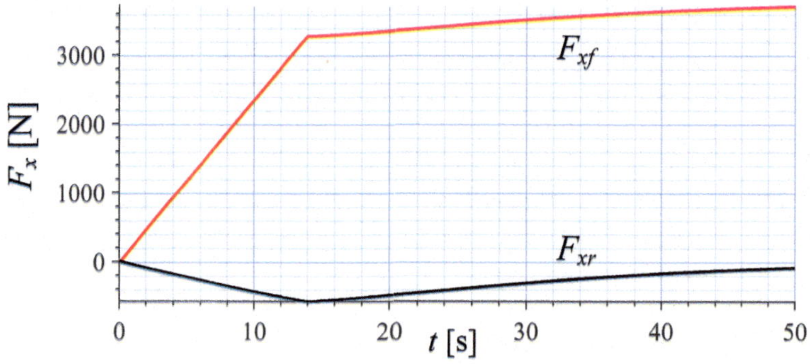

Fig. 3.13 The longitudinal forces F_{x_f} and F_{x_r} of a bicycle roll vehicle model under front torque and steer angle

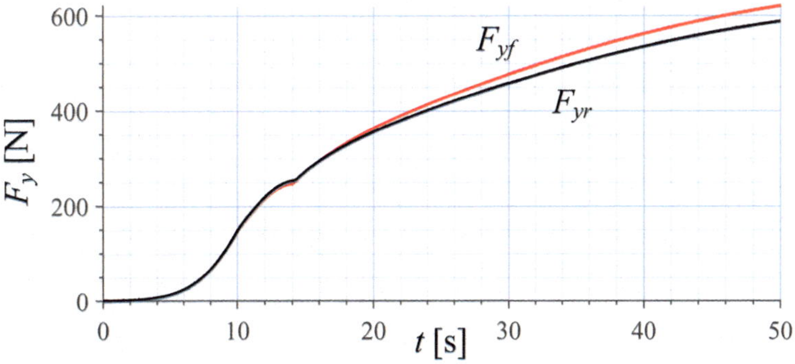

Fig. 3.14 The lateral forces F_{y_f} and F_{y_r} of a bicycle roll vehicle model under front torque and steer angle

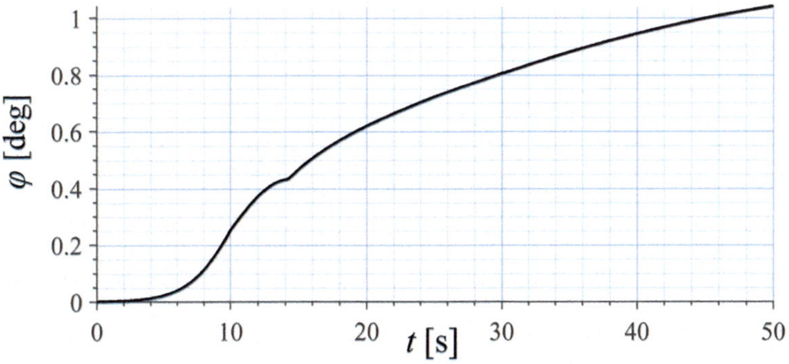

Fig. 3.15 The roll angle φ of a bicycle roll vehicle model under front torque and steer angle

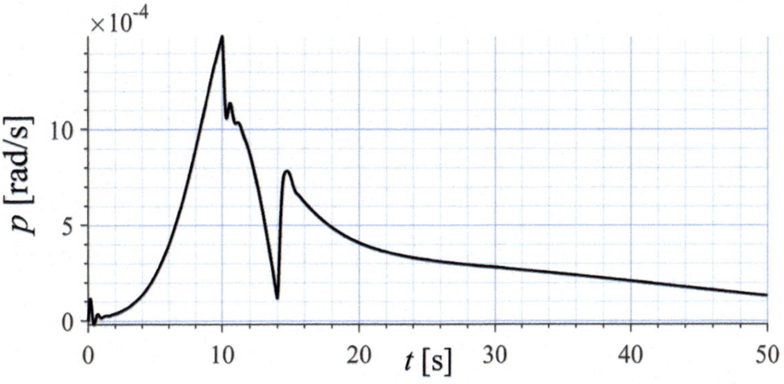

Fig. 3.16 The roll rate p of a bicycle roll vehicle model under front torque and steer angle

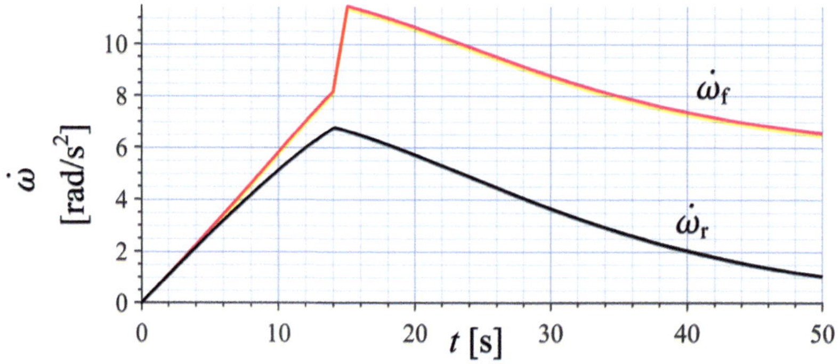

Fig. 3.17 The angular acceleration components $\dot{\omega}_f$ and $\dot{\omega}_r$ of a bicycle roll vehicle model under front torque and steer angle

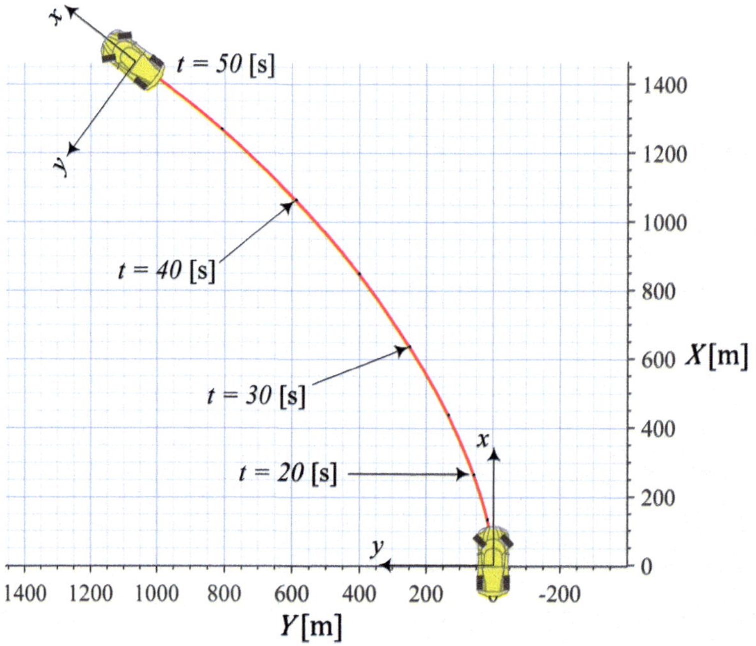

Fig. 3.18 Path of motion of a bicycle roll vehicle model under front torque and steer angle

Example 87 Two-wheel roll vehicle, increasing steer angle and front torque, no combined forces.

To compare the effect of the combined tire force model with no combined forces, let us consider the same Example 86 with $C_{\alpha s} = 0$, $C_{s\alpha} = 0$. Ignoring the interaction between s and α makes them to go higher than real and therefore their associated forces to be higher as well. The vehicle of this example will have saturated s_f and unsaturated other slips, similar to Example 86.

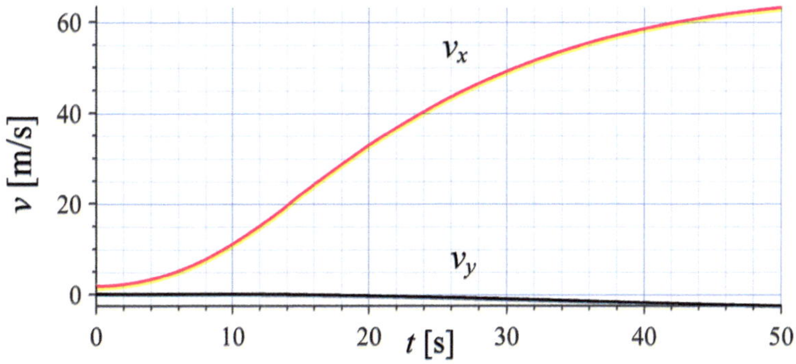

Fig. 3.19 The forward velocity components v_x and v_y of a bicycle roll vehicle model under front torque and steer angle and no combined tire forces

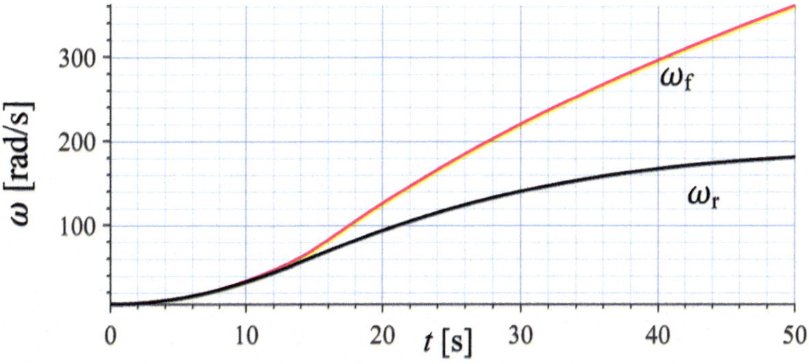

Fig. 3.20 The angular velocity components ω_f and ω_r of a bicycle roll vehicle model under front torque and steer angle and no combined tire forces

Figure 3.19 depicts the velocity components of the vehicle v_x and v_y, measured in body coordinate frame B. Figure 3.20 illustrates the angular velocities of the front and rear wheels ω_f and ω_r. Figure 3.21 illustrates the sideslip angles of the front and rear wheels α_f and α_r. Figure 3.22 depicts the longitudinal slip ratios s_f and s_r for the front and rear tires. The acceleration components a_X and a_Y of the vehicle are plotted in Fig. 3.23. Figure 3.24 depicts the forces F_x and F_y on the vehicle at its mass center. Figure 3.25 illustrates variation of the vertical loads F_{zf} and F_{zr}. Figure 3.26 illustrates the longitudinal forces F_{xf} and F_{xr} on front and rear tires. Figure 3.27 shows the lateral forces F_{yf} and F_{yr} on front and rear tires. The roll angle φ and roll rate p are shown in Figs. 3.28 and 3.29 respectively. The yaw rate r is shown in Fig. 3.30. Figure 3.31 depicts the angular accelerations $\dot{\omega}_f$ and $\dot{\omega}_r$ of the front and rear wheels. Figure 3.32 illustrates the path of the vehicle.

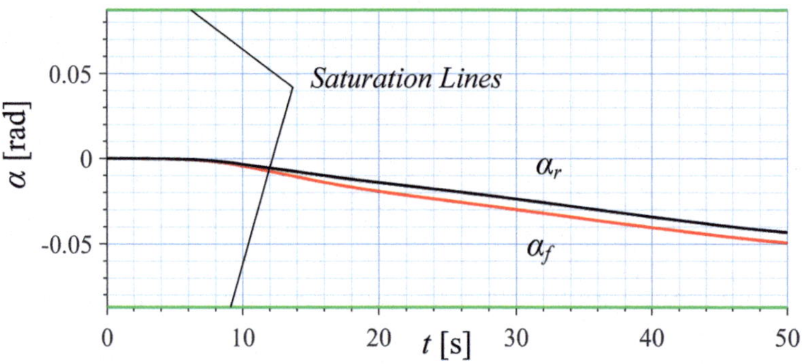

Fig. 3.21 The sideslip angles α_f and α_r of a bicycle roll vehicle model under front torque and steer angle and no combined tire forces

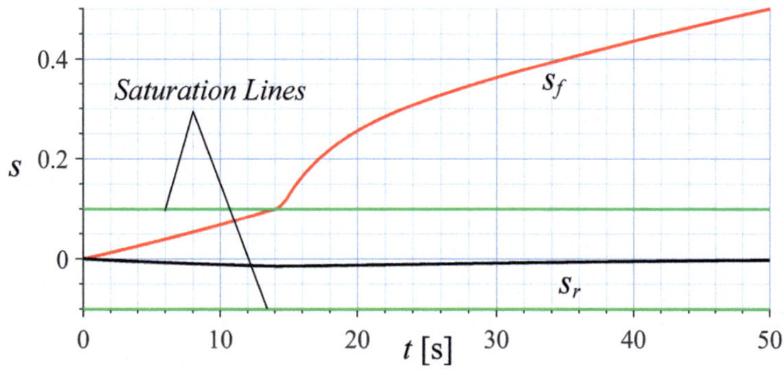

Fig. 3.22 The slip ratios s_f and s_r of a bicycle roll vehicle model under front torque and steer angle and no combined tire forces

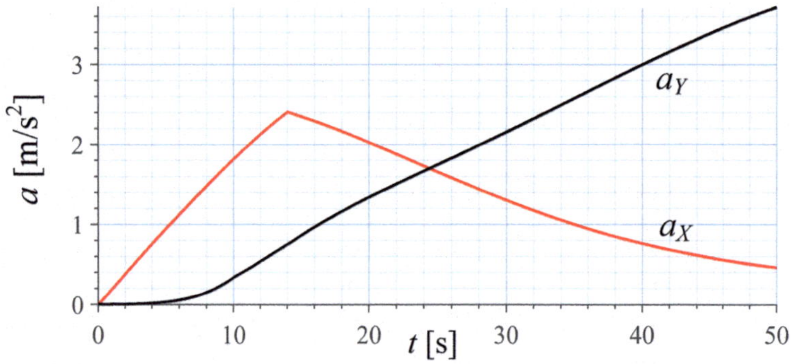

Fig. 3.23 The global acceleration components a_X and a_Y of a bicycle roll vehicle model under front torque and steer angle and no combined tire forces

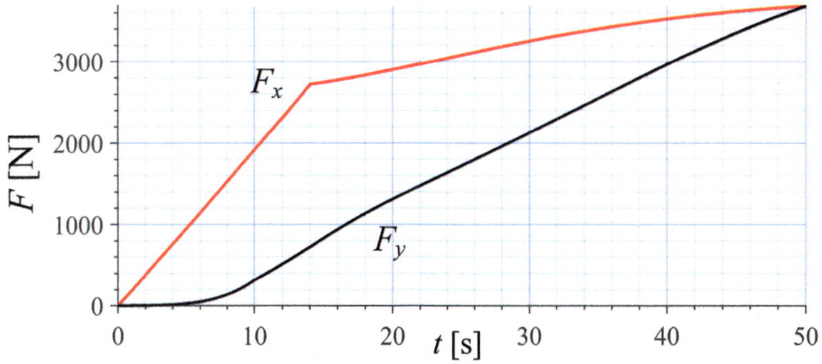

Fig. 3.24 The force components F_x and F_y of a bicycle roll vehicle model under front torque and steer angle and no combined tire forces

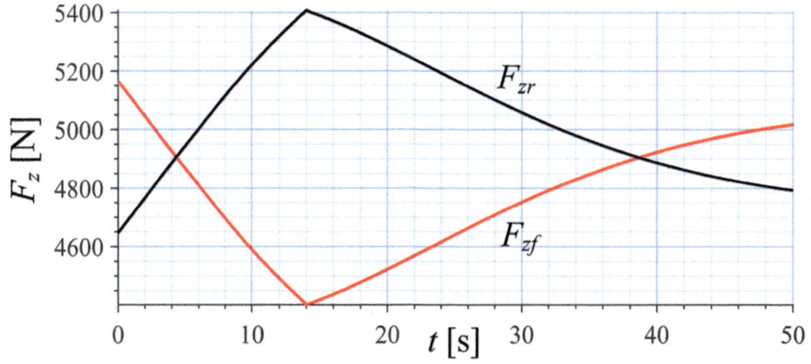

Fig. 3.25 The vertical forces F_{z_f} and F_{z_r} of a bicycle roll vehicle model under front torque and steer angle and no combined tire forces

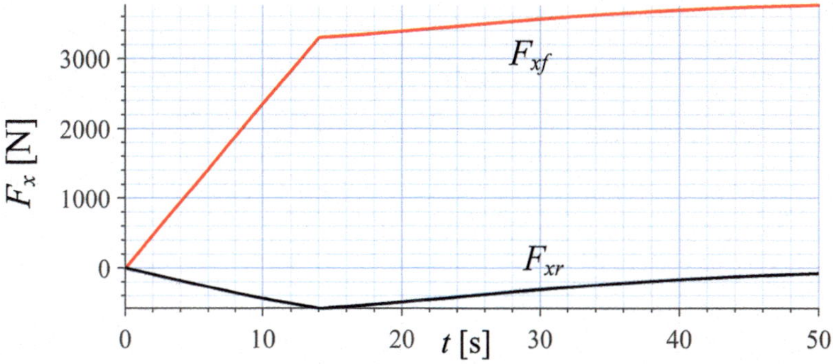

Fig. 3.26 The longitudinal forces F_{x_f} and F_{x_r} of a bicycle roll vehicle model under front torque and steer angle and no combined tire forces

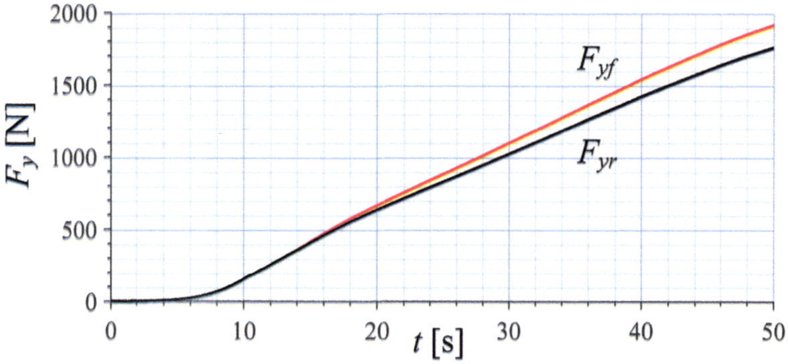

Fig. 3.27 The lateral forces F_{y_f} and F_{y_r} of a bicycle roll vehicle model under front torque and steer angle and no combined tire forces

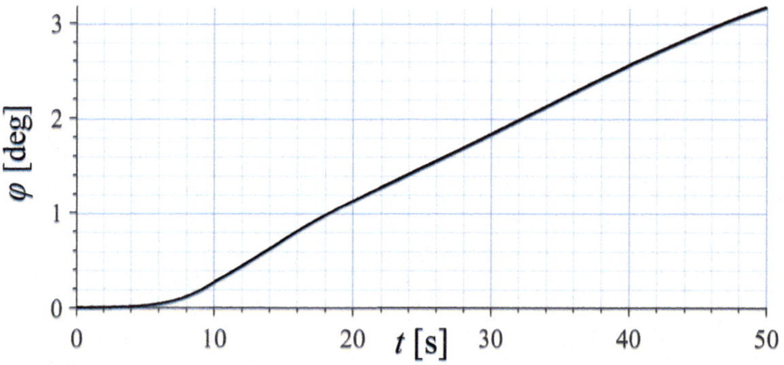

Fig. 3.28 The roll angle φ of a bicycle roll vehicle model under front torque and steer angle and no combined tire forces

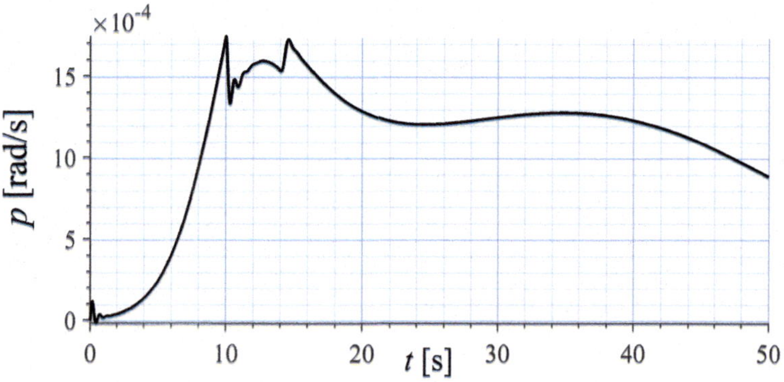

Fig. 3.29 The roll rate p of a bicycle roll vehicle model under front torque and steer angle and no combined tire forces

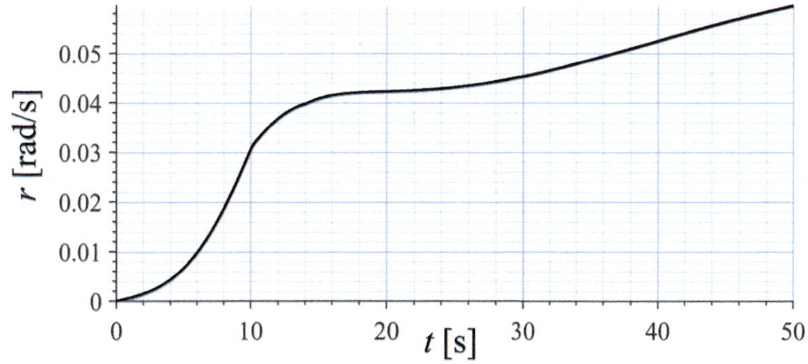

Fig. 3.30 The yaw rate r of a bicycle roll vehicle model under front torque and steer angle and no combined tire forces

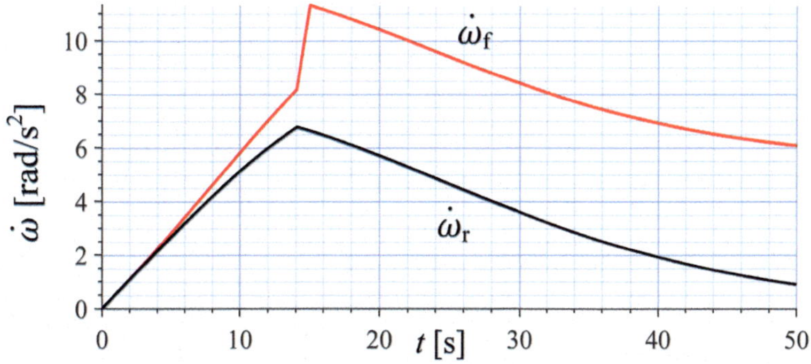

Fig. 3.31 The angular acceleration components $\dot{\omega}_f$ and $\dot{\omega}_r$ of a bicycle roll vehicle model under front torque and steer angle and no combined tire forces

Example 88 Two-wheel roll vehicle, increasing steer angle and rear torque.

Consider a vehicle with the following data

$$\begin{aligned}
&m = 1000\,\mathrm{kg} \qquad I_f = I_r = 30\,\mathrm{kg\,m^2} \qquad C_A = 0.8 \\
&I_x = 300\,\mathrm{kg\,m^2} \qquad I_z = 2000\,\mathrm{kg\,m^2} \qquad R_g = 0.35\,\mathrm{m} \\
&C_{\alpha f} = C_{\alpha r} = 8.5 \qquad \alpha_s = 5\,\mathrm{deg} \qquad C_{sf} = C_{sr} = 7.5 \\
&C_{\alpha s} = C_{s\alpha} = 0.5 \qquad s_s = 0.1 \qquad w_f = w_r = 1.8\,\mathrm{m} \\
&C_{\beta_f} = -0.4 \qquad C_{\beta_r} = -0.1 \qquad C_{T_f} = C_{T_r} = 0.4 \\
&C_{\delta\varphi_f} = C_{\delta\varphi_r} = 0.01 \qquad C_{\varphi_f} = 3200 \qquad C_{\varphi_r} = 0 \\
&a_1 = 1.35\,\mathrm{m} \qquad a_2 = 1.5\,\mathrm{m} \qquad h = 0.9\,\mathrm{m} \\
&k_\varphi = 26{,}612\,\mathrm{N\,m/rad} \qquad c_\varphi = 1700\,\mathrm{N\,m\,s/rad}
\end{aligned} \tag{3.218}$$

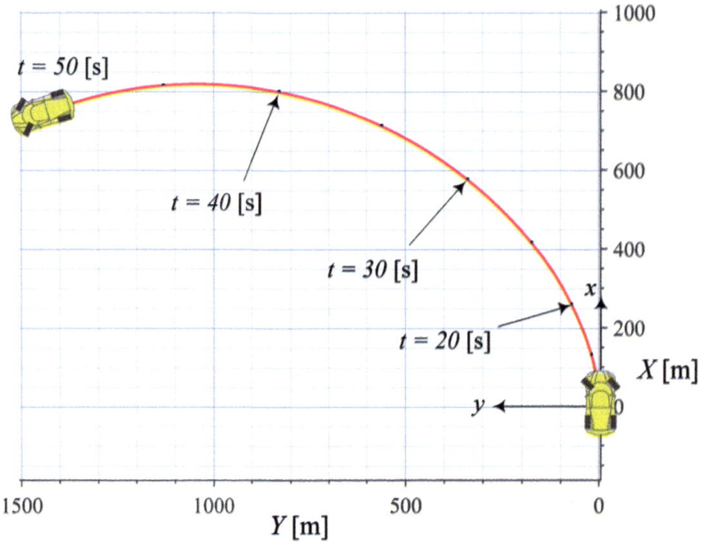

Fig. 3.32 Path of motion of a bicycle roll vehicle model under front torque and steer angle and no combined tire forces

and assume the vehicle is moving slowly straight.

$$v_x = 2\,\mathrm{m/s} \qquad \omega_f = \omega_r = \frac{v_x}{R_g} = \frac{2}{0.35} = 5.714\,\mathrm{rad/s} \tag{3.219}$$

At time $t = 0$, we apply a linearly increasing torque on the rear wheel up to $T_f = 300\,\mathrm{N\,m}$ and keep it constant after that. The steer angle is also linearly increasing at a very low rate up to $\delta = 1.5\,\mathrm{deg}$ and remains constant after that.

$$T_r = \begin{cases} 30t\,\mathrm{N\,m} & 0 < t < 10\,\mathrm{s} \\ 300\,\mathrm{N\,m} & 10\,\mathrm{s} \le t \end{cases} \qquad T_f = 0 \tag{3.220}$$

$$\delta = \begin{cases} 0.1t\,\mathrm{deg} = \frac{0.1\pi}{180}t\,\mathrm{rad} & 0 < t < 15\,\mathrm{s} \\ 1.5\,\mathrm{deg} = \frac{1.5\pi}{180}\,\mathrm{rad} & 15\,\mathrm{s} \le t \end{cases} \tag{3.221}$$

We solve the equations of motion numerically and plot the vehicle dynamic variables. Figure 3.33 illustrates the applied torque history. Figure 3.34 depicts the forward velocity components of the vehicle v_x and v_y, measured in body coordinate frame B. Figure 3.35 illustrates the angular velocities of the front and rear wheels ω_f and ω_r. Figure 3.36 illustrates the sideslip angles of the front and rear wheels α_f and α_r. Figure 3.37 depicts the longitudinal slip ratios s_f and s_r for the front and rear tires. The acceleration components a_X and a_Y of the vehicle are plotted in Fig. 3.38. Figure 3.39 depicts the forces F_x and F_y on the vehicle at its mass

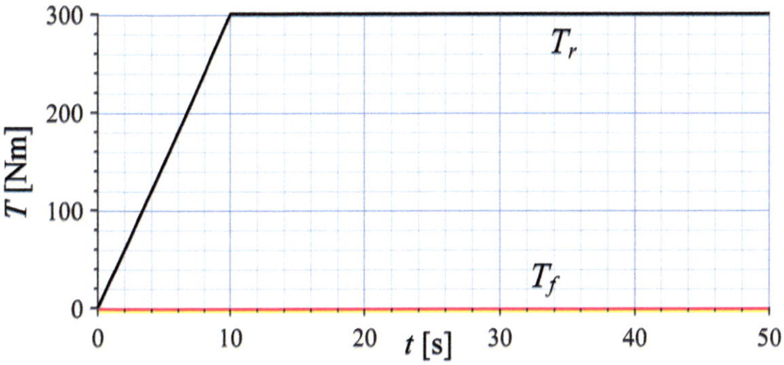

Fig. 3.33 The applied torque history T_f and T_r of a two-wheel roll vehicle model under rear torque and steer angle

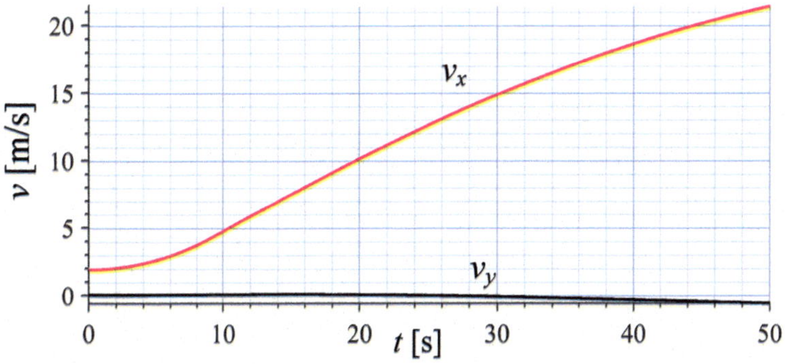

Fig. 3.34 The velocity components v_x and v_y of a two-wheel roll vehicle model under rear torque and steer angle

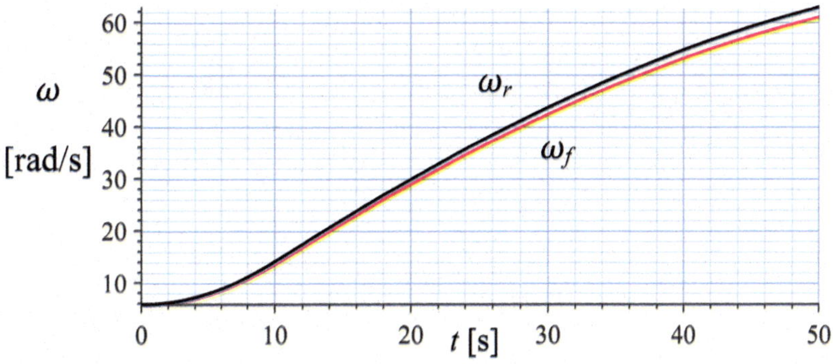

Fig. 3.35 The angular velocity ω_f and ω_r of a two-wheel roll vehicle model under rear torque and steer angle

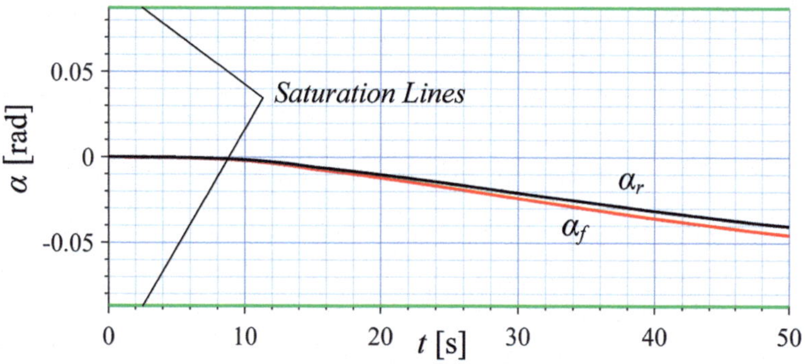

Fig. 3.36 The angular acceleration α_f and α_r of a two-wheel roll vehicle model under rear torque and steer angle

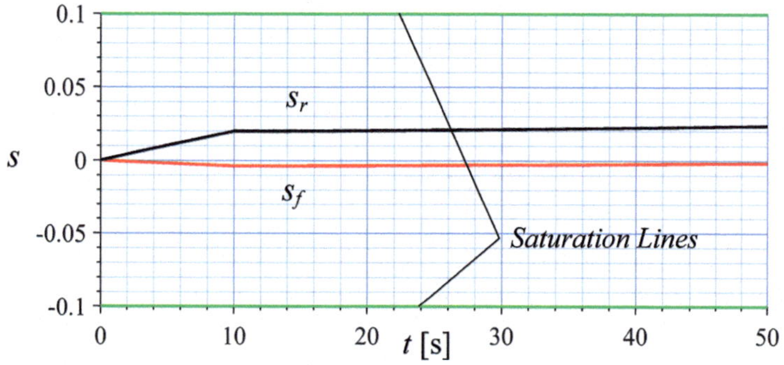

Fig. 3.37 The slip ratios s_f and s_r of a two-wheel roll vehicle model under rear torque and steer angle

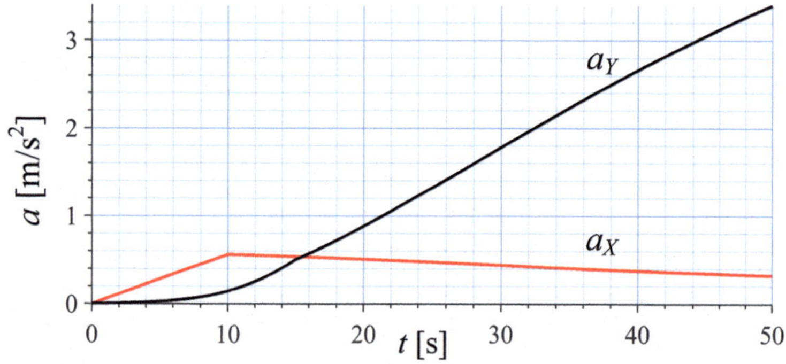

Fig. 3.38 The global acceleration a_X and a_Y of a two-wheel roll vehicle model under rear torque and steer angle

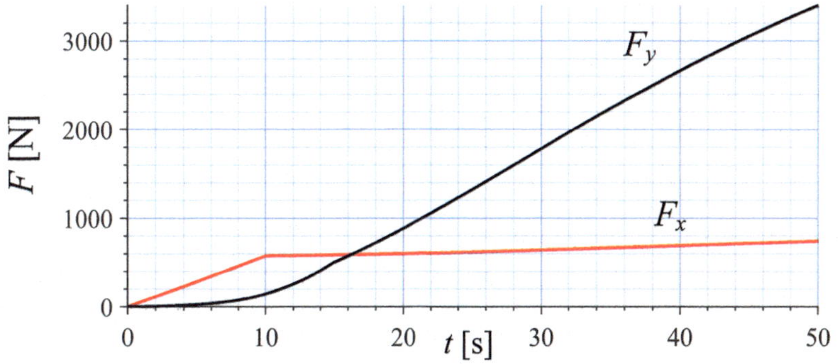

Fig. 3.39 The force components F_x and F_y of a two-wheel roll vehicle model under rear torque and steer angle

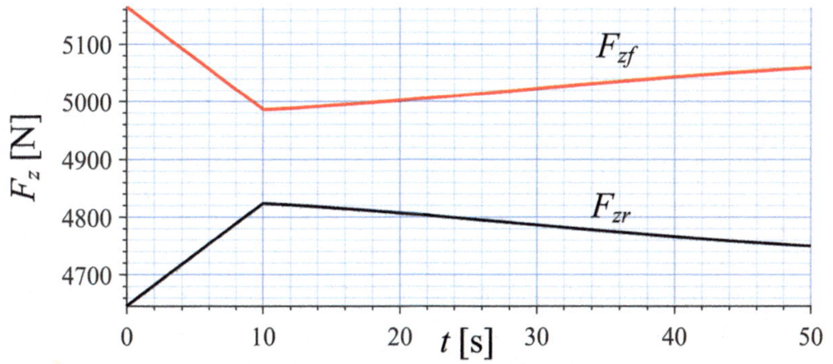

Fig. 3.40 The vertical forces F_{zf} and F_{zr} of a two-wheel roll vehicle model under rear torque and steer angle

center. Figure 3.40 illustrates variation of the vertical loads F_{zf} and F_{zr}. Figure 3.41 illustrates the longitudinal forces F_{xf} and F_{xr} on front and rear tires. Figure 3.42 shows the lateral forces F_{yf} and F_{yr} on front and rear tires. The roll angle φ and roll rate p are shown in Figs. 3.43 and 3.44 respectively. Figure 3.45 depicts the yaw rate r of vehicle. Figure 3.46 illustrates the path of the vehicle.

3.5 Four-Wheel Roll Vehicle Dynamics

The four-wheel roll vehicle model is the best practical vehicle mathematical model. This model provides us with in-wheel torques T_i, tire slips α_i, s_i, β_i, tire and vehicle forces F_x, F_y, F_{x_i}, F_{y_i}, F_{z_i}, velocity components of the vehicle v_x, v_y, ω_i, as well as yaw and roll angular variables φ, ψ, p, r. This model is an extension to the two-wheel roll vehicle model to include the lateral weight transfer as well as roll effects on vehicle dynamics. The four-wheel roll vehicle model is an excellent model to simulate drifting of vehicles.

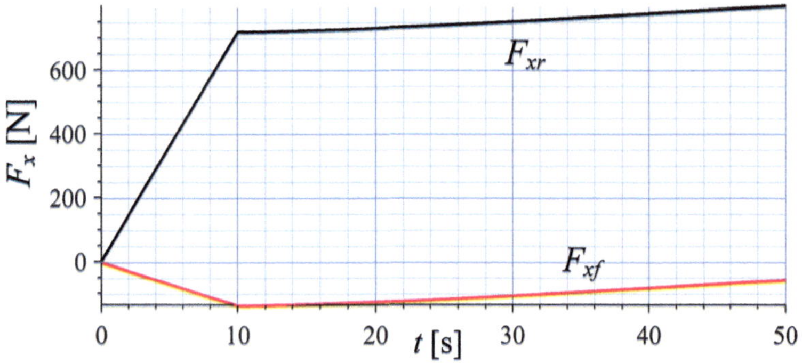

Fig. 3.41 The longitudinal forces F_{xf} and F_{xr} of a two-wheel roll vehicle model under rear torque and steer angle

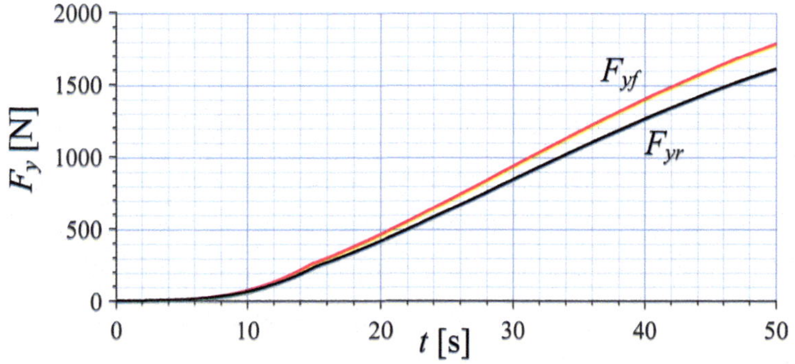

Fig. 3.42 The lateral forces F_{yf} and F_{yr} of a two-wheel roll vehicle model under rear torque and steer angle

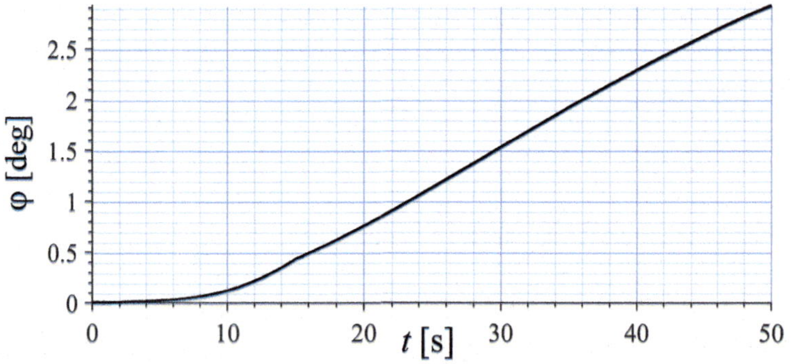

Fig. 3.43 The roll angle φ of a two-wheel roll vehicle model under rear torque and steer angle

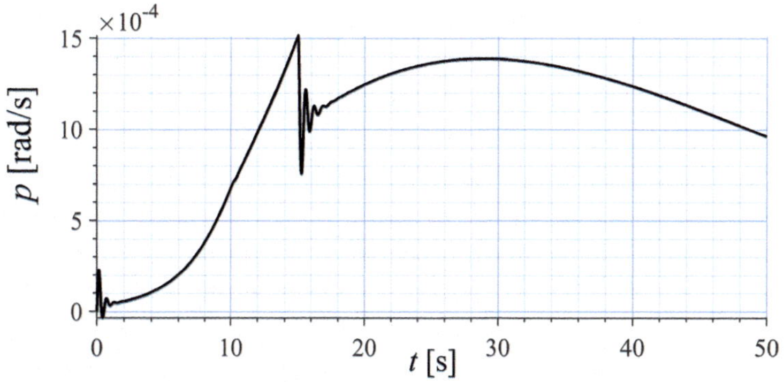

Fig. 3.44 The roll rate p of a two-wheel roll vehicle model under rear torque and steer angle

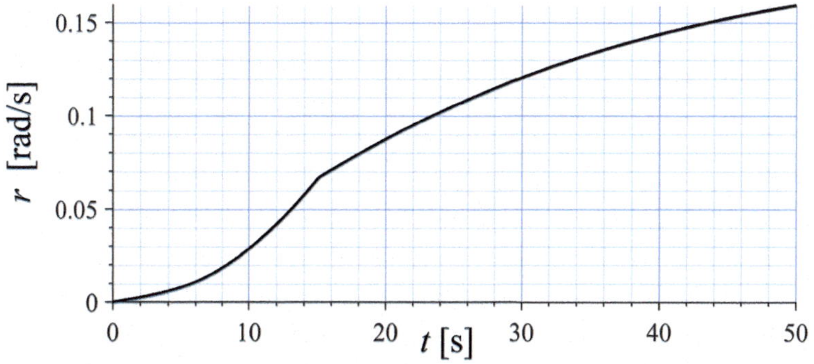

Fig. 3.45 The yaw rate r of a two-wheel roll vehicle model under rear torque and steer angle

Let us assume that the front and rear tracks are different.

$$b_{1f} + b_{2f} = w_f \tag{3.222}$$

$$b_{1r} + b_{2r} = w_r \tag{3.223}$$

The differential equations of motion of the four-wheel roll vehicle are:

$$\dot{v}_x = \frac{1}{m}\left(F_x - F_A\right) + r\, v_y \tag{3.224}$$

$$\dot{v}_y = \frac{1}{m} F_y - r\, v_x \tag{3.225}$$

$$\dot{r} = \frac{1}{I_z} M_z \tag{3.226}$$

$$\dot{\omega}_1 = \frac{1}{I_1} T_1 - \frac{R_g}{I_1} F_{x_1} \tag{3.227}$$

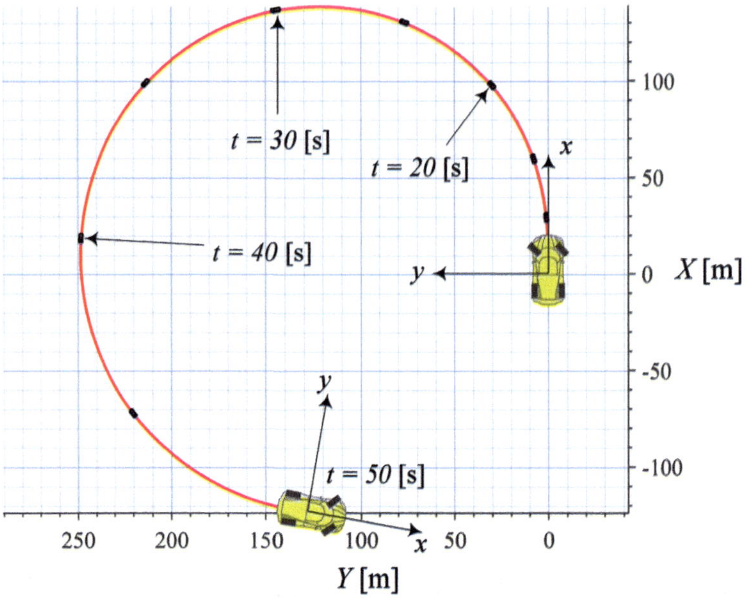

Fig. 3.46 Path of motion of a two-wheel roll vehicle model under rear torque and steer angle

$$\dot{\omega}_2 = \frac{1}{I_2} T_2 - \frac{R_g}{I_2} F_{x_2} \tag{3.228}$$

$$\dot{\omega}_1 = \frac{1}{I_3} T_3 - \frac{R_g}{I_3} F_{x_3} \tag{3.229}$$

$$\dot{\omega}_1 = \frac{1}{I_4} T_4 - \frac{R_g}{I_4} F_{x_4} \tag{3.230}$$

The resultant force system on the vehicle is:

$$\begin{aligned} F_x = {} & F_{x_1} \cos\left(\delta_1 + \delta_{\varphi_1}\right) + F_{x_2} \cos\left(\delta_2 + \delta_{\varphi_2}\right) \\ & + F_{x_3} \cos\delta_{\varphi_3} + F_{x_4} \cos\delta_{\varphi_4} - F_{y_3} \sin\delta_{\varphi_3} - F_{y_4} \sin\delta_{\varphi_4} \\ & - F_{y_1} \sin\left(\delta_1 + \delta_{\varphi_1}\right) - F_{y_2} \sin\left(\delta_2 + \delta_{\varphi_2}\right) \end{aligned} \tag{3.231}$$

$$\begin{aligned} F_y = {} & F_{y_1} \cos\left(\delta_1 + \delta_{\varphi_1}\right) + F_{y_2} \cos\left(\delta_2 + \delta_{\varphi_2}\right) \\ & + F_{y_3} \cos\delta_{\varphi_3} + F_{y_4} \cos\delta_{\varphi_4} + F_{x_3} \sin\delta_{\varphi_3} + F_{x_4} \sin\delta_{\varphi_4} \\ & + F_{x_1} \sin\left(\delta_1 + \delta_{\varphi_1}\right) + F_{x_2} \sin\left(\delta_2 + \delta_{\varphi_2}\right) \end{aligned} \tag{3.232}$$

$$F_A = C_A\, v_x^2 \tag{3.233}$$

$$M_x = C_{T_1} F_{y_1} + C_{T_2} F_{y_2} + C_{T_3} F_{y_3} + C_{T_4} F_{y_4} - k_\varphi \varphi - c_\varphi \dot{\varphi} \tag{3.234}$$

$$\begin{aligned} M_z = & a_1 F_{x_1} \sin\left(\delta_1 + \delta_{\varphi_1}\right) + a_1 F_{y_1} \cos\left(\delta_1 + \delta_{\varphi_1}\right) \\ & + a_1 F_{x_2} \sin\left(\delta_2 + \delta_{\varphi_2}\right) + a_1 F_{y_2} \cos\left(\delta_2 + \delta_{\varphi_2}\right) \\ & - b_{1f} F_{x_1} \cos\left(\delta_1 + \delta_{\varphi_1}\right) + b_{1f} F_{y_1} \sin\left(\delta_1 + \delta_{\varphi_1}\right) \\ & - b_{2f} F_{y_2} \sin\left(\delta_2 + \delta_{\varphi_2}\right) + b_{2f} F_{x_2} \cos\left(\delta_2 + \delta_{\varphi_2}\right) \\ & - a_2 F_{y_3} - a_2 F_{y_4} - b_{1r} F_{x_3} + b_{2r} F_{x_4} \end{aligned} \tag{3.235}$$

where

$$\delta_{\varphi_i} = C_{\delta\varphi_i} \varphi \tag{3.236}$$

The tire force components are:

$$F_{x_i} = F_{z_i} C_{si} \, S\left(s_i - s_s\right) \sqrt{1 - C_{s\alpha} \left(\frac{S\left(\alpha_i - \alpha_s\right)}{\alpha_s}\right)^2} \tag{3.237}$$

$$F_{y_i} = -F_{z_i} C_{\alpha i} \, S\left(\alpha_i - \alpha_s\right) \sqrt{1 - C_{\alpha s} \left(\frac{S\left(s_i - s_s\right)}{s_s}\right)^2} - C_{\varphi_i} \varphi \tag{3.238}$$

where S is the $Saturation$ function (1.60).

$$S(x - x_0) = \begin{cases} x_0 & x_0 < x \\ x & -x_0 < x < x_0 \\ -x_0 & x < -x_0 \end{cases} \tag{3.239}$$

The vertical force of tires is:

$$F_{z_1} = \frac{m}{l w_f g} \left(a_2 g - h\left(\dot{v}_x - r\, v_y\right)\right)\left(b_{2f} g - h\left(\dot{v}_y + r\, v_x\right)\right) \tag{3.240}$$

$$F_{z_2} = \frac{m}{l w_f g} \left(a_2 g - h\left(\dot{v}_x - r\, v_y\right)\right)\left(b_{1f} g + h\left(\dot{v}_y + r\, v_x\right)\right) \tag{3.241}$$

$$F_{z_3} = \frac{m}{l w_r g} \left(a_1 g + h\left(\dot{v}_x - r\, v_y\right)\right)\left(b_{2r} g - h\left(\dot{v}_y + r\, v_x\right)\right) \tag{3.242}$$

$$F_{z_4} = \frac{m}{l w_r g} \left(a_1 g + h\left(\dot{v}_x - r\, v_y\right)\right)\left(b_{1r} g + h\left(\dot{v}_y + r\, v_x\right)\right) \tag{3.243}$$

and tire slips as

$$\alpha_i = \beta_i - \delta_i - \delta_{\varphi_i} = \arctan\left(\frac{v_y + x_i\, r - C_{\beta_i}\, p}{v_x - y_i\, r}\right) - \delta_i - C_{\delta\varphi_i} \varphi \tag{3.244}$$

$$\beta_i = \arctan\left(\frac{v_{y_i}}{v_{x_i}}\right) = \arctan\left(\frac{v_y + x_i\, r - C_{\beta_i}\, p}{v_x - y_i\, r}\right) \tag{3.245}$$

$$\beta = \arctan\frac{v_y}{v_x} \tag{3.246}$$

$$x_1 = x_2 = a_1 \qquad y_1 = b_{1f} \qquad y_3 = b_{1r} \tag{3.247}$$

$$x_3 = x_4 = -a_2 \qquad y_2 = -b_{2f} \qquad y_4 = -b_{2r} \tag{3.248}$$

$$s_i = \frac{R_g\,\omega_i - v_{xTi}}{R_g\,\omega_i\, H(R_g\,\omega_w - v_{xT_i}) + v_{xT_i}\, H(v_{xTi} - R_g\,\omega_i)} \tag{3.249}$$

$$v_{xTi} = v_{x_i}\cos\left(\delta_i + \delta_{\varphi_i}\right) + v_{y_i}\sin\left(\delta_i + \delta_{\varphi_i}\right) \tag{3.250}$$

$$v_{xT1} = \left(v_x - rb_{1f}\right)\cos\left(\delta_1 + \delta_{\varphi_1}\right) + \left(v_y + ra_1 - C_{\beta_1}p\right)\sin\left(\delta_1 + \delta_{\varphi_1}\right) \tag{3.251}$$

$$v_{xT2} = \left(v_x + rb_{1f}\right)\cos\left(\delta_2 + \delta_{\varphi_2}\right) + \left(v_y + ra_1 - C_{\beta_2}p\right)\sin\left(\delta_2 + \delta_{\varphi_2}\right) \tag{3.252}$$

$$v_{xT3} = (v_x - rb_{2r})\cos\delta_{\varphi_3} + \left(v_y - ra_2 - C_{\beta_3}p\right)\sin\delta_{\varphi_3} \tag{3.253}$$

$$v_{xT4} = (v_x + rb_{2r})\cos\delta_{\varphi_4} + \left(v_y - ra_2 - C_{\beta_4}p\right)\sin\delta_{\varphi_4} \tag{3.254}$$

where H is the *Heaviside* function (1.61).

The front left and right steer angles δ_1, δ_2 are related due to steering mechanism.

$$\delta_2 = f\,(\delta_1) \tag{3.255}$$

Proof The equations of motion for a roll vehicle model in the local coordinate frame B, at its mass center C, are given in Eqs. (3.1)–(3.5). The equations for a four-wheel roll vehicle model will be:

$$F_x = m\,\dot{v}_x - mr\,v_y + F_A \tag{3.256}$$

$$F_y = m\,\dot{v}_y + mr\,v_x \tag{3.257}$$

$$M_z = I_z\,\dot{\omega}_z = I_z\,\dot{r} \tag{3.258}$$

$$M_x = I_x\,\dot{\omega}_x = I_x\,\dot{p} \tag{3.259}$$

$$T_i = I_{w_i}\,\dot{\omega}_{w_i} + R_w\,F_{x_i} \tag{3.260}$$

$$T_1 = I_{w_1}\,\dot{\omega}_1 + R_w\,F_{x_1} \tag{3.261}$$

$$T_2 = I_{w_2}\,\dot{\omega}_2 + R_w\,F_{x_2} \tag{3.262}$$

$$T_3 = I_{w_3}\,\dot{\omega}_r + R_w\,F_{x_3} \tag{3.263}$$

$$T_4 = I_{w_4}\,\dot{\omega}_r + R_w\,F_{x_4} \tag{3.264}$$

The aerodynamic force F_A is the environment resistance forces on the vehicle. The F_A is assumed to be effective only in the x-direction.

$$F_A = \frac{1}{2}\rho\, C_D\, A_f\, v_x^2 = C_A\, v_x^2 \tag{3.265}$$

We combine the coefficients ρ, C_D, and A_f into a single coefficient C_A.

Suspension of vehicles are supposed to provide flexibility to the sprung mass with respect to unsprung mass. Although there are several members connected to both sprung and unsprung masses, the vehicle body is the main sprung mass and wheels including in-wheel motors are the main unsprung masses. Besides suspending the vehicle body, suspensions need to provide some motion freedom to the wheels with respect to the vehicle body and lock some other motions. Wheels with respect to the vehicle body must be locked for translation in the x and y directions and be able to move in the z-direction. They also must be able to rotate about the x, y, and z directions. Rotation about the y-direction is the spin of the wheels. Rotation about the z-direction is the steer angle that is controlled by actuators or mechanisms connected to the steer wheel. Rotation about the x-direction is camber of the wheel. Because of these flexibility, the roll angle φ of the body is different than the camber angle γ, although we may assume $\varphi = \gamma$ as long as we are not separating sprung and unsprung masses in dynamic analysis. Due to steering and suspension mechanism interaction, the body roll may generate some steer angles. Such a roll-steer action may be designed on purpose or be an unavoidable mechanism function. Due to the roll-steer phenomenon, the actual steer angle δ_a of such a tire will be:

$$\delta_a = \delta + \delta_\varphi \tag{3.266}$$

The wheel roll steering is assumed to be proportional to the vehicle roll angle φ.

$$\delta_\varphi = C_{\delta\varphi}\varphi \tag{3.267}$$

$$C_{\delta\varphi} = \lim_{\varphi \longrightarrow 0} \frac{d\delta}{d\varphi} \tag{3.268}$$

Depending on the design of suspension and steering mechanism, the roll-steering coefficient $C_{\delta\varphi}$ may be positive or negative, although a positive $C_{\delta\varphi}$ is desirable.

From (3.47) to (3.49), the applied forces on the vehicle are:

$$\begin{aligned} {}^B F_x &= \sum_i F_{xw_i} \cos\left(\delta_i + \delta_{\varphi_i}\right) - \sum_i F_{yw_i} \sin\left(\delta_i + \delta_{\varphi_i}\right) \\ &= F_{x_1} \cos\left(\delta_1 + \delta_{\varphi_1}\right) + F_{x_2} \cos\left(\delta_2 + \delta_{\varphi_2}\right) \\ &\quad + F_{x_3} \cos \delta_{\varphi_3} + F_{x_4} \cos \delta_{\varphi_4} - F_{y_3} \sin \delta_{\varphi_3} - F_{y_4} \sin \delta_{\varphi_4} \\ &\quad - F_{y_1} \sin\left(\delta_1 + \delta_{\varphi_1}\right) - F_{y_2} \sin\left(\delta_2 + \delta_{\varphi_2}\right) \end{aligned} \tag{3.269}$$

$$
\begin{aligned}
{}^{B}F_y &= \sum_i F_{yw_i} \cos\left(\delta_i + \delta_{\varphi_i}\right) + \sum_i F_{xw_i} \sin\left(\delta_i + \delta_{\varphi_i}\right) \\
&= F_{y_1} \cos\left(\delta_1 + \delta_{\varphi_1}\right) + F_{y_2} \cos\left(\delta_2 + \delta_{\varphi_2}\right) \\
&\quad + F_{y_3} \cos\delta_{\varphi_3} + F_{y_4} \cos\delta_{\varphi_4} + F_{x_3} \sin\delta_{\varphi_3} + F_{x_4} \sin\delta_{\varphi_4} \\
&\quad + F_{x_1} \sin\left(\delta_1 + \delta_{\varphi_1}\right) + F_{x_2} \sin\left(\delta_2 + \delta_{\varphi_2}\right) \qquad (3.270)
\end{aligned}
$$

$$
F_A = C_A\, v_x^2 \qquad (3.271)
$$

$$
{}^{B}M_x = \sum_i M_{x_i} + \sum_i y_i F_{z_i} - \sum_i z_i F_{y_i} \qquad (3.272)
$$

$$
= C_{T_1} F_{y_1} + C_{T_2} F_{y_2} + C_{T_3} F_{y_3} + C_{T_4} F_{y_4} - k_\varphi \varphi - c_\varphi \dot{\varphi} \qquad (3.273)
$$

$$
\begin{aligned}
{}^{B}M_z &= \sum_i M_{z_i} + \sum_i x_i F_{y_i} - \sum_i y_i F_{x_i} \\
&= a_1 F_{x_1} \sin\left(\delta_1 + \delta_{\varphi_1}\right) + a_1 F_{y_1} \cos\left(\delta_1 + \delta_{\varphi_1}\right) \qquad (3.274) \\
&\quad + a_1 F_{x_2} \sin\left(\delta_2 + \delta_{\varphi_2}\right) + a_1 F_{y_2} \cos\left(\delta_2 + \delta_{\varphi_2}\right) \\
&\quad - b_{1f} F_{x_1} \cos\left(\delta_1 + \delta_{\varphi_1}\right) + b_{1f} F_{y_1} \sin\left(\delta_1 + \delta_{\varphi_1}\right) \\
&\quad - b_{2f} F_{y_2} \sin\left(\delta_2 + \delta_{\varphi_2}\right) + b_{2f} F_{x_2} \cos\left(\delta_2 + \delta_{\varphi_2}\right) \\
&\quad - a_2 F_{y_3} - a_2 F_{y_4} - b_{1r} F_{x_3} + b_{2r} F_{x_4} \qquad (3.275)
\end{aligned}
$$

A lateral reaction force will be generated under tires or a rolled vehicle called tire roll-thrust. Tire roll-thrust is assumed to be proportional to the vehicle roll angle φ.

$$
F_{y\varphi} = -C_\varphi \varphi \qquad (3.276)
$$

$$
C_\varphi = \lim_{\varphi \longrightarrow 0} \frac{d\left(-F_y\right)}{d\varphi} \qquad (3.277)
$$

Therefore, employing the combined slip elliptical model, the tire tangential forces F_{x_i} and F_{y_i} are:

$$
F_{x_i} = F_{z_i} C_{si}\, S\left(s_i - s_s\right) \sqrt{1 - C_{s\alpha} \left(\frac{S\left(\alpha_i - \alpha_s\right)}{\alpha_s}\right)^2} \qquad (3.278)
$$

$$
F_{y_i} = -F_{z_i} C_{\alpha i}\, S\left(\alpha_i - \alpha_s\right) \sqrt{1 - C_{\alpha s} \left(\frac{S\left(s_i - s_s\right)}{s_s}\right)^2} - C_{\varphi_i} \varphi_i \qquad (3.279)
$$

The vertical forces on front and rear wheels F_{z_i} are calculated in (1.431)–(1.434) in terms of global accelerations a_X, a_Y. To make the equations suitable to be substituted in (3.278) and (3.279) we need to express a_X and a_Y in body coordinate frame B.

$$a_X = \dot{v}_x - r\,v_y \tag{3.280}$$

$$a_Y = \dot{v}_y + r\,v_x \tag{3.281}$$

Therefore, the vertical force of tires will be:

$$F_{z_1} = \frac{m}{l w_f g}\left(g a_2 - h\dot{v}_x + h r v_y\right)\left(h\dot{v}_y + g b_{2f} - h r v_x\right) \tag{3.282}$$

$$F_{z_2} = \frac{m}{l w_f g}\left(g a_2 - h\dot{v}_x + h r v_y\right)\left(h\dot{v}_y + g b_{1f} + h r v_x\right) \tag{3.283}$$

$$F_{z_3} = \frac{m}{l w_r g}\left(h\dot{v}_x + g a_1 - h r v_y\right)\left(h\dot{v}_y + g b_{2r} - h r v_x\right) \tag{3.284}$$

$$F_{z_4} = \frac{m}{l w_r g}\left(h\dot{v}_x + g a_1 - h r v_y\right)\left(h\dot{v}_y + g b_{1r} + h r v_x\right) \tag{3.285}$$

where

$$w_f = b_{1f} + b_{2f} \tag{3.286}$$

$$w_r = b_{1r} + b_{2r} \tag{3.287}$$

$$l = a_1 + a_2 \tag{3.288}$$

The tire side slip angle α_i is defined as:

$$\alpha_i = \beta_i - \delta_i - C_{\delta\varphi_i}\varphi \tag{3.289}$$

$$\beta_i = \arctan\left(\frac{v_{y_i}}{v_{x_i}}\right) \tag{3.290}$$

We determine the velocity components v_{x_i} and v_{y_i} of wheels, using velocity of the vehicle, ${}^B\mathbf{v}$.

$${}^B\mathbf{v}_i = {}^B\mathbf{v} + {}^B_G\boldsymbol{\omega}_B \times {}^B\mathbf{r}_i \tag{3.291}$$

Substituting the position vector of wheel number i, and the angular velocity of the vehicle ${}^B_G\boldsymbol{\omega}_B$,

$${}^B\mathbf{r}_i = \begin{bmatrix} x_i \\ y_i \\ z_i \end{bmatrix} \qquad {}^B_G\boldsymbol{\omega}_B = \begin{bmatrix} p \\ 0 \\ r \end{bmatrix}$$

we find

$$v_{x_i} = v_x - y_i r \tag{3.292}$$

$$v_{y_i} = v_y + x_i r - z_i p \tag{3.293}$$

To compensate the assumption of having no roll difference between sprung and unsprung masses, we substitute the term $z_i p$ with $C_{\beta_i} p$ in which the *tire roll rate coefficient* C_{β_i} is determined by experiments. The z_i is vertical distance of the wheel center number i from the xy-plane. Although z_i is very small, it may vary because of $\psi \neq \gamma$. Considering the variation of $z_i p$ due to roll rate, we simplify it by $C_{\beta_i} p$ and determine C_{β_i} by experiment. Depending on the design of suspension mechanisms, C_{β_i} may be positive or negative. Therefore, the lateral velocity component because of roll rate p is assumed to be proportional to p.

$$C_{\beta_i} = \lim_{p \longrightarrow 0} \frac{dv_{y_i}}{dp} \tag{3.294}$$

Hence, the velocity components of wheel number i will be

$$v_{x_i} = v_x - y_i r \tag{3.295}$$

$$v_{y_i} = v_y + r x_i - C_{\beta_i} p \tag{3.296}$$

and therefore,

$$\begin{aligned} \alpha_i &= \beta_i - \delta_i - \delta_{\varphi_i} \\ &= \arctan\left(\frac{v_y + x_i\, r - C_{\beta_i}\, p}{v_x - y_i\, r}\right) - \delta_i - C_{\delta\varphi_i}\varphi \end{aligned} \tag{3.297}$$

$$\beta_i = \arctan\left(\frac{v_{y_i}}{v_{x_i}}\right) = \arctan\left(\frac{v_y + x_i\, r - C_{\beta_i}\, p}{v_x - y_i\, r}\right) \tag{3.298}$$

$$\beta = \arctan\frac{v_y}{v_x} \tag{3.299}$$

From Eq. (1.57), the longitudinal slip ratio s_i is calculated as a function of the tire forward velocity v_{xT_i} and its angular velocity ω_i.

$$s_i = \frac{R_g\, \omega_i - v_{xTi}}{R_g\, \omega_i\, H(R_g\, \omega_w - v_{xTi}) + v_{xT_i}\, H(v_{xT_i} - R_g\, \omega_i)} \tag{3.300}$$

$$v_{xTi} = v_{x_i} \cos\left(\delta_i + \delta_{\varphi_i}\right) + v_{y_i} \sin\left(\delta_i + \delta_{\varphi_i}\right) \tag{3.301}$$

The forward velocity v_{xT_i} of the four-wheels in the tire forward direction, ${}^{T}\hat{\imath}_i$, will be

$$v_{xT1} = \left(v_x - rb_{1f}\right)\cos\left(\delta_1 + \delta_{\varphi_1}\right) + \left(v_y + ra_1 - C_{\beta_1} p\right)\sin\left(\delta_1 + \delta_{\varphi_1}\right) \tag{3.302}$$

$$v_{xT2} = \left(v_x + rb_{1f}\right)\cos\left(\delta_2 + \delta_{\varphi_2}\right) + \left(v_y + ra_1 - C_{\beta_2} p\right)\sin\left(\delta_2 + \delta_{\varphi_2}\right) \tag{3.303}$$

$$v_{xT3} = \left(v_x - rb_{2r}\right)\cos\delta_{\varphi_3} + \left(v_y - ra_2 - C_{\beta_3} p\right)\sin\delta_{\varphi_3} \tag{3.304}$$

$$v_{xT4} = \left(v_x + rb_{2r}\right)\cos\delta_{\varphi_4} + \left(v_y - ra_2 - C_{\beta_4} p\right)\sin\delta_{\varphi_{43}} \tag{3.305}$$

because

$$^{T}\mathbf{v}_i = {}^{C}R_T^{T}\,{}^{B}\mathbf{v}_i \tag{3.306}$$

$$= \begin{bmatrix} \cos\left(\delta_i + \delta_{\varphi_i}\right) & -\sin\left(\delta_i + \delta_{\varphi_i}\right) & 0 \\ \sin\left(\delta_i + \delta_{\varphi_i}\right) & \cos\left(\delta_i + \delta_{\varphi_i}\right) & 0 \\ 0 & 0 & 1 \end{bmatrix}^T \begin{bmatrix} v_x - ry_i \\ v_y + rx_i - C_{\beta_i} p \\ 0 \end{bmatrix}$$

■

Example 89 Four-wheel roll vehicle model, steer angle and front torque.
Consider a vehicle with the following characteristics

$$\begin{aligned}
m &= 1000\,\text{kg} \qquad I_1 = I_r = I_3 = I_4 = 30\,\text{kg m}^2 \\
I_x &= 300\,\text{kg m}^2 \qquad I_z = 2000\,\text{kg m}^2 \qquad C_A = 0.8 \\
C_{\alpha_1} &= C_{\alpha_2} = C_{\alpha_3} = C_{\alpha_4} = 8.5 \qquad \alpha_s = 5\,\text{deg} \\
C_{s_1} &= C_{s_2} = C_{s_3} = C_{s_4} = 7.5 \qquad s_s = 0.1 \\
R_g &= 0.35\,\text{m} \qquad C_{\alpha s} = 0.5 \qquad C_{s\alpha} = 0.5 \\
a_1 &= 1.35\,\text{m} \qquad a_2 = 1.5\,\text{m} \qquad h = 0.9\,\text{m} \\
C_{\varphi_1} &= C_{\varphi_2} = 1600 \qquad C_{\varphi_3} = C_{\varphi_4} = 0 \\
k_\varphi &= 26{,}612\,\text{N m/rad} \qquad c_\varphi = 1700\,\text{N m s/rad} \\
C_{\beta_1} &= C_{\beta_2} = -0.4 \qquad C_{\beta_3} = C_{\beta_4} = -0.1 \\
C_{\delta\varphi_1} &= C_{\delta\varphi_2} = C_{\delta\varphi_3} = C_{\delta\varphi_4} = 0.01 \\
C_{T_1} &= C_{T_2} = C_{T_3} = C_{T_4} = 0.4 \qquad w_f = w_r = 1.8\,\text{m}
\end{aligned} \tag{3.307}$$

The vehicle is moving very slowly straight.

$$v_x = 2\,\text{m/s} \qquad \omega_i = \frac{v_x}{R_g} = \frac{2}{0.35} = 5.714\,\text{rad/s} \tag{3.308}$$

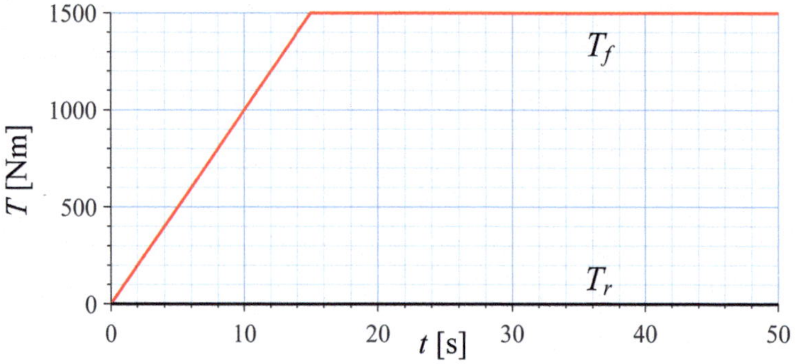

Fig. 3.47 The torques T_i of a four-wheel roll vehicle with active front torques and steer angle

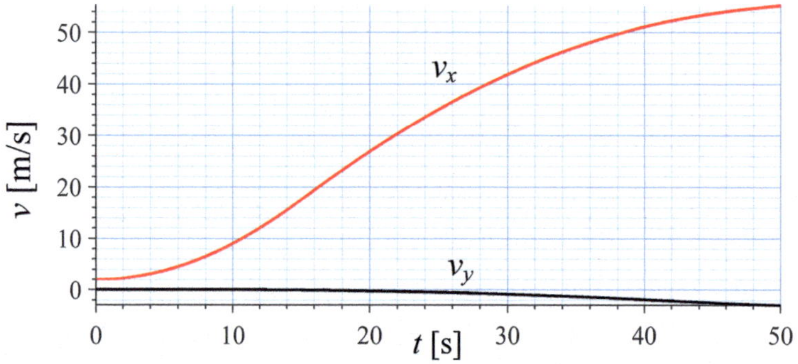

Fig. 3.48 The vehicle velocity v_x and v_y of a four-wheel roll vehicle with active front torques and steer angle

At time $t = 0$, an increasing torque will be applied on the front wheels up to $T_1 = T_2 = 1500\,\mathrm{N\,m}$ and constant after that. The steer angle is also increasing at a very low rate up to $\delta = 0.5\,\mathrm{deg}$ and remains constant after that.

$$T_1 = T_2 = \begin{cases} 50t\,\mathrm{N\,m} & 0 < t < 15\,\mathrm{s} \\ 750\,\mathrm{N\,m} & 15\,\mathrm{s} \le t \end{cases} \qquad T_3 = T_4 = 0 \tag{3.309}$$

$$\delta = \begin{cases} 0.05t\,\mathrm{deg} = \frac{0.05\pi}{180}t\,\mathrm{rad} & 0 < t < 10\,\mathrm{s} \\ 0.5\,\mathrm{deg} = \frac{0.5\pi}{180}\,\mathrm{rad} & 10\,\mathrm{s} \le t \end{cases} \tag{3.310}$$

The torques T_i of the four-wheel roll vehicle are shown in Fig. 3.47. Figure 3.48 depicts the velocity components of the vehicle v_x and v_y, measured in body coordinate B-frame. Figure 3.49 illustrates the angular velocities of the wheels ω_i. Figure 3.50 illustrates the sideslip angles of the wheels α_i. Figure 3.51 depicts the

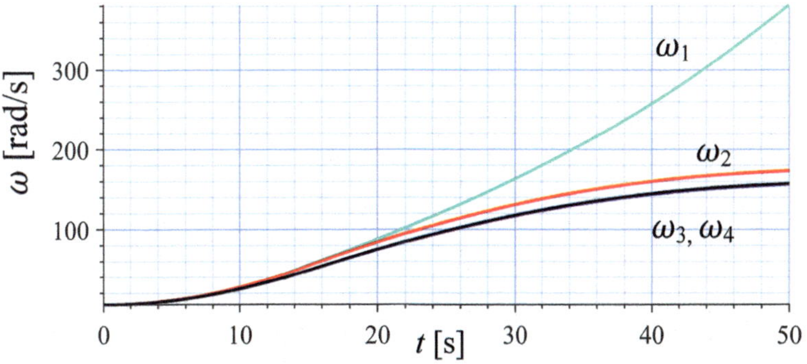

Fig. 3.49 The wheels angular velocity ω_i of a four-wheel roll vehicle with active front torques and steer angle

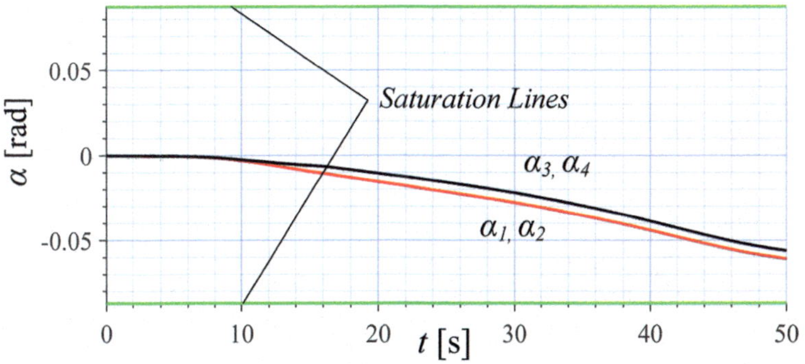

Fig. 3.50 The sideslip angles of the wheels α_i of a four-wheel roll vehicle with active front torques and steer angle

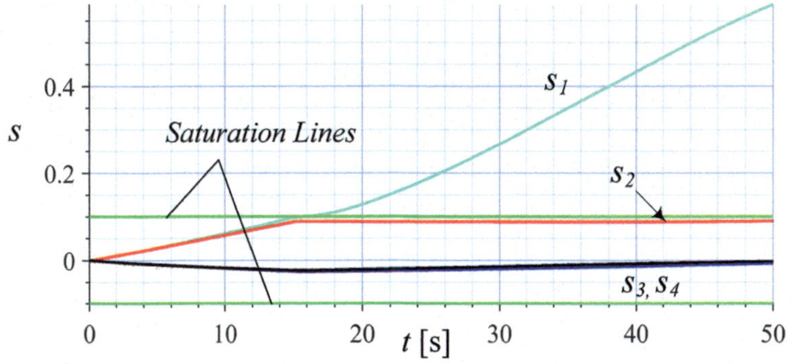

Fig. 3.51 The slip ratios of the wheels s_i of a four-wheel roll vehicle with active front torques and steer angle

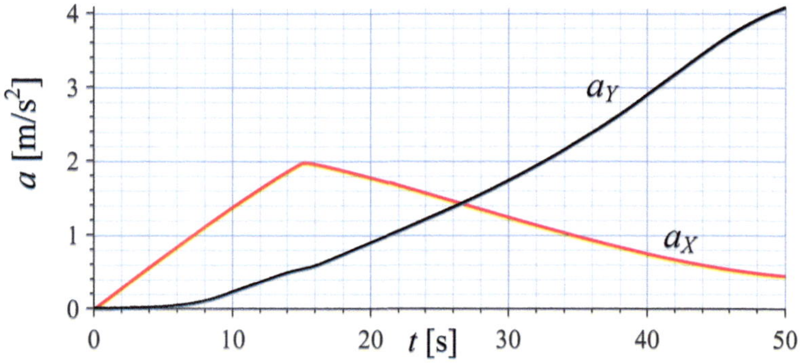

Fig. 3.52 The vehicle acceleration a_X and a_Y/g of a four-wheel roll vehicle with active front torques and steer angle

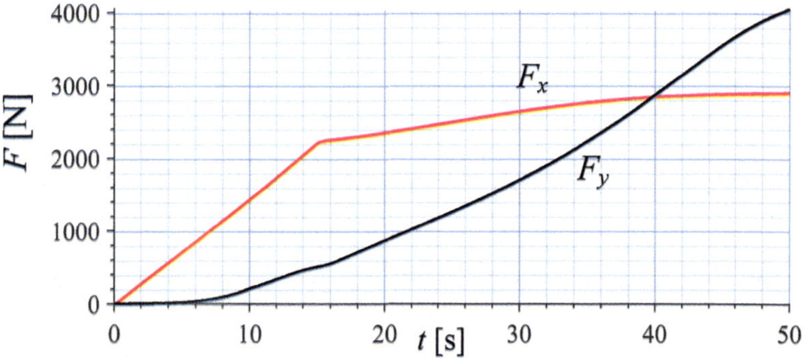

Fig. 3.53 The force components F_x and F_y of a four-wheel roll vehicle with active front torques and steer angle

longitudinal slip ratios s_i for the tires. The acceleration components a_X and a_Y of the vehicle are plotted in Fig. 3.52. Figure 3.53 depicts the forces F_x and F_y on the vehicle at its mass center. Figure 3.54 illustrates variation of the vertical loads F_{z_i}. Figure 3.55 illustrates the longitudinal forces F_{x_i} on tires. Figure 3.56 shows the lateral forces F_{y_i} on tires. The roll angle φ and roll rate p are shown in Figs. 3.57 and 3.58, respectively. Figure 3.59 depicts the angular accelerations $\dot{\omega}_f$ and $\dot{\omega}_r$ of the front and rear wheels. Figure 3.60 depicts the yaw rate r of vehicle. Figure 3.61 illustrates the path of the vehicle.

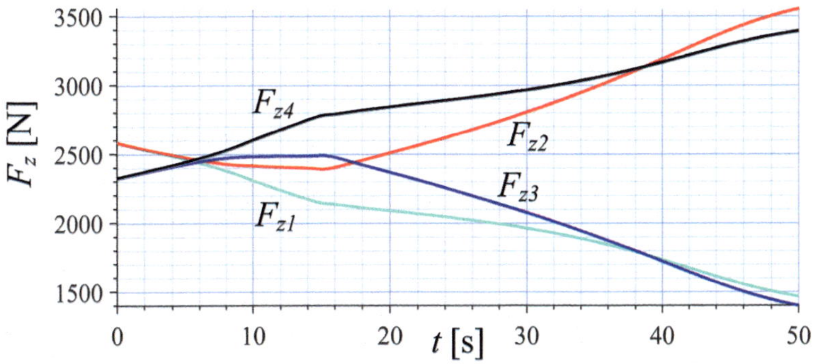

Fig. 3.54 The vertical forces of the wheels F_{zi} of a four-wheel roll vehicle with active front torques and steer angle

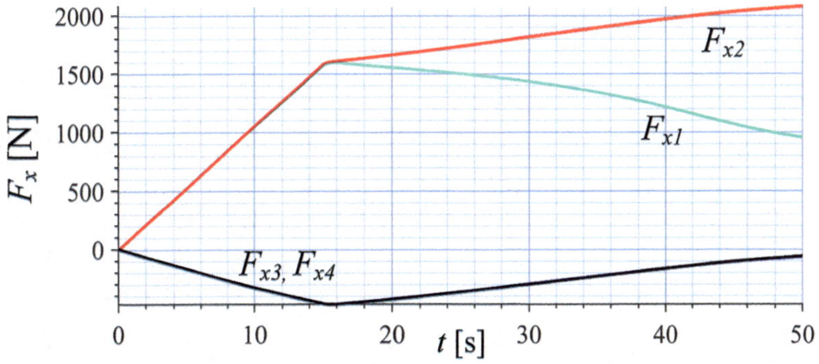

Fig. 3.55 The longitudinal forces of the wheels F_{xi} of a four-wheel roll vehicle with active front torques and steer angle

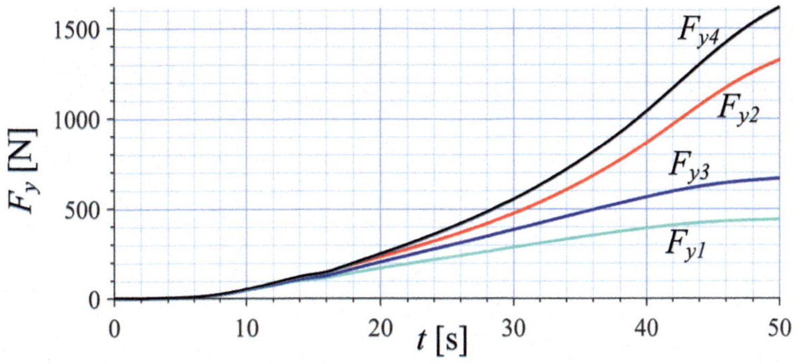

Fig. 3.56 The lateral forces of the wheels F_{yi} of a four-wheel roll vehicle with active front torques and steer angle

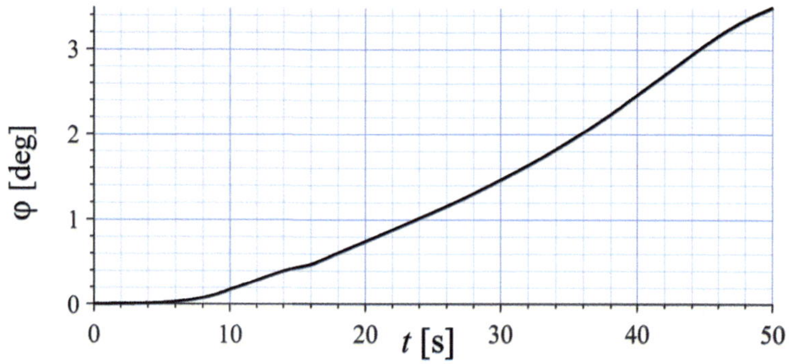

Fig. 3.57 The roll angle φ of a four-wheel roll vehicle with active front torques and steer angle

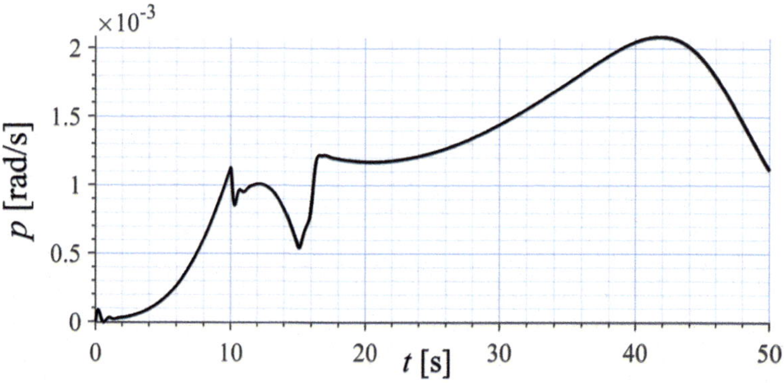

Fig. 3.58 The roll rate p of a four-wheel roll vehicle with active front torques and steer angle

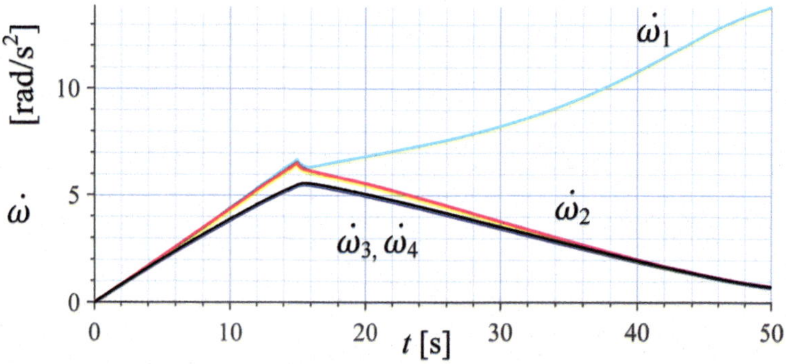

Fig. 3.59 The wheels angular accelerations $\dot{\omega}_i$ of a four-wheel roll vehicle with active front torques and steer angle

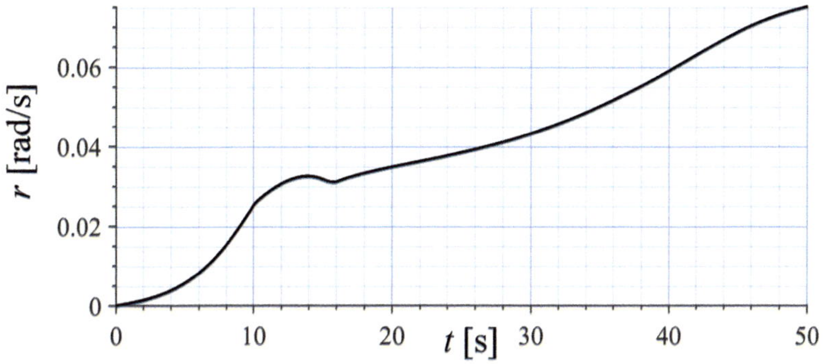

Fig. 3.60 The yaw rate r of a four-wheel roll vehicle with active front torques and steer angle

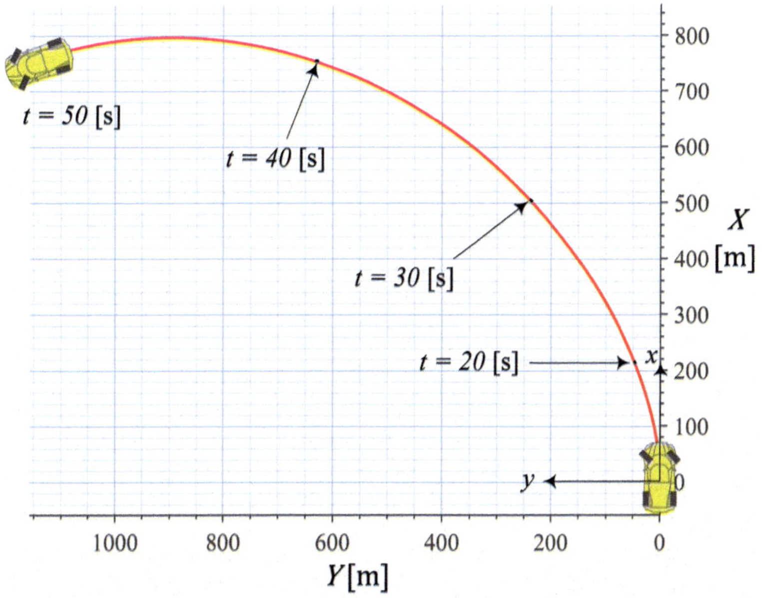

Fig. 3.61 Path of motion of a four-wheel roll vehicle with active front torques and steer angle

Example 90 Four-wheel roll vehicle model, steer angle and rear torque.

Consider a vehicle with the following data

$$m = 1000\,\text{kg} \qquad I_1 = I_r = I_3 = I_4 = 30\,\text{kg m}^2$$

$$I_x = 300\,\text{kg m}^2 \qquad I_z = 2000\,\text{kg m}^2 \qquad C_A = 0.8$$

$$C_{\alpha_1} = C_{\alpha_2} = C_{\alpha_3} = C_{\alpha_4} = 8.5 \qquad \alpha_s = 5\,\text{deg}$$

$$C_{s_1} = C_{s_2} = C_{s_3} = C_{s_4} = 7.5 \qquad s_s = 0.1$$

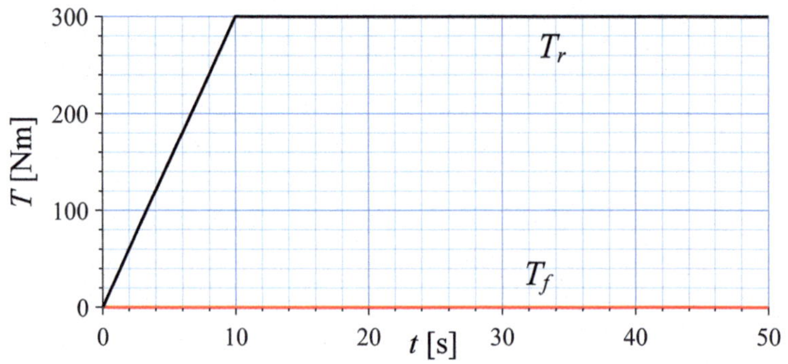

Fig. 3.62 The torques T_i of a four-wheel roll vehicle with active rear torques and steer angle

$$\begin{aligned}
R_g &= 0.35\,\mathrm{m} \qquad C_{\alpha s} = 0.5 \qquad C_{s\alpha} = 0.5 \\
a_1 &= 1.35\,\mathrm{m} \qquad a_2 = 1.5\,\mathrm{m} \qquad h = 0.9\,\mathrm{m} \\
C_{\varphi_1} &= C_{\varphi_2} = 1600 \qquad C_{\varphi_3} = C_{\varphi_4} = 0 \\
k_\varphi &= 26{,}612\,\mathrm{N\,m/rad} \qquad c_\varphi = 1700\,\mathrm{N\,m\,s/rad} \\
C_{\beta_1} &= C_{\beta_2} = -0.4 \qquad C_{\beta_3} = C_{\beta_4} = -0.1 \\
C_{\delta\varphi_1} &= C_{\delta\varphi_2} = C_{\delta\varphi_3} = C_{\delta\varphi_4} = 0.01 \\
C_{T_1} &= C_{T_2} = C_{T_3} = C_{T_4} = 0.4 \qquad w_f = w_r = 1.8\,\mathrm{m}
\end{aligned} \tag{3.311}$$

and assume the vehicle is moving slowly straight.

$$v_x = 2\,\mathrm{m/s} \qquad \omega_f = \omega_r = \frac{v_x}{R_g} = \frac{2}{0.35} = 5.714\,\mathrm{rad/s} \tag{3.312}$$

At time $t = 0$, we apply an increasing torque on the rear wheels up to $T_r = 300\,\mathrm{N\,m}$ and keep it constant after that. The steer angle is also increasing at a very low rate up to $\delta = 1.5$ deg and remains constant after that.

$$T_3 = T_4 = \begin{cases} 15t\,\mathrm{N\,m} & 0 < t < 10\,\mathrm{s} \\ 150\,\mathrm{N\,m} & 10\,\mathrm{s} \le t \end{cases} \qquad T_1 = T_2 = 0 \tag{3.313}$$

$$\delta = \begin{cases} 0.1t\,\mathrm{deg} = \frac{0.1\pi}{180} t\,\mathrm{rad} & 0 < t < 10\,\mathrm{s} \\ 1.5\,\mathrm{deg} = \frac{1.5\pi}{180}\,\mathrm{rad} & 10\,\mathrm{s} \le t \end{cases} \tag{3.314}$$

Figure 3.62 illustrates the applied torque history. Figure 3.63 depicts the forward velocity components of the vehicle v_x and v_y, measured in body coordinate frame B. Figure 3.64 illustrates the angular velocities of the wheels ω_i. Figure 3.65 illustrates the sideslip angles of the wheels α_i. Figure 3.66 depicts the longitudinal slip ratios

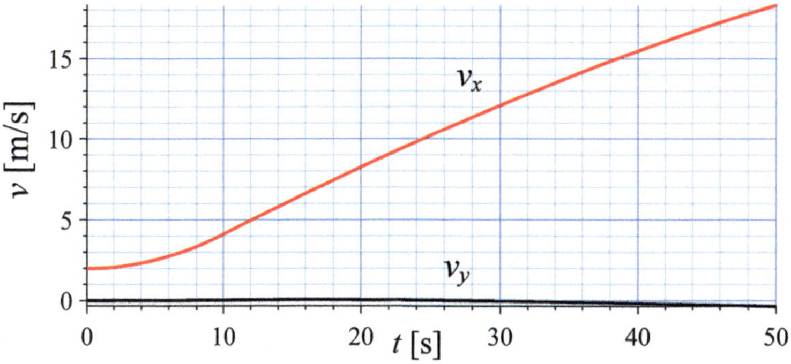

Fig. 3.63 The vehicle velocity v_x and v_y of a four-wheel roll vehicle with active rear torques and steer angle

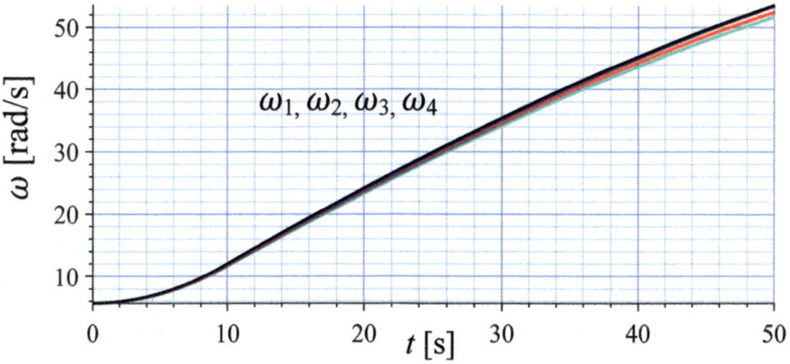

Fig. 3.64 The wheels angular velocity ω_i of a four-wheel roll vehicle with active rear torques and steer angle

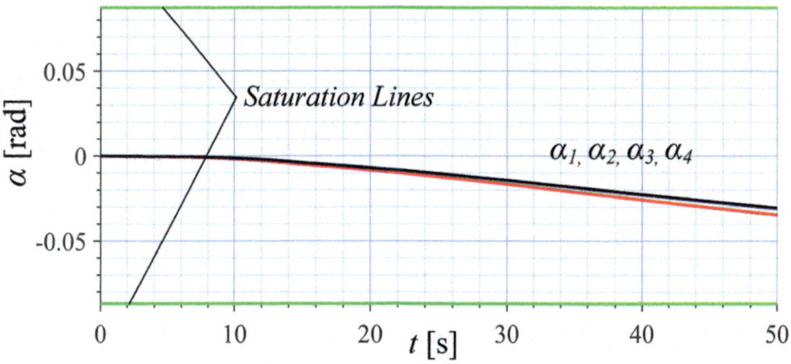

Fig. 3.65 The sideslip angles of the wheels α_i of a four-wheel roll vehicle with active rear torques and steer angle

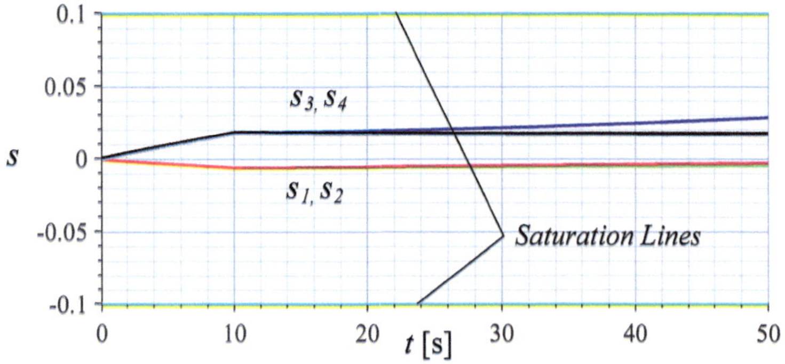

Fig. 3.66 The slip ratios of the wheels s_i of a four-wheel roll vehicle with active rear torques and steer angle

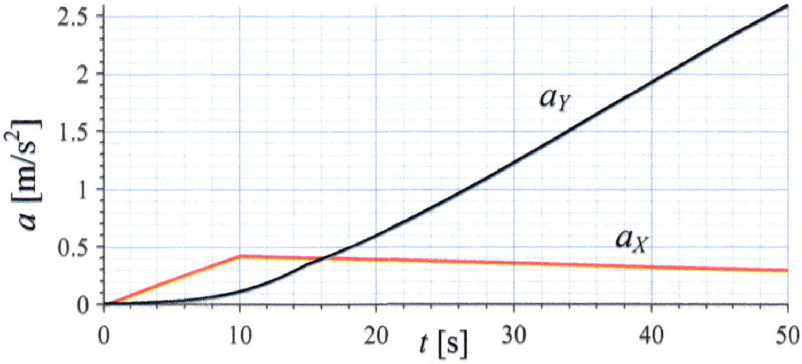

Fig. 3.67 The vehicle acceleration a_X and a_Y of a four-wheel roll vehicle with active rear torques and steer angle

s_i for the tires. The acceleration components a_X and a_Y of the vehicle are plotted in Fig. 3.67. Figure 3.68 depicts the forces F_x and F_y on the vehicle at its mass center. Figure 3.69 illustrates variation of the vertical loads F_{zi}. Figure 3.70 illustrates the longitudinal forces F_{xi} on front and rear tires. Figure 3.71 shows the lateral forces F_{yi} on front and rear tires. The roll angle φ and roll rate p are shown in Figs. 3.72 and 3.73, respectively. Figure 3.74 depicts the yaw rate r of vehicle. Figure 3.75 illustrates the path of the vehicle.

Example 91 Four-wheel roll model, no friction on left side, rear torque.

Let us assume that left side of the vehicle is moving on a no friction pavement while an increasing torque is applied on the rear wheels. The friction coefficients of the tires number 1 and 3 would be zero.

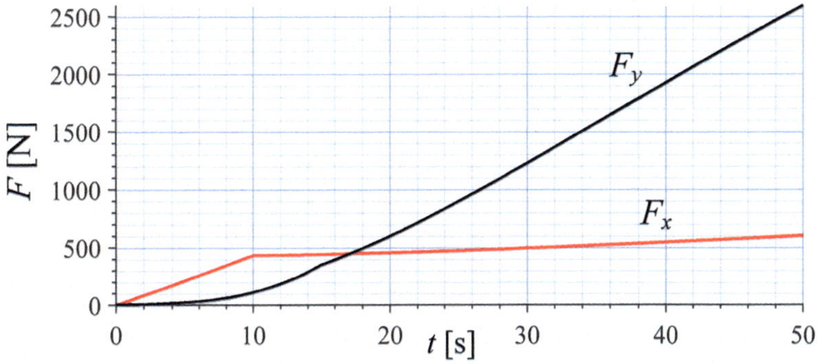

Fig. 3.68 The force components F_x and F_y of a four-wheel roll vehicle with active rear torques and steer angle

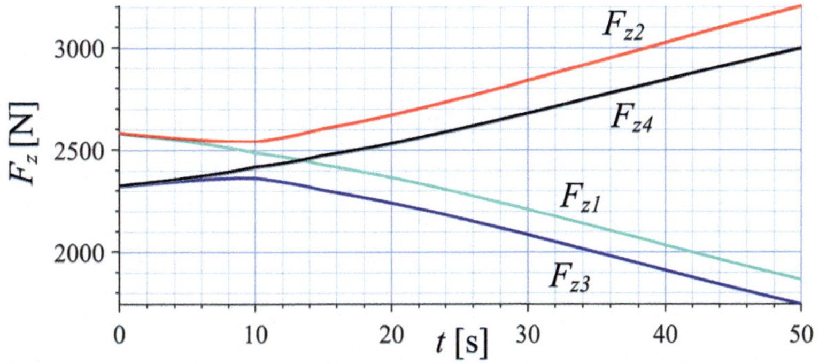

Fig. 3.69 The vertical forces of the wheels F_{zi} of a four-wheel roll vehicle with active rear torques and steer angle

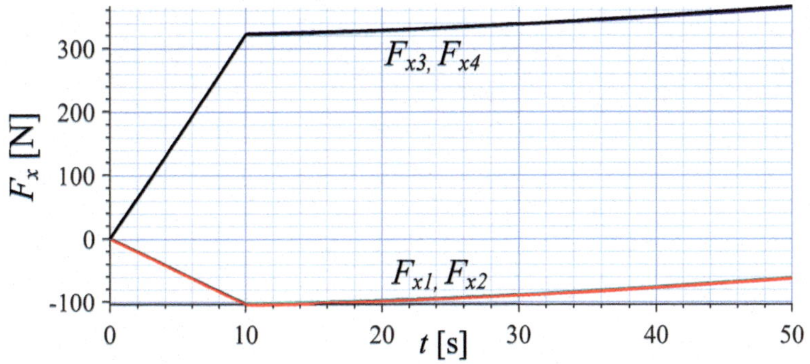

Fig. 3.70 The longitudinal forces of the wheels F_{xi} of a four-wheel roll vehicle with active rear torques and steer angle

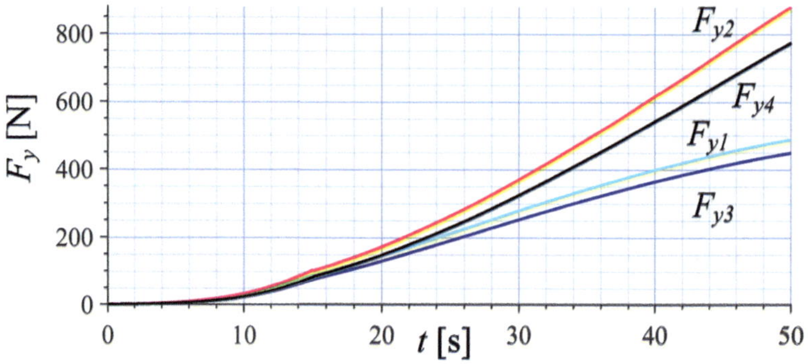

Fig. 3.71 The lateral forces of the wheels F_{yi} of a four-wheel roll vehicle with active rear torques and steer angle

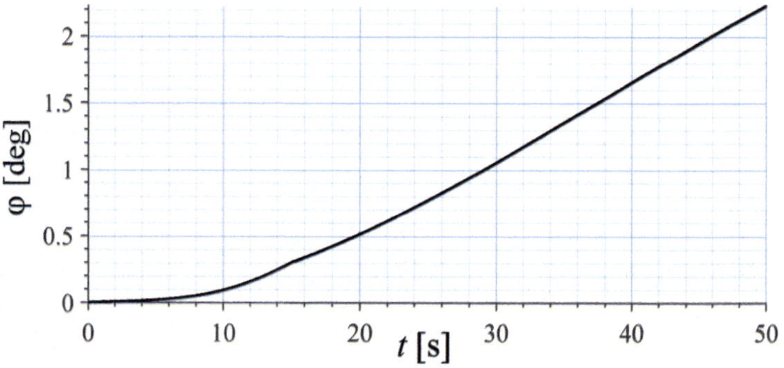

Fig. 3.72 The roll angle φ of a four-wheel roll vehicle with active rear torques and steer angle

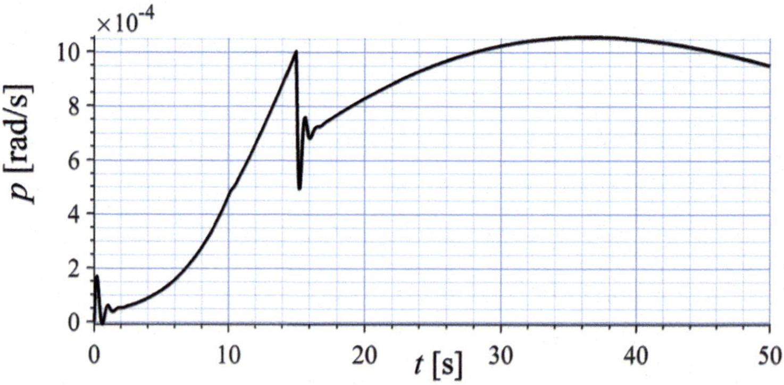

Fig. 3.73 The roll rate p of a four-wheel roll vehicle with active rear torques and steer angle

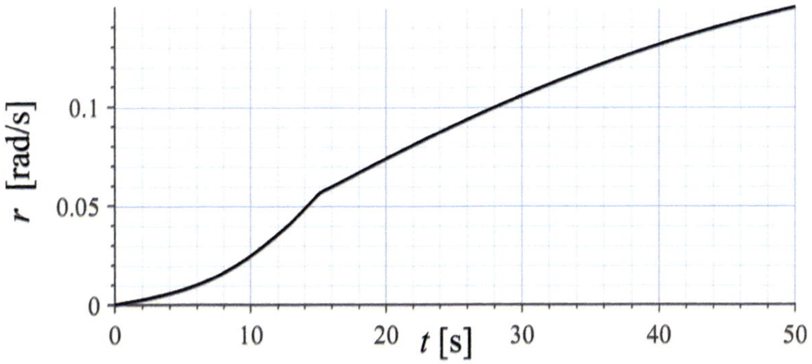

Fig. 3.74 The yaw rate r of a four-wheel roll vehicle with active rear torques and steer angle

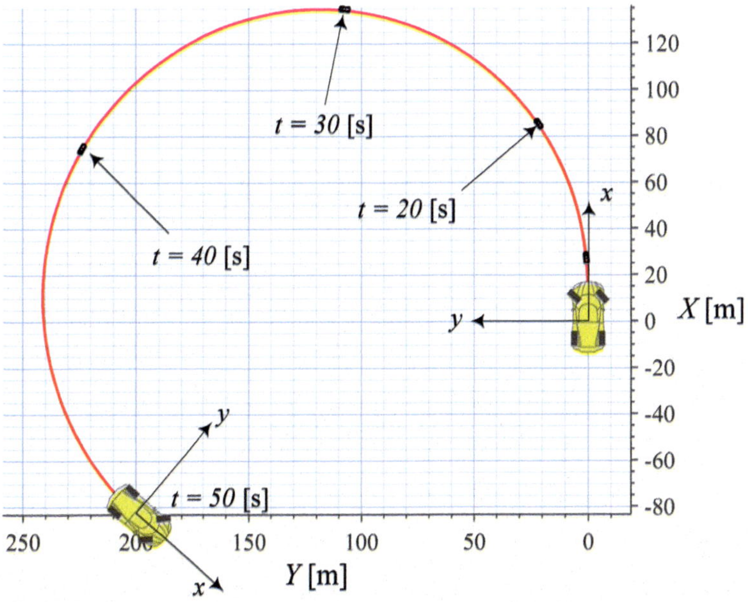

Fig. 3.75 Path of motion of a four-wheel roll vehicle with active rear torques and steer angle

Consider a vehicle with the following data

$$m = 1000\,\text{kg} \qquad I_1 = I_r = I_3 = I_4 = 30\,\text{kg m}^2$$

$$I_x = 300\,\text{kg m}^2 \qquad I_z = 2000\,\text{kg m}^2 \qquad C_A = 0$$

$$C_{\alpha_1} = C_{\alpha_3} = 0 \qquad C_{\alpha_2} = C_{\alpha_4} = 8.5 \qquad \alpha_s = 5\,\text{deg}$$

$$C_{s_1} = C_{s_3} = 0 \qquad C_{s_2} = C_{s_4} = 7.5 \qquad s_s = 0.1$$

$$R_g = 0.35\,\text{m} \qquad C_{\alpha s} = 0.5 \qquad C_{s\alpha} = 0.5$$

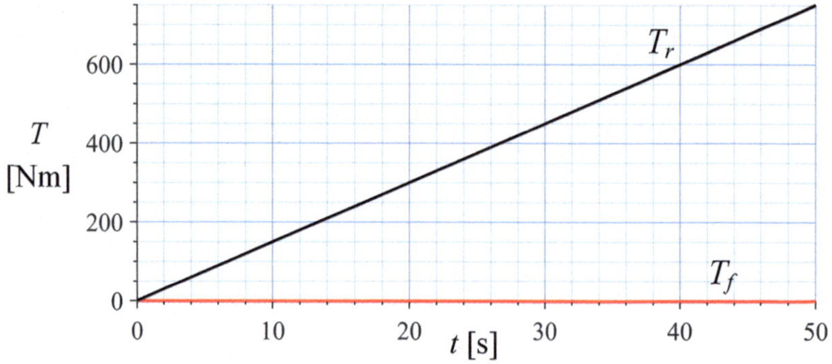

Fig. 3.76 The torques T_i of a four-wheel roll vehicle with rear torques and no friction on left side

$$a_1 = 1.35\,\mathrm{m} \qquad a_2 = 1.5\,\mathrm{m} \qquad h = 0.9\,\mathrm{m} \tag{3.315}$$
$$C_{\varphi_2} = 1600 \qquad C_{\varphi_1} = C_{\varphi_3} = C_{\varphi_4} = 0$$
$$k_\varphi = 26{,}612\,\mathrm{N\,m/rad} \qquad c_\varphi = 1700\,\mathrm{N\,m\,s/rad}$$
$$C_{\beta_1} = C_{\beta_2} = -0.4 \qquad C_{\beta_3} = C_{\beta_4} = -0.1$$
$$C_{\delta\varphi_i} = 0.01 \qquad C_{T_i} = 0.4 \qquad w_f = w_r = 1.8\,\mathrm{m}$$

and assume the vehicle is moving slowly straight. The vehicle is moving slowly straight with

$$v_x = 2\,\mathrm{m/s} \qquad \omega_i = \frac{v_x}{R_g} = \frac{2}{0.35} = 5.714\,\mathrm{rad/s} \tag{3.316}$$
$$\delta = 0 \tag{3.317}$$

At time $t = 0$, we apply an increasing torque on the rear wheels as expressed below.

$$T_1 = T_2 = 0 \qquad T_4 = T_3 = 15t\,\mathrm{N\,m} \tag{3.318}$$

Figure 3.76 illustrates the applied torques. Figure 3.77 depicts the forward velocity components of the vehicle v_x and v_y, measured in body coordinate frame B. Figure 3.78 illustrates the angular velocities of the wheels ω_i. Figure 3.79 illustrates the sideslip angles of the wheels α_i. Figure 3.80 depicts the longitudinal slip ratios s_i for the tires. The acceleration components a_X and a_Y of the vehicle are plotted in Fig. 3.81. Figure 3.82 depicts the forces F_x and F_y on the vehicle at its mass center. Figure 3.83 illustrates variation of the vertical loads F_{zi}. Figure 3.84 illustrates the longitudinal forces F_{xi} on front and rear tires. Figure 3.85 shows the lateral forces F_{yi} on front and rear tires. The roll angle φ and roll rate p are shown in Figs. 3.86 and 3.87, respectively. Figure 3.88 depicts the yaw rate r of vehicle. The angular accelerations $\dot{\omega}_i$ of the wheels are shown in Fig. 3.89. Figure 3.90 illustrates the path of motion of the vehicle.

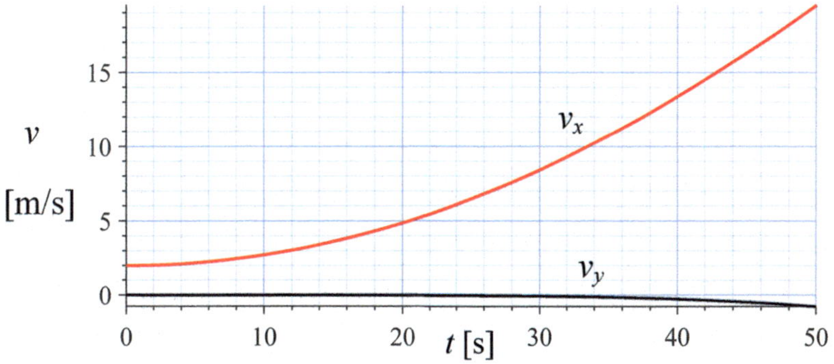

Fig. 3.77 The vehicle velocity v_x and v_y of a four-wheel roll vehicle with rear torques and no friction on left side

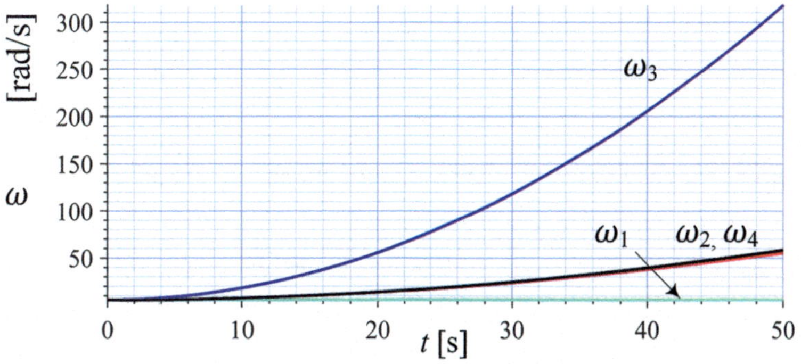

Fig. 3.78 The angular velocity ω_i of a four-wheel roll vehicle with rear torques and no friction on left side

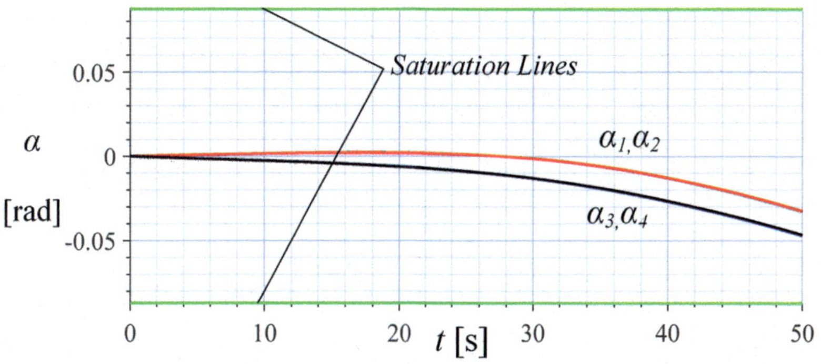

Fig. 3.79 The sideslip angles α_i of a four-wheel roll vehicle with rear torques and no friction on left side

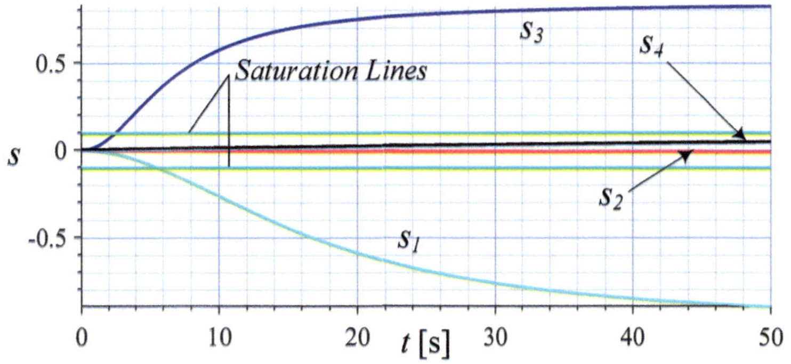

Fig. 3.80 The slip ratios s_i of a four-wheel roll vehicle with rear torques and no friction on left side

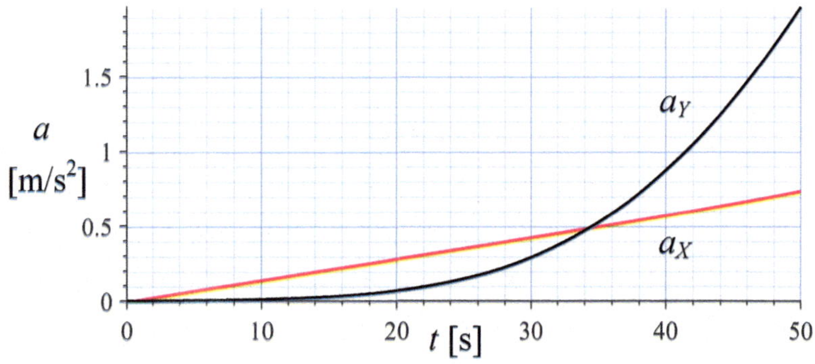

Fig. 3.81 The vehicle accelerations a_X and a_Y of a four-wheel roll vehicle with rear torques and no friction on left side

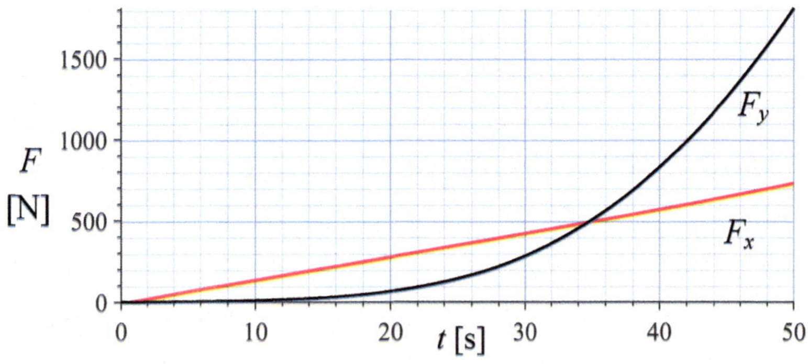

Fig. 3.82 The vehicle forces F_x and F_y of a four-wheel roll vehicle with rear torques and no friction on left side

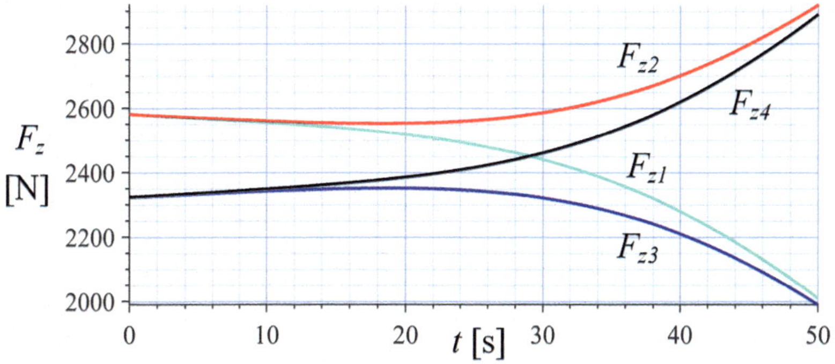

Fig. 3.83 The vertical forces of the wheels F_{zi} of a four-wheel roll vehicle with rear torques and no friction on left side

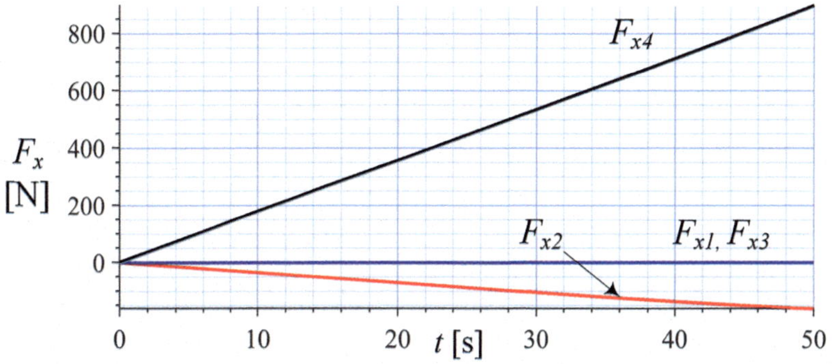

Fig. 3.84 The longitudinal forces of the wheels F_{xi} of a four-wheel roll vehicle with rear torques and no friction on left side

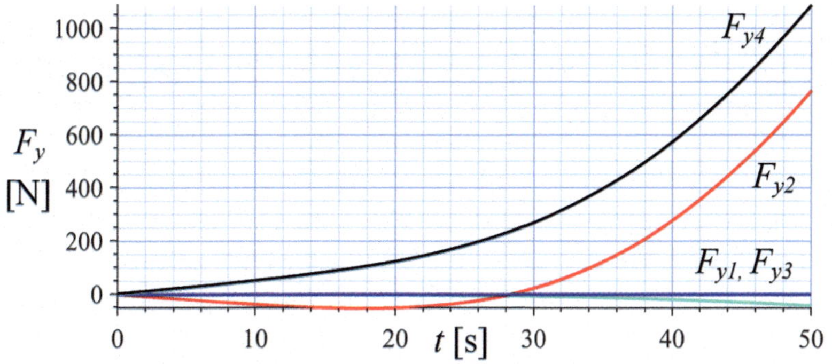

Fig. 3.85 The lateral forces of the wheels F_{yi} of a four-wheel roll vehicle with rear torques and no friction on left side

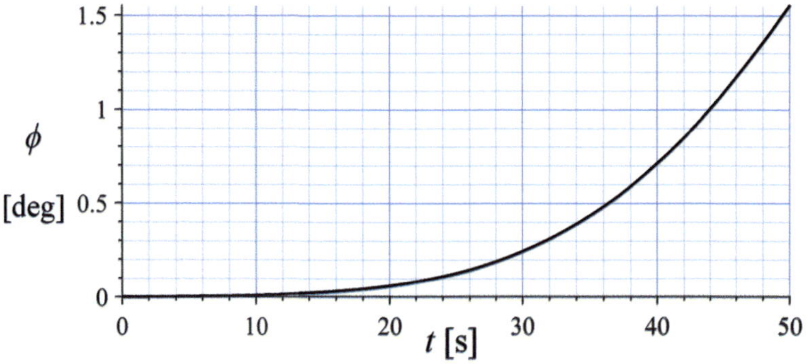

Fig. 3.86 The roll angle φ of a four-wheel roll vehicle with rear torques and no friction on left side

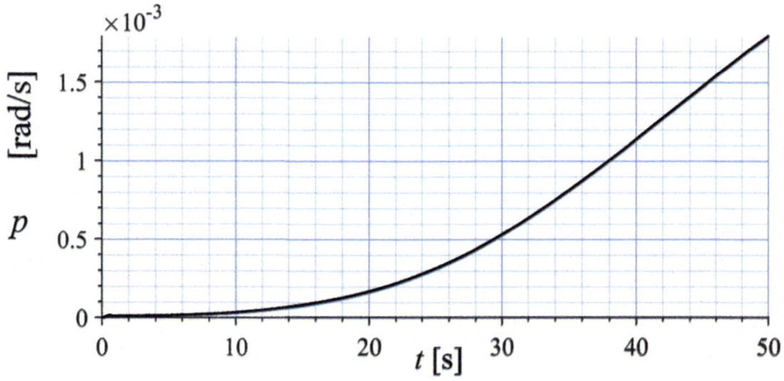

Fig. 3.87 The roll rate p of a four-wheel roll vehicle with rear torques and no friction on left side

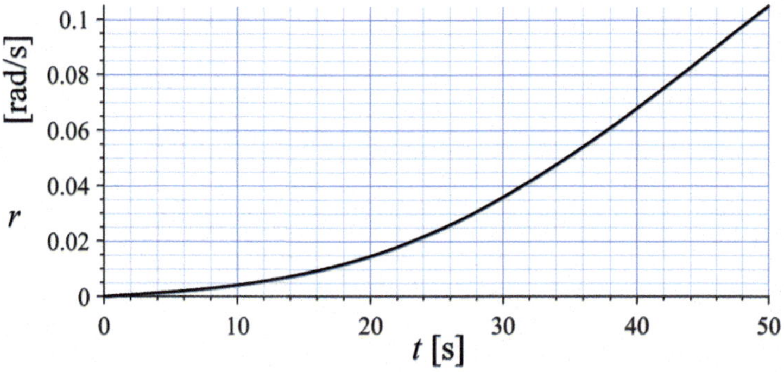

Fig. 3.88 The yaw rate r of a four-wheel roll vehicle with rear torques and no friction on left side

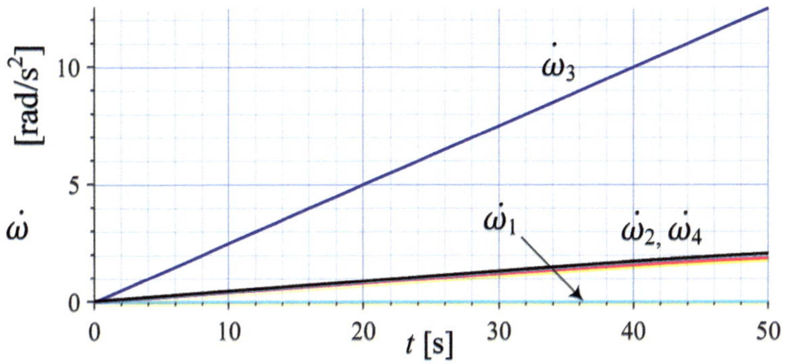

Fig. 3.89 The angular acceleration $\dot{\omega}_i$ of a four-wheel roll vehicle with rear torques and no friction on left side

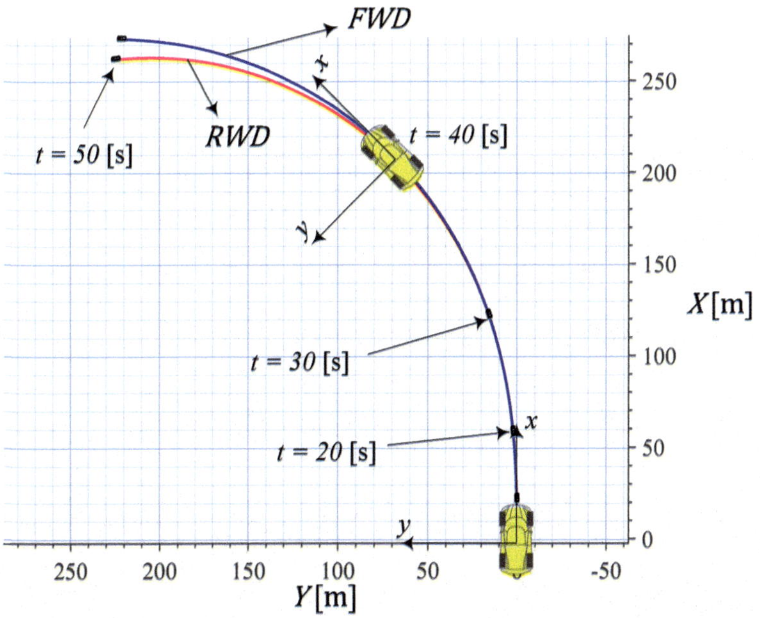

Fig. 3.90 Path of motion of a four-wheel roll vehicle with rear torques and no friction on left side

Example 92 Only one-wheel torque.

We examine a vehicle that due to some reasons, only one of the wheel actuators is working. Consider a vehicle with the following data

$$m = 1000\,\text{kg} \qquad I_1 = I_r = I_3 = I_4 = 30\,\text{kg}\,\text{m}^2$$

$$I_x = 300\,\text{kg}\,\text{m}^2 \qquad I_z = 2000\,\text{kg}\,\text{m}^2 \qquad C_A = 0.8$$

$$C_{\alpha_i} = 8.5 \qquad \alpha_s = 5\,\text{deg} \qquad C_{s_i} = 7.5 \qquad s_s = 0.1$$

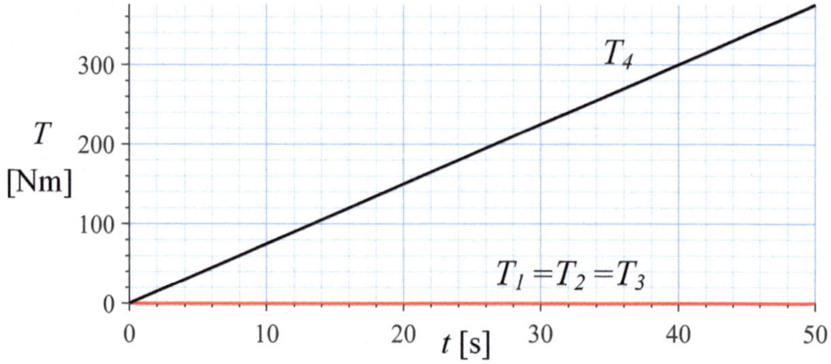

Fig. 3.91 The torques T_i of a four-wheel roll vehicle with only torque T_4

$$R_g = 0.35\,\text{m} \qquad C_{\alpha s} = 0.5 \qquad C_{s\alpha} = 0.5$$
$$a_1 = 1.35\,\text{m} \qquad a_2 = 1.5\,\text{m} \qquad h = 0.9\,\text{m} \tag{3.319}$$

$$C_{\varphi_1} = C_{\varphi_2} = 1600 \qquad C_{\varphi_3} = C_{\varphi_4} = 0$$
$$k_\varphi = 26{,}612\,\text{N m/rad} \qquad c_\varphi = 1700\,\text{N m s/rad}$$
$$C_{\beta_1} = C_{\beta_2} = -0.4 \qquad C_{\beta_3} = C_{\beta_4} = -0.1$$
$$C_{T_i} = 0.4 \qquad C_{\delta\varphi_i} = 0.01 \qquad w_f = w_r = 1.8\,\text{m} \tag{3.320}$$

and assume the vehicle is moving slowly straight.

$$v_x = 2\,\text{m/s} \qquad \omega_f = \omega_r = \frac{v_x}{R_g} = \frac{2}{0.35} = 5.714\,\text{rad/s} \tag{3.321}$$

$$\delta = 0 \tag{3.322}$$

At time $t = 0$, we apply an increasing torque on the wheel number 4.

$$T_1 = T_2 = T_3 = 0 \qquad T_4 = 7.5t\,\text{N m} \tag{3.323}$$

Figure 3.91 illustrates the applied torques. Figure 3.92 depicts the forward velocity components of the vehicle v_x and v_y, measured in body coordinate frame B. Figure 3.93 illustrates the angular velocities of the wheels ω_i. Figure 3.94 illustrates the sideslip angles of the wheels α_i. Figure 3.95 depicts the longitudinal slip ratios s_i for the tires. The acceleration components a_X and a_Y of the vehicle are plotted in Fig. 3.96. Figure 3.97 depicts the forces F_x and F_y on the vehicle at its mass center. Figure 3.98 illustrates variation of the vertical loads F_{zi}. Figure 3.99 illustrates the longitudinal forces F_{xi} on front and rear tires. Figure 3.100 shows the lateral forces F_{yi} on front and rear tires. The roll angle φ and roll rate p are shown in Figs. 3.101 and 3.102, respectively. Figure 3.103 depicts the yaw rate r of vehicle. The angular accelerations $\dot{\omega}_i$ of the wheels are shown in Fig. 3.104. Figure 3.105 illustrates the path of motion of the vehicle.

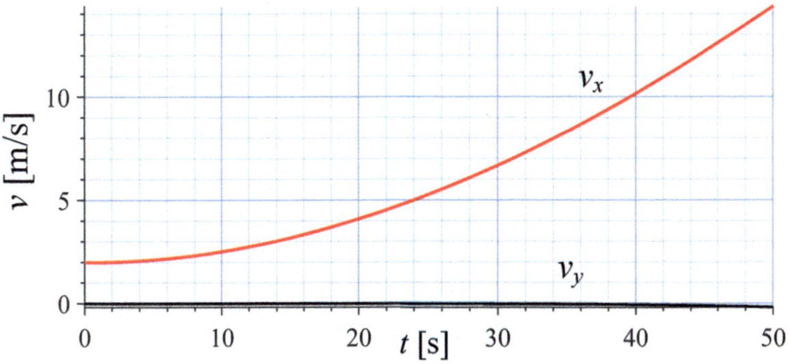

Fig. 3.92 The vehicle velocity v_x and v_y of a four-wheel roll vehicle with only torque T_4

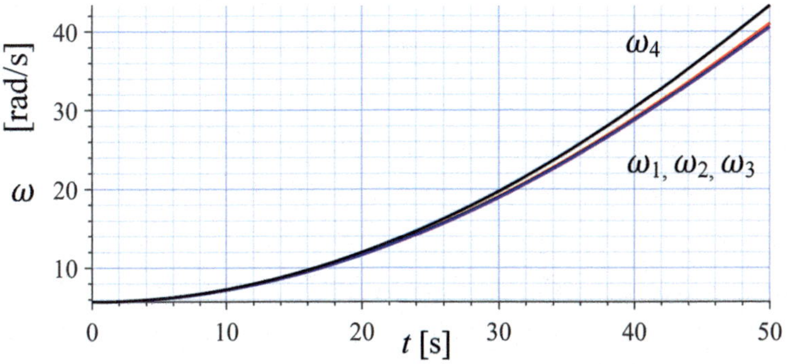

Fig. 3.93 The angular velocity ω_i of a four-wheel roll vehicle with only torque T_4

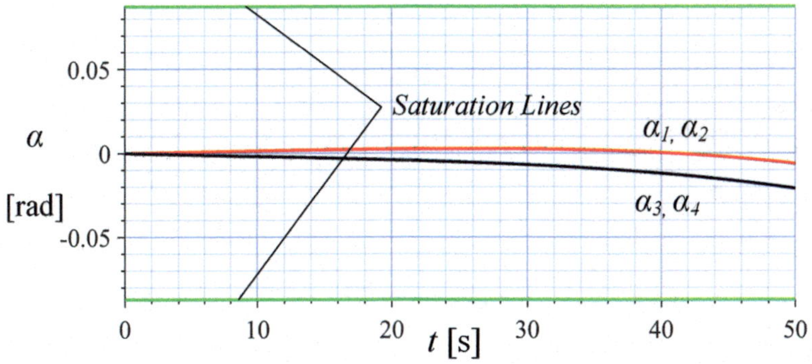

Fig. 3.94 The sideslip angles α_i of a four-wheel roll vehicle with only torque T_4

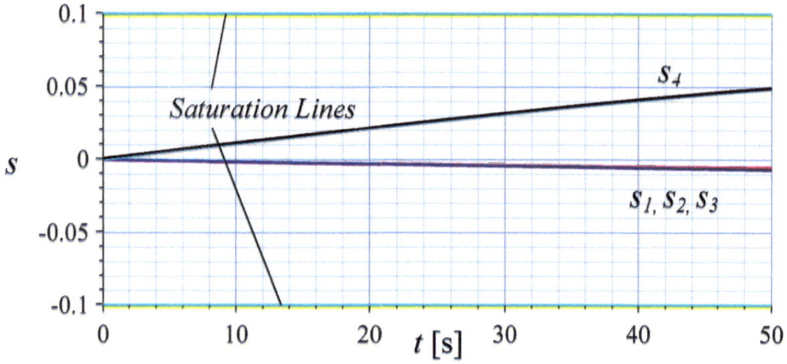

Fig. 3.95 The slip ratios s_i of a four-wheel roll vehicle with only torque T_4

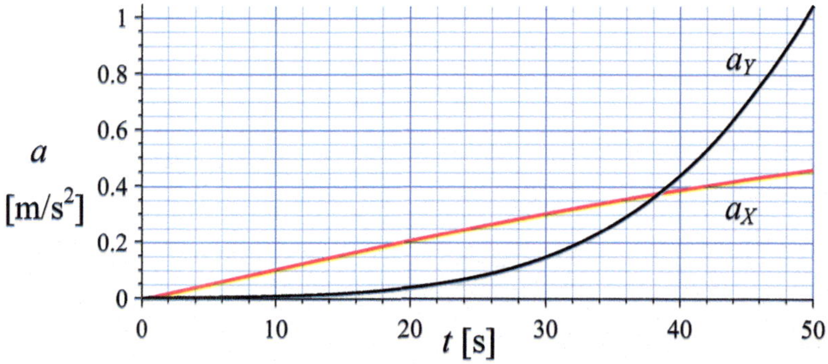

Fig. 3.96 The vehicle acceleration a_X and a_Y of a four-wheel roll vehicle with only torque T_4

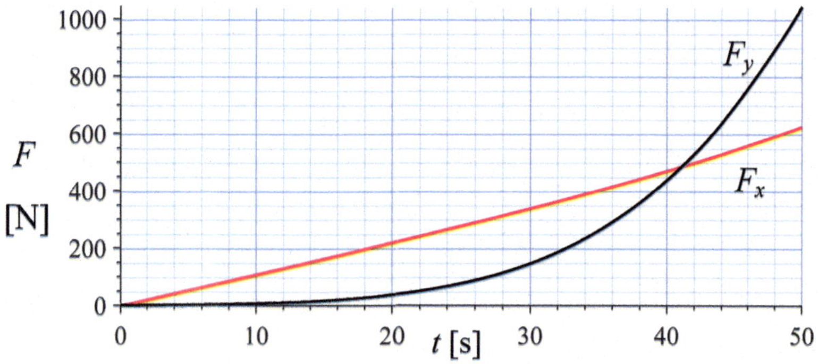

Fig. 3.97 The angular velocity ω_i of a four-wheel roll vehicle with only torque T_4

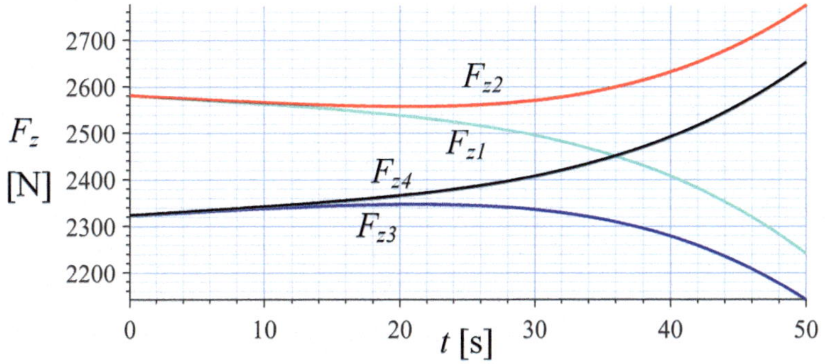

Fig. 3.98 The vertical forces of the wheels F_{zi} of a four-wheel roll vehicle with only torque T_4

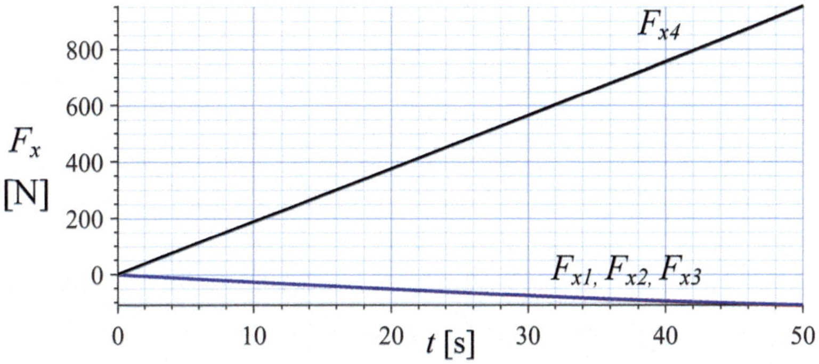

Fig. 3.99 The longitudinal forces of the wheels F_{xi} of a four-wheel roll vehicle with only torque T_4

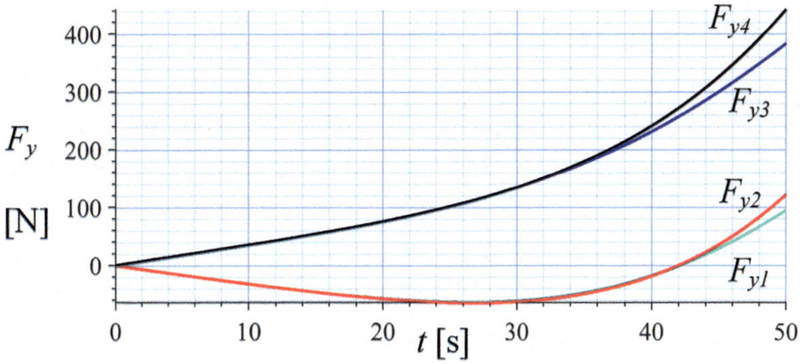

Fig. 3.100 The lateral forces of the wheels F_{yi} of a four-wheel roll vehicle with only torque T_4

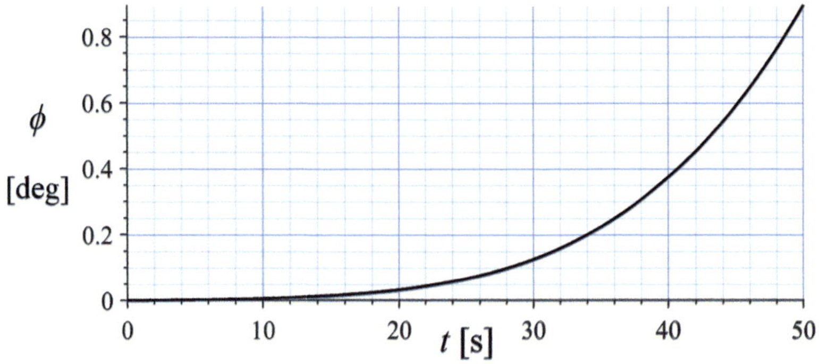

Fig. 3.101 The roll angle φ of a four-wheel roll vehicle with only torque T_4

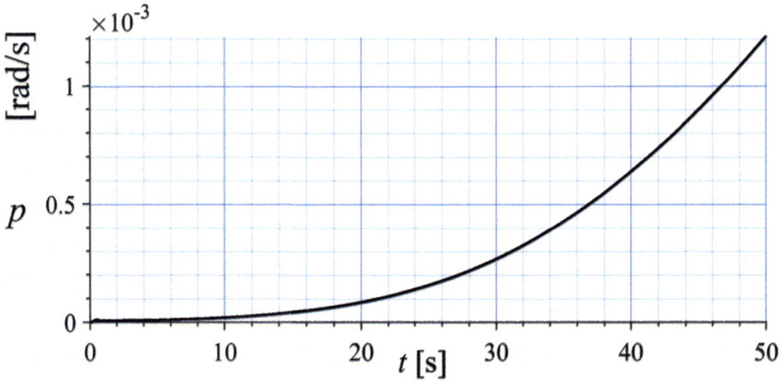

Fig. 3.102 The roll rate p of a four-wheel roll vehicle with only torque T_4

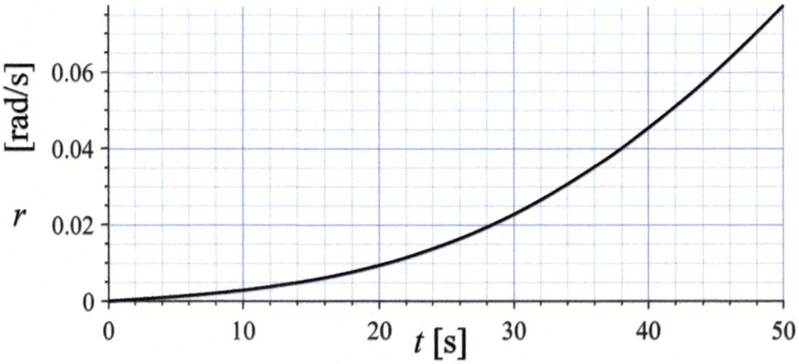

Fig. 3.103 The yaw rate r of a four-wheel roll vehicle with only torque T_4

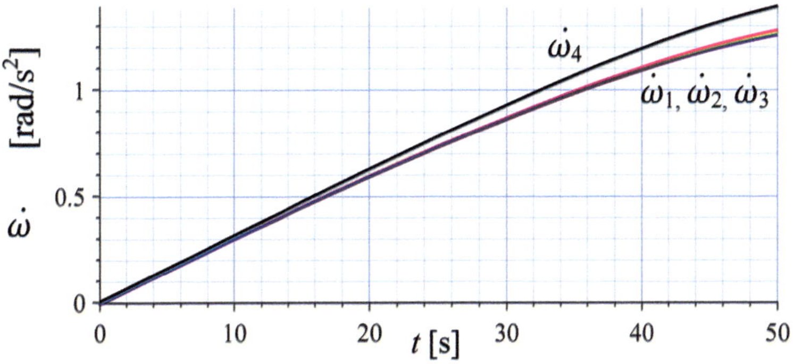

Fig. 3.104 The angular acceleration $\dot{\omega}_i$ of a four-wheel roll vehicle with only torque T_4

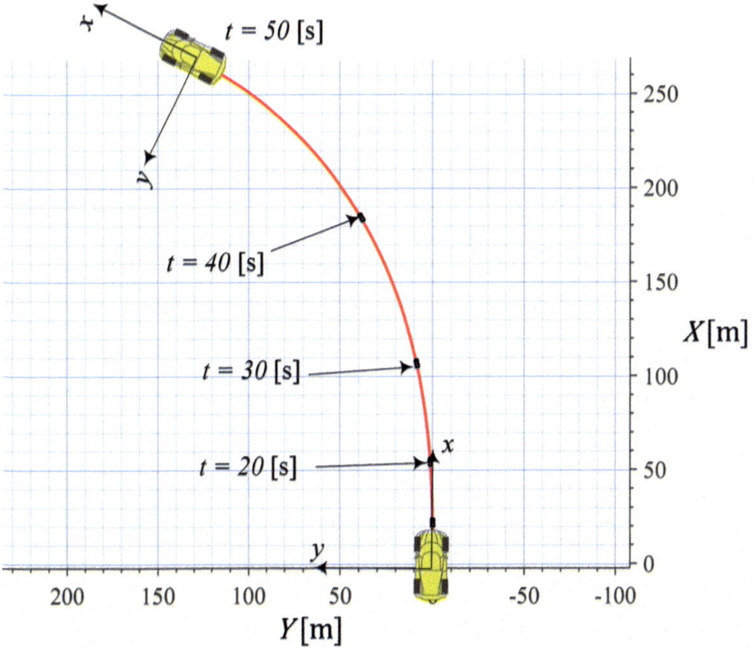

Fig. 3.105 Path of motion of a four-wheel roll vehicle with only torque T_4

3.6 Chapter Summary

The roll vehicle dynamic model is well expressed by four kinematic variables: the forward motion x, the lateral motion y, the roll angle φ, and the yaw angle ψ, plus four equations for the dynamics of each wheel. In the roll model of vehicle dynamics, we do not consider vertical movement z and pitch motion θ. The model of a rollable rigid vehicle is more exact and more realistic compared to the vehicle planar model. Using roll dynamic model, we are able to analyze the roll behavior of a vehicle in maneuvers. Angular orientation of the vehicle is expressed by three angles: *roll* φ, *pitch* θ, and *yaw* ψ, and the vehicle angular velocities are expressed by their rates: *roll rate* p, *pitch rate* q, and *yaw rate* r.

A rolled vehicle introduces new reactions in the tires of the vehicle that must be considered in development of the dynamic equations of motion. The most important reactions are:

1. Roll-thrust $F_{y\varphi}$.
   An extra lateral force appears because of the vehicle roll. Tire roll-thrust is assumed to be proportional to the vehicle roll angle φ.
2. Roll-steer angle δ_φ.
   An extra steer angle appears because of the wheel roll. The roll steer is a result of suspension mechanisms that provide some steer angle when the vehicle rolls and the mechanism deflects. The wheel roll steering is assumed to be proportional to the vehicle roll angle φ. Therefore, the actual steer angle δ_a of such a tire will be $\delta_a = \delta + \delta_\varphi$.

In this chapter we introduce bicycle as well as four-wheel roll models with independent in-wheel motors. The four-wheel roll vehicle model is the best practical vehicle mathematical model. This model provides us with in-wheel torques T_i, tire slips α_i, s_i, β_i, tire and vehicle forces F_x, F_y, F_{x_i}, F_{y_i}, F_{z_i}, velocity components of the vehicle v_x, v_y, ω_i, as well as yaw and roll angular variables φ, ψ, p, r. This model is an extension to the two-wheel roll vehicle model to include the lateral weight transfer as well as roll effects on vehicle dynamics. The four-wheel roll vehicle model is an excellent model to simulate drifting of vehicles.

3.7 Key Symbols

$a \equiv \ddot{x}$	Acceleration
a_i	Longitudinal distance of the axle number i from lateral y-axis
A_f	Frontal area of vehicle
b_i	Lateral distance of tire number i from longitudinal x-axis
$B(Cxyz)$	Vehicle coordinate frame
c_φ	Roll damping of vehicle
C	Mass center
C_A	Aerodynamic coefficient
C_α	Tire sideslip coefficient
$C_{\alpha f}$	Front sideslip coefficient
$C_{\alpha i}$	Sideslip coefficient of tire number i
$C_{\alpha s}$	Tire lateral force drop factor
$C_{\alpha r}$	Rear sideslip coefficient
C_γ	Camber coefficient, camber stiffness
C_s	Tire slip ratio coefficient

C_{sf}	Front slip ratio coefficient
C_{sr}	Rear slip ratio coefficient
$C_{s\alpha}$	Tire longitudinal force drop factor
C_D	Drag coefficient
C_{sf}	Front slip ratio coefficient
C_{si}	Slip ratio coefficient of tire number i
C_{sr}	Rear slip ratio coefficient
C_T	Torque coefficient
C_{T_f}	Torque coefficient of front wheel
C_{T_r}	Torque coefficient of rear wheel
C_{T_i}	Torque coefficient of wheel number i
C_β	Coefficient between F_y and β at steady-state
C_{β_i}	Wheel slip coefficient of number i
C_δ	Coefficient between F_y and δ at steady-state
C_κ	Coefficient between F_y and κ at steady-state
$C_{\delta\varphi}$	Roll-steering coefficient
$C_{\delta\varphi_i}$	Roll-steering coefficient of wheel number i
C_φ	Roll-thrust coefficient
$\mathbf{d}$	Location vector
D_β	Coefficient between M_z and β at steady-state
D_δ	Coefficient between M_z and δ at steady-state
D_κ	Coefficient between M_z and κ at steady-state
F_A	Aerodynamic force
F_x	Longitudinal force, forward force, traction force
F_{xi}	Longitudinal force of tire number i
F_y	Lateral force of vehicle
F_{yi}	Lateral force of tire number i
F_{yf}	Front lateral force
F_{yr}	Rear lateral force
$F_{y\varphi}$	Roll-thrust
F_{zi}	Normal force, vertical force of tire number i
$\mathbf{F}_i$	Force vector of tire number i
$\mathbf{F}, \mathbf{M}$	Force system
$\mathbf{g}, g$	Gravitational acceleration
$G(OXYZ)$	Global coordinate frame
h	Height of mass center from the ground
H	Heaviside function
I	Mass moment
I_i, I_{wi}	Wheel number i mass moment
k_φ	Roll stiffness of vehicle

K	Stability factor
l	Wheel base
$\mathbf{L}$	Moment of momentum
m	Mass
M_x	Roll moment, bank moment, tilting torque
M_y	Pitch moment
M_z	Yaw moment, aligning moment
o, O	Origin of a coordinate frame
$p = \dot{\varphi}$	Roll rate
$\mathbf{p}$	Momentum
$q = \dot{\theta}$	Pitch rate
q_i	Generalized coordinate
$r = \dot{\psi}$	Yaw rate
$\mathbf{r}$	Position vector
R, R_g	Tire radius
${}^G R_B$	Rotation matrix to go from B frame to G frame
s	Longitudinal slip
s_s	Saturation value of longitudinal slip
S	Saturation function
$S_\kappa = \kappa/\delta$	Curvature response
$S_\beta = \beta/\delta$	Sideslip response
t	Time
T	Tire coordinate frame
T_i	Wheel torque
$v \equiv \dot{x},\ \mathbf{v}$	Velocity
$\mathbf{v}_{wind}$	Wind velocity
v_{x_i}	Longitudinal velocity of wheel number i
v_{y_i}	Lateral velocity of wheel number i
w	Wheelbase
w_f	Front wheelbase
w_r	Rear wheelbase
$x, y, z,\ \mathbf{x}$	Displacement
X, Y, Z	Global displacement
α	Tire sideslip angle between $\mathbf{v}_w$ and x_w-axis
α_s	Sideslip angle saturation
$\beta = v_y/v_x$	Vehicle slip angle between $\mathbf{v}$ and x-axis
β	Attitude angle
β_f	Sideslip angle of front wheel
β_i	Sideslip angle of wheel number i
β_r	Sideslip angle of rear wheel
$\beta + \psi$	Cruise angle
δ	Steer angle

δ_a	Actual steer angle
δ_f	Front steer angle
δ_i	Front steer angle of wheel number i
δ_r	Rear steer angle
δ_φ	Roll-steer angle
θ	Pitch angle
$\dot{\theta} = q$	Pitch rate
$\kappa = 1/\rho$	Curvature
λ	Eigenvalue
ρ	Radius of rotation, air density
φ	Roll angle
$\dot{\varphi} = p$	Roll rate
ψ	Yaw angle
$\dot{\psi} = r$	Yaw rate
ψ	Heading angle
ω, $\boldsymbol{\omega}$	Angular velocity
ω_i	Angular velocity of wheel number i
$\dot{\omega}$, $\dot{\boldsymbol{\omega}}$	Angular acceleration
$\dot{\omega}_i$	Angular acceleration of wheel number i

Exercises

1. Global equations of motion
   The equation of motion of a vehicle, expressed in the global coordinate frame, is called the G-expression.

$$F_X = m \frac{d}{dt} \dot{X} = m \dot{v}_X \tag{3.324}$$

$$F_Y = m \frac{d}{dt} \dot{Y} = m \dot{v}_Y \tag{3.325}$$

$$M_Z = I_Z \frac{d}{dt} \dot{\psi} = I_{Zz} \dot{\omega}_Z \tag{3.326}$$

$$M_X = I_X \frac{d}{dt} \dot{\varphi} = I_X \dot{\omega}_X \tag{3.327}$$

$$\dot{\omega}_x = \dot{p} \hat{I} \tag{3.328}$$

$$\dot{\omega}_z = \dot{r} \hat{K} \tag{3.329}$$

   These are not practical because the force systems **F**, **M** are dependent on the orientation of the vehicle. Transform these equations into the B-frame and derive the vehicle roll model equations of motion.

2. Lane-change maneuver.
   Passing and lane-change maneuvers are two standard tests to examine a vehicle's dynamic responses. Lane-change may be expressed by a half-sine or a sine-squared function for steering input. Two examples of such functions are

$$\delta(t) = \begin{cases} \delta_0 \sin^2 \omega t & t_1 < t < \dfrac{\pi}{\omega} \\ 0 & \dfrac{\pi}{\omega} < t < t_1 \end{cases} \text{ rad} \tag{3.330}$$

$$\delta(t) = \begin{cases} \delta_0 \sin \omega t & t_1 < t < \dfrac{\pi}{\omega} \\ 0 & \dfrac{\pi}{\omega} < t < t_1 \end{cases} \text{ rad} \tag{3.331}$$

$$\omega = \frac{\pi L}{v_x} \tag{3.332}$$

   where L is the moving length during the lane-change and v_x is the forward speed of the vehicle.
   Examine a vehicle with the characteristics given below

$$\begin{aligned} m &= 1000\,\text{kg} \qquad I_1 = I_r = I_3 = I_4 = 30\,\text{kg m}^2 \\ I_x &= 300\,\text{kg m}^2 \qquad I_z = 2000\,\text{kg m}^2 \qquad C_A = 0.8 \\ C_{\alpha_1} &= C_{\alpha_2} = C_{\alpha_3} = C_{\alpha_4} = 8.5 \qquad \alpha_s = 5\,\text{deg} \\ C_{s_1} &= C_{s_2} = C_{s_3} = C_{s_4} = 7.5 \qquad s_s = 0.1 \\ R_g &= 0.35\,\text{m} \qquad C_{\alpha s} = 0.5 \qquad C_{s\alpha} = 0.5 \\ a_1 &= 1.35\,\text{m} \qquad a_2 = 1.5\,\text{m} \qquad h = 0.9\,\text{m} \\ C_{\varphi_1} &= C_{\varphi_2} = 1600 \qquad C_{\varphi_3} = C_{\varphi_4} = 0 \\ k_\varphi &= 26{,}612\,\text{N m/rad} \qquad c_\varphi = 1700\,\text{N m s/rad} \\ C_{\beta_1} &= C_{\beta_2} = -0.4 \qquad C_{\beta_3} = C_{\beta_4} = -0.1 \\ C_{\delta\varphi_1} &= C_{\delta\varphi_2} = C_{\delta\varphi_3} = C_{\delta\varphi_4} = 0.01 \qquad w_f = w_r = 1.8\,\text{m} \end{aligned} \tag{3.333}$$

$$v_x = 2\,\text{m/s} \qquad \omega_i = \frac{v_x}{R_g} = \frac{2}{0.35} = 5.714\,\text{rad/s} \tag{3.334}$$

   and a change in half-sine steering input $\delta(t)$.

$$\delta(t) = \begin{cases} 0.2 \sin \dfrac{\pi L}{v_x} t & 0 < t < \dfrac{v_x}{L} \\ 0 & \dfrac{v_x}{L} < t < 0 \end{cases} \text{ rad} \tag{3.335}$$

$$L = 100\,\text{m} \qquad v_x = 40\,\text{m/s} \tag{3.336}$$

Solve the equations of motion numerically and plot kinematic variables, v_x, v_y, α_i, s_i, a_x, a_y, ω_i, $\dot{\omega}_i$, φ, ψ, p, r, and forces F_x, F_y, F_{zi}, F_{xi}, F_{yi}, and path of motion of the vehicle.

3. ★Front locked wheel.
   Consider a two-wheel roll model vehicle with the data given in (3.214). Assume the vehicle is moving slowly straight with

$$v_x = 2\,\mathrm{m/s} \qquad \omega_r = \frac{v_x}{R_g} = \frac{2}{0.35} = 5.714\,\mathrm{rad/s} \tag{3.337}$$

   At time $t = 0$, we apply an increasing torque on the rear wheel

$$T_f = 0 \qquad T_r = 100t\,\mathrm{N\,m} \tag{3.338}$$

   while the steer angle is kept at zero $\delta = 0$ and the front wheel is locked at $\omega_f = 0$.
   Solve the equations of motion numerically and plot $T_f, T_r, s_f, s_r, \alpha_f, \alpha_r, F_f, F_r, F_z, v_x, a_x, \omega_r, \dot{\omega}_r$, for $0 < t < 50$ s.
4. ★Rear locked wheel.
   Consider a two-wheel roll model vehicle with the data given in (3.214). Assume the vehicle is moving slowly straight with

$$v_x = 2\,\mathrm{m/s} \qquad \omega_r = \frac{v_x}{R_g} = \frac{2}{0.35} = 5.714\,\mathrm{rad/s} \tag{3.339}$$

   At time $t = 0$, we apply an increasing torque on the rear wheel

$$T_r = 0 \qquad T_f = 100t\,\mathrm{N\,m} \tag{3.340}$$

   while the steer angle is kept at zero $\delta = 0$ and the rear wheel is locked at $\omega_r = 0$.
   Solve the equations of motion numerically and plot $T_f, T_r, s_f, s_r, \alpha_f, \alpha_r, F_f, F_r, F_z, v_x, a_x, \omega_r, \dot{\omega}_r$, for $0 < t < 50$ s.
5. Increasing steer angle and front torque.
   Repeat Example 86 for,

   (a) $C_{\alpha s} = 0.2, C_{s\alpha} = 0.8$
   (b) $C_{\alpha s} = 0.8, C_{s\alpha} = 0.2$

6. Increasing steer angle and rear torque.
   Repeat Example 87 for,

   (a) $C_\alpha = 0.2, C_s = 0.8$
   (b) $C_\alpha = 0.8, C_s = 0.2$

7. Four-wheel roll vehicle, increasing rear torque, straight driving.
   This exercise simulates changing one tire with a new or different tire. Repeat Example 89 assuming $C_{\alpha_1} = 10$, $C_{\alpha_2} = 8.5$, $C_{\alpha_3} = 8.5$, $C_{\alpha_4} = 8.5$.
8. Four-wheel roll vehicle, a different tire.
   Repeat Example 90 assuming $C_{\alpha_1} = 10$, $C_{\alpha_2} = 8.5$, $C_{\alpha_3} = 8.5$, $C_{\alpha_4} = 8.5$.
9. No friction on one tire and applying front torque.
   In Example 89 assume the wheel number 1 has no friction.

$$\begin{array}{cccc} C_{\alpha_1} = 0 & C_{\alpha_2} = 8.5 & C_{\alpha_3} = 8.5 & C_{\alpha_4} = 8.5 \\ C_{s_1} = 0 & C_{s_2} = 7.5 & C_{s_3} = 7.5 & C_{s_4} = 7.5 \end{array} \tag{3.341}$$

   Repeat the example and plot the same graphs to compare.

Chapter 4
Road Dynamics

Passenger cars are developed to move on smooth paved pre-designed roads. To keep vehicles on road, we need a steering mechanism to provide steer angle as an input to the vehicle dynamic system. Ideally, all wheels of a vehicle should be able to steer independently such that the vehicle follows the desired path at the given speed. In this chapter we review steer and road dynamics.

4.1 Road Design

Roads are made by continuously connecting straight and circular paths by proper transition turning sections. Having a continuous and well-behaved curvature is a necessary criterion in road design. The *clothoid spiral* is the best smooth transition connecting curve in road design which is expressed by parametric equations called *Fresnel Integrals*:

$$X(t) = a \int_0^t \cos\left(\frac{\pi}{2}u^2\right) du \tag{4.1}$$

$$Y(t) = a \int_0^t \sin\left(\frac{\pi}{2}u^2\right) du \tag{4.2}$$

The curvature of the clothoid curve varies linearly with arc length and this linearity makes clothoid the smoothest driving transition curve.

Figure 4.1 illustrates the clothoid curve for the *scaling parameter* $a = 1$ and *variable* $-\pi \le t \le \pi$. The scaling parameter a is a magnification factor that shrinks or magnifies the curve. The range of t determines the variation of curvature within the clothoid, as well as the initial and final tangent angles of the clothoid curve.

The *arc length*, s, of a clothoid for a given value of t is

R. N. Jazar, *Advanced Vehicle Dynamics*,
https://doi.org/10.1007/978-3-030-13062-6_4

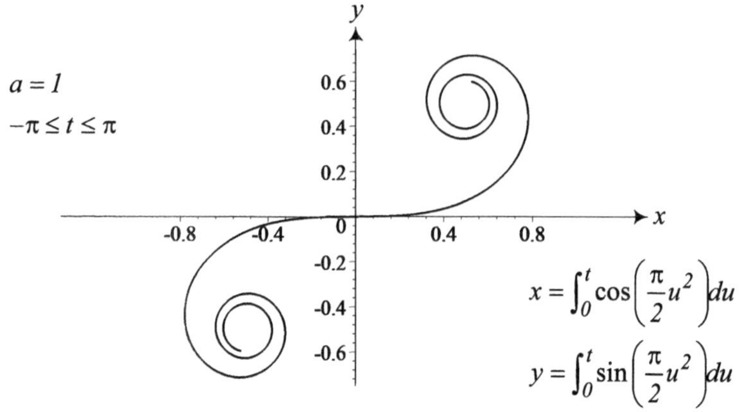

Fig. 4.1 The clothoid curve for $a = 1$ and $-\pi \le t \le \pi$

$$s = at \tag{4.3}$$

If the variable t indicates time, then a would be the speed of motion along the path. The *curvature* κ and *radius of curvature* ρ of a clothoid at a given t are:

$$\kappa = \frac{\pi t}{a} \tag{4.4}$$

$$\rho = \frac{1}{\kappa} = \frac{a}{\pi t} \tag{4.5}$$

The *tangent angle* θ of a clothoid at a given value of t is:

$$\theta = \frac{\pi}{2} t^2 \tag{4.6}$$

Having a road with linearly increasing curvature is equivalent to entering the path with a steering wheel at the neutral position and turning the steering wheel with a constant angular velocity. This is a desirable and natural driving action (Jazar 2017; Marzbani et al. 2015a).

Proof Arc length s of a parametric planar curve $X = X(t)$, $Y = Y(t)$ between t_1 and t_2 is calculated by

$$s = \int_{t_1}^{t_2} \sqrt{\left(\frac{dX}{dt}\right)^2 + \left(\frac{dY}{dt}\right)^2} \, dt \tag{4.7}$$

Substituting the clothoid spiral parametric expression (4.1)–(4.2), we have

$$s = \int_{t_1}^{t_2} \sqrt{\left(\frac{dX}{du}\right)^2 + \left(\frac{dY}{du}\right)^2} dt$$

$$= a \int_0^t \sqrt{\cos^2\left(\frac{\pi}{2}u^2\right) + \sin^2\left(\frac{\pi}{2}u^2\right)} dt = a \int_0^t dt = at \tag{4.8}$$

Curvature κ of a planar curve $X = X(s)$, $Y = Y(s)$ that is expressed parametrically by its arc length s is

$$\kappa = \sqrt{\left(\frac{d^2X}{ds^2}\right)^2 + \left(\frac{d^2Y}{ds^2}\right)^2} \tag{4.9}$$

Using the result of (4.8), we can replace the variable t with arc length s

$$t = \frac{s}{a} \tag{4.10}$$

and define the clothoid spiral parametric equations (4.1)–(4.2) as:

$$X(s) = a \int_0^{s/a} \cos\left(\frac{\pi}{2}u^2\right) du \tag{4.11}$$

$$Y(s) = a \int_0^{s/a} \sin\left(\frac{\pi}{2}u^2\right) du \tag{4.12}$$

Therefore, the curvature of clothoid spiral is

$$\kappa = \frac{\pi s}{a^2} \sqrt{\cos^2\left(\frac{\pi}{2}\frac{s^2}{a^2}\right) + \sin^2\left(\frac{\pi}{2}\frac{s^2}{a^2}\right)} = \frac{\pi s}{a^2} = \frac{\pi t}{a} \tag{4.13}$$

The slope of the tangent to clothoid spiral $\tan\theta$ at a point t is

$$\tan\theta = \frac{dY}{dX} = \frac{dY/dt}{dX/dt} = \frac{a \sin\left(\frac{\pi}{2}t^2\right)}{a \cos\left(\frac{\pi}{2}t^2\right)} = \tan\left(\frac{\pi}{2}t^2\right) \tag{4.14}$$

and therefore, the slope angle θ of the tangent line is

$$\theta = \frac{\pi}{2}t^2 = \frac{\pi}{2}\frac{s^2}{a^2} \tag{4.15}$$

The clothoid curve approaches the point $(a/2, a/2)$ at infinity, $t \to \infty$,

$$X(t) = \lim_{t\to\infty}\left(a\int_0^t \cos\left(\frac{\pi}{2}u^2\right)du\right) = \frac{a}{2} \tag{4.16}$$

$$Y(t) = \lim_{t\to\infty}\left(a\int_0^t \sin\left(\frac{\pi}{2}u^2\right)du\right) = \frac{a}{2} \tag{4.17}$$

Combining the equations for arc length (4.8) and curvature (4.13), we can express the curvature κ of a clothoid curve such that it varies linearly with its arc-length s. The curvature of such a curve is

$$\kappa(s) = \frac{\pi}{a^2}s = ks \tag{4.18}$$

$$k = \frac{\pi}{a^2} \tag{4.19}$$

where s is the arc length and k is the *sharpness* or the rate of change of curvature.

Using

$$t = \frac{\kappa a}{\pi} \tag{4.20}$$

we can also define the parametric equations of the transition clothoid road as:

$$X(\kappa) = a\int_0^{\kappa a/\pi} \cos\left(\frac{\pi}{2}u^2\right)du \tag{4.21}$$

$$Y(\kappa) = a\int_0^{\kappa a/\pi} \sin\left(\frac{\pi}{2}u^2\right)du \tag{4.22}$$

Figure 4.2 illustrates a design graph of the relationship between the clothoid and parameters of scaling or magnification factor a, curvature κ, and slope θ. The higher the magnification factor a the larger the clothoid. The clothoid curves of different a are intersecting with the constant slope lines of θ. The curves for constant curvature k intersect both the constant a and constant θ curves.

The clothoid transition equation is a proper solution for any required change in any parameter of a road. As an example the change of the bank angle from a flat straight road to a tilted road on a circular path needs a clothoid transition bank angle. ■

Example 93 Derivative of a clothoid spiral.

Differentiation of a definite integral is based on the Leibniz formula

$$\frac{d}{dt}\int_{a(t)}^{b(t)} f(u,t)\,du = \int_{a(t)}^{b(t)} \frac{df}{dt}du + f(b(t),t)\frac{db}{dt} - f(a(t),t)\frac{da}{dt} \tag{4.23}$$

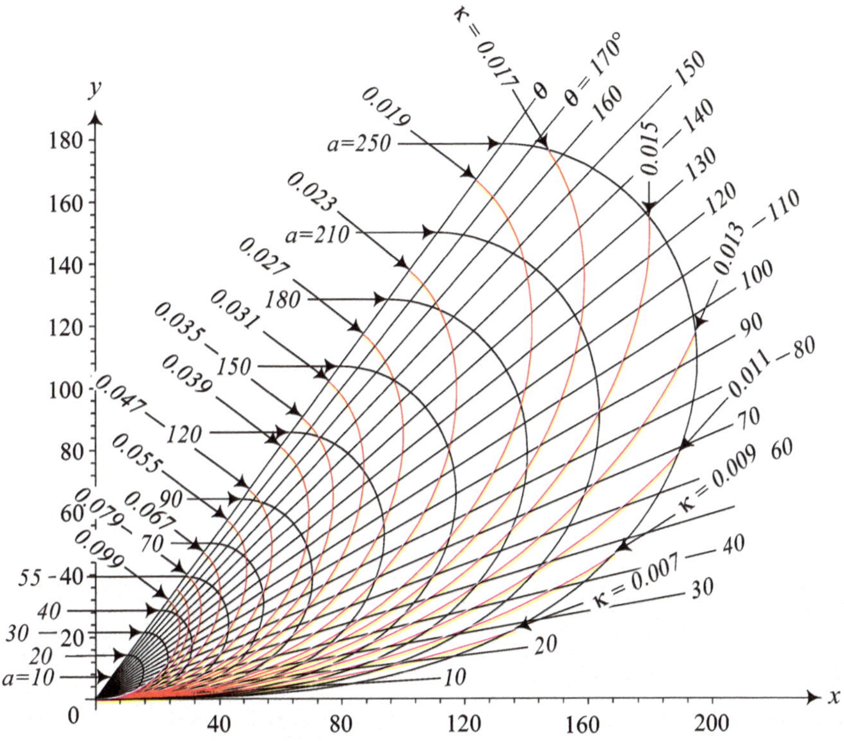

Fig. 4.2 A design graph of the relationship between the clothoid and parameters of magnification factor a, curvature κ, and slope θ

Taking derivative of the clothoid spiral parametric equations in calculating the arc length s of (4.7) is based on the Leibniz formula.

$$\frac{dX}{dt} = a\frac{d}{dt}\int_0^t \cos\left(\frac{\pi}{2}u^2\right) du = a\cos\left(\frac{\pi}{2}t^2\right) \tag{4.24}$$

$$\frac{dY}{dt} = a\frac{d}{dt}\int_0^t \sin\left(\frac{\pi}{2}u^2\right) du = a\sin\left(\frac{\pi}{2}t^2\right) \tag{4.25}$$

The calculation of the curvature κ of (4.9) is also based on the Leibniz formula (Jazar 2010b, 2012).

$$\frac{dX}{ds} = a\frac{d}{dt}\int_0^{s/a} \cos\left(\frac{\pi}{2}u^2\right) du = \cos\left(\frac{\pi}{2}\frac{s^2}{a^2}\right) \tag{4.26}$$

$$\frac{dY}{ds} = a\frac{d}{dt}\int_0^{s/a} \sin\left(\frac{\pi}{2}u^2\right) du = \sin\left(\frac{\pi}{2}\frac{s^2}{a^2}\right) \tag{4.27}$$

$$\frac{d^2X}{ds^2} = \frac{d}{ds}\cos\left(\frac{\pi}{2}\frac{s^2}{a^2}\right) = -\frac{\pi s}{a^2}\sin\left(\frac{\pi}{2}\frac{s^2}{a^2}\right) \tag{4.28}$$

$$\frac{d^2Y}{ds^2} = \frac{d}{ds}\sin\left(\frac{\pi}{2}\frac{s^2}{a^2}\right) = \frac{\pi s}{a^2}\cos\left(\frac{\pi}{2}\frac{s^2}{a^2}\right) \tag{4.29}$$

Therefore,

$$\sqrt{\left(\frac{d^2X}{ds^2}\right)^2 + \left(\frac{d^2Y}{ds^2}\right)^2} = \frac{\pi}{a^2} = \kappa(s) \tag{4.30}$$

Example 94 A connecting road with given a and κ.

Let us set $a = 200$ and plot a clothoid road starting from $(0, 0)$ and end up at a point with a given curvature of $\kappa = 0.01$ equal to a radius of curvature of $\rho = 100$ m.

Using

$$t = \frac{\kappa a}{\pi} \tag{4.31}$$

we can define the parametric equations of the transition road (4.1)–(4.2) as

$$X(\kappa) = a\int_0^{\kappa a/\pi}\cos\left(\frac{\pi}{2}u^2\right)du \tag{4.32}$$

$$Y(\kappa) = a\int_0^{\kappa a/\pi}\sin\left(\frac{\pi}{2}u^2\right)du \tag{4.33}$$

The coordinates of the clothoid road at $\kappa = 0.01$ and $a = 200$ are

$$X_0 = 122.2596310 \qquad Y_0 = 26.24682756 \tag{4.34}$$

The slope of the road at the point is

$$\theta = \frac{1}{2\pi}a^2\kappa^2 = 0.6366197722\,\text{rad} = 36.475\,\text{deg} \tag{4.35}$$

and therefore, the tangent line to the road is

$$Y = -64.14007833 + 0.7393029502X \tag{4.36}$$

and the normal line to the road is

$$Y = 191.6183183 - 1.352625469X \tag{4.37}$$

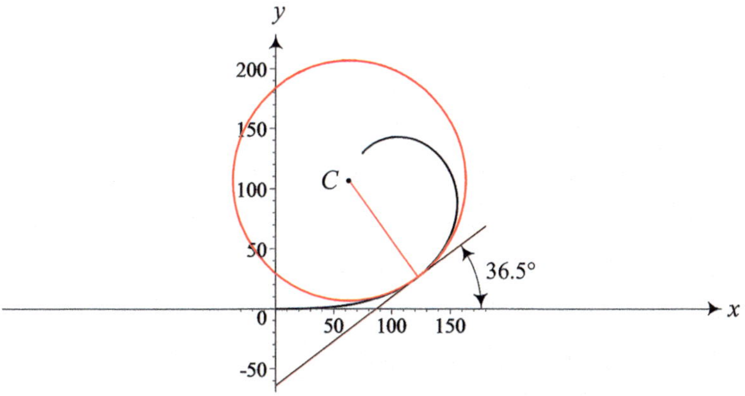

Fig. 4.3 The tangent line, normal line, and the tangent circle to the clothoid at the point where $\kappa = 0.01$ for a given $a = 200$. The clothoid is plotted up to $\kappa = 0.025$

Having the radius of the tangent curvature circle, $\rho = 1/\kappa = 100$ m, we are able to find the coordinates of the center C of the tangent circle on the line (4.37).

$$X_C = 62.81155414 \qquad Y_C = 106.6578104 \tag{4.38}$$

Figure 4.3 illustrates the tangent line, normal line, and the tangent circle to the clothoid at the point where $\kappa = 0.01$. The clothoid is plotted up to $\kappa = 0.025$.

Example 95 Connecting a straight road to a circle.

Assume that we need to define a clothoid road to begin with zero curvature and meet a given circular curve. Let us consider the road to be on the X-axis and the circle of $\rho = 100$ m at center $C\,(62.811, 106.658)$.

$$(X - 62.811)^2 + (Y - 106.658)^2 = 100^2 \tag{4.39}$$

Therefore, the transition road must begin with $\kappa = 0$ on the X-axis and touch the circle at a point when its curvature is $\kappa = 1/100$. Because of

$$\kappa = \frac{\pi s}{a^2} = \frac{\pi t}{a} \tag{4.40}$$

we can define the parametric equations of the transition road (4.1)–(4.2) by κ.

$$X(\kappa) = a \int_0^{\kappa a/\pi} \cos\left(\frac{\pi}{2} u^2\right) du$$

$$Y(\kappa) = a \int_0^{\kappa a/\pi} \sin\left(\frac{\pi}{2} u^2\right) du \tag{4.41}$$

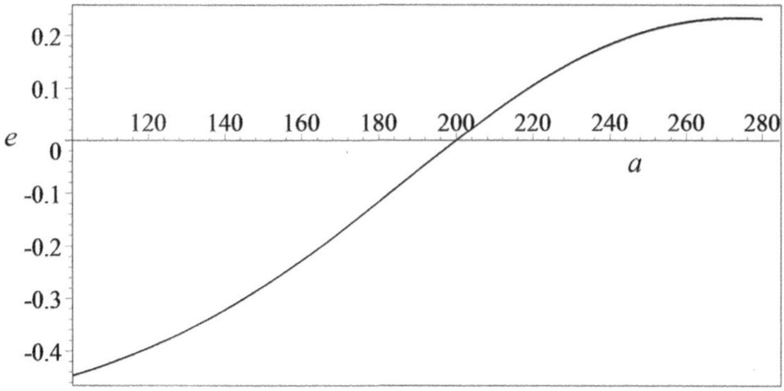

Fig. 4.4 The error $e = \arctan \dfrac{-(X - 62.811)}{Y - 106.658} - 1.5915 \times 10^{-5} a^2$ as a function of a

Having $\kappa = 0.01$ at the destination point, we find the coordinates of the end point of the clothoid as functions of a.

$$X(a) = a \int_0^{0.01a/\pi} \cos\left(\frac{\pi}{2} u^2\right) du$$
$$Y(a) = a \int_0^{0.01a/\pi} \sin\left(\frac{\pi}{2} u^2\right) du \tag{4.42}$$

We need to find the magnifying factor a, such that the clothoid (4.41) touches the circle (4.39) with the same slope. The slope of the circle at (X, Y) is

$$Y' = \tan\theta = -\frac{(X - 62.811)}{(Y - 106.658)} \tag{4.43}$$

and the slope angle of the clothoid is

$$\theta = \frac{\pi}{2} t^2 = \frac{1}{2\pi} a^2 \kappa^2 = 1.5915 \times 10^{-5} a^2 \tag{4.44}$$

To make the clothoid have the same slope, we derive an equation that relates the magnification factor a to the components of the final point of the clothoid.

$$\arctan\theta = \arctan\frac{-(X - 62.811)}{Y - 106.658} = 1.5915 \times 10^{-5} a^2 \tag{4.45}$$

Equations (4.45), (4.42), and (4.39) provide us with an equation to find a. Let us define and plot an error equation e versus a in Fig. 4.4

$$e = \arctan\theta - \theta = \arctan\frac{-(X-62.811)}{Y-106.658} - 1.5915\times10^{-5}a^2$$
$$= \arctan\frac{-(X-62.811)}{\sqrt{100^2-(X-62.811)^2}} - 1.5915\times10^{-5}a^2$$
$$= \arctan\frac{-\left(a\int_0^{0.01a/\pi}\cos\left(\frac{\pi}{2}u^2\right)du - 62.811\right)}{\sqrt{100^2-\left(a\int_0^{0.01a/\pi}\cos\left(\frac{\pi}{2}u^2\right)du-62.811\right)^2}}$$
$$-1.5915\times10^{-5}a^2 \tag{4.46}$$

and solve the equation for a that makes $e=0$.

$$a = 200 \tag{4.47}$$

Therefore, the clothoid equation is

$$X(\kappa) = a\int_0^{\kappa a/\pi}\cos\left(\frac{\pi}{2}u^2\right)du$$
$$Y(\kappa) = a\int_0^{\kappa a/\pi}\sin\left(\frac{\pi}{2}u^2\right)du \tag{4.48}$$

where at $\kappa = 0.01$ reaches to:

$$X_0 = 122.2596310 \qquad Y_0 = 26.24682756 \tag{4.49}$$

The slope of the road at the point, the tangent line to the road, and the normal line to the road are

$$\theta = \frac{1}{2\pi}a^2\kappa^2 = 0.6366197722\,\text{rad} = 36.475\,\text{deg} \tag{4.50}$$
$$Y = -64.14007833 + 0.7393029502X \tag{4.51}$$
$$Y = 191.6183183 - 1.352625469X \tag{4.52}$$

Figure 4.3 illustrates the clothoid, tangent line, normal line, and the tangent circle to the clothoid at the point where $\kappa = 0.01$. The clothoid is plotted up to $\kappa = 0.025$.

Example 96 Connecting a straight road to another circle.

Assume that we need to determine a transition clothoid road to begin with zero curvature on the X-axis and meet a given circular curve of $R = 80$ m at center $C(100, 100)$.

$$(X-100)^2 + (Y-100)^2 = 80^2 \tag{4.53}$$

The expression of the transition road (4.1)–(4.2) by κ is:

$$X(\kappa) = a \int_0^{\kappa a/\pi} \cos\left(\frac{\pi}{2}u^2\right) du$$

$$Y(\kappa) = a \int_0^{\kappa a/\pi} \sin\left(\frac{\pi}{2}u^2\right) du \tag{4.54}$$

Using $\kappa = 1/R = 0.0125$ at the destination point determines the coordinates of the end of the clothoid as functions of a.

$$X(a) = a \int_0^{0.0125a/\pi} \cos\left(\frac{\pi}{2}u^2\right) du$$

$$Y(a) = a \int_0^{0.0125a/\pi} \sin\left(\frac{\pi}{2}u^2\right) du \tag{4.55}$$

The slope of the tangent to the circle (4.53) at a point (X, Y) is

$$Y' = \tan\theta = -\frac{(X - 100)}{Y - 100} \tag{4.56}$$

and the slope angle of the clothoid as a function of a is

$$\theta = \frac{\pi}{2}t^2 = \frac{1}{2\pi}a^2\kappa^2 = 2.4868 \times 10^{-5}a^2 \tag{4.57}$$

The clothoid should have the same slope, therefore,

$$\arctan\frac{-(X - 100)}{Y - 100} = 2.4868 \times 10^{-5}a^2 \tag{4.58}$$

Equations (4.58) and (4.53) along with (4.55) provide us with an equation to find a. However, substituting $Y = Y(X)$ and replacing tan and arctan generate four equations to be solved for possible a. To visualize the possible solutions, let us define two error equations (4.59)–(4.60).

$$e = \arctan\frac{-(X - 100)}{\pm\sqrt{80^2 - (X - 100)^2}} - 2.4868 \times 10^{-5}a^2 \tag{4.59}$$

$$e = \frac{-(X - 100)}{\pm\sqrt{80^2 - (X - 100)^2}} - \tan\left(2.4868 \times 10^{-5}a^2\right) \tag{4.60}$$

Figure 4.5 depicts Eq. (4.59) and Fig. 4.6 shows Eq. (4.60). Equation (4.59) provides the solutions of

$$a = 230.7098693 \qquad a = 130.8889343 \tag{4.61}$$

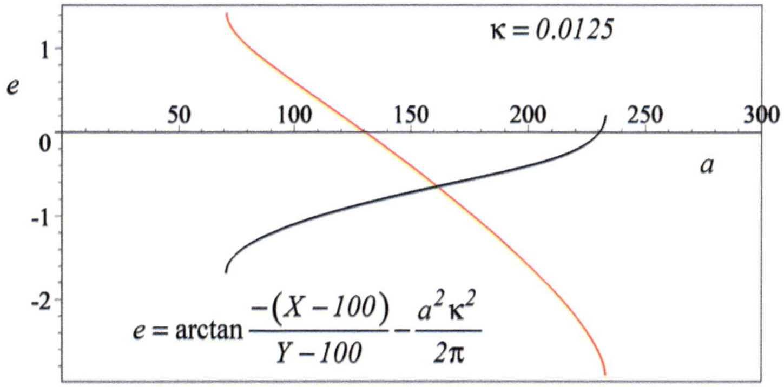

Fig. 4.5 Plot of $e = \arctan \dfrac{-(X-100)}{\pm\sqrt{80^2-(X-100)^2}} - 2.4868 \times 10^{-5} a^2$ versus a

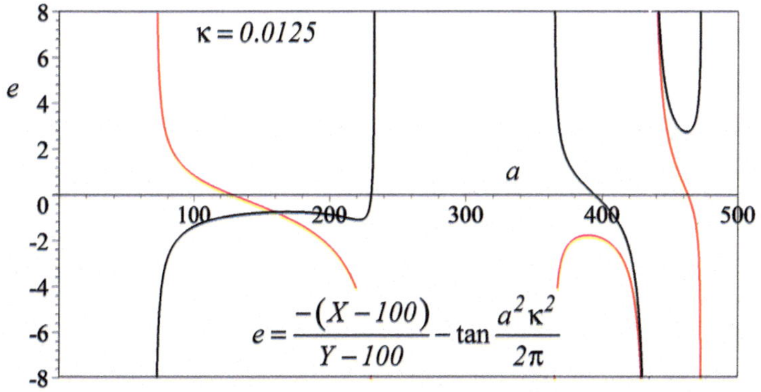

Fig. 4.6 Plot of $e = \dfrac{-(X-100)}{\pm\sqrt{80^2-(X-100)^2}} - \tan\left(2.4868 \times 10^{-5} a^2\right)$ versus a

and Eq. (4.60) provides the solutions of

$$a = 230.7098693 \qquad a = 130.8889343$$
$$a = 394.0940573 \qquad a = 463.5589702 \tag{4.62}$$

The correct answer is $a = 230.7098693$ and Fig. 4.7 depicts the circle and the proper clothoid. Using a, we define the clothoid equation

$$X(\kappa) = a\int_0^{\kappa a/\pi} \cos\left(\frac{\pi}{2}u^2\right) du$$
$$Y(\kappa) = a\int_0^{\kappa a/\pi} \sin\left(\frac{\pi}{2}u^2\right) du \tag{4.63}$$

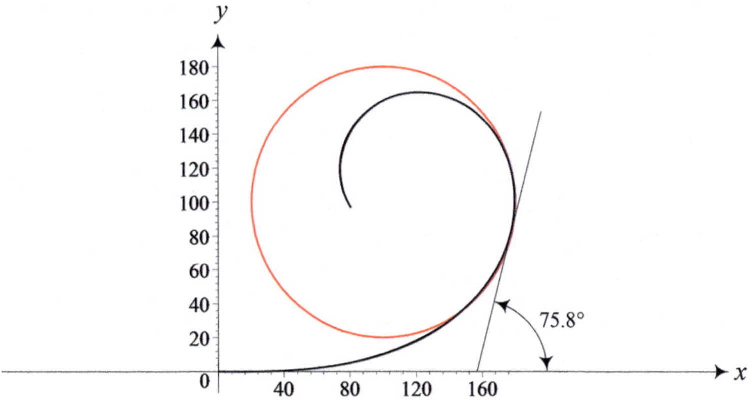

Fig. 4.7 The transition road starting on the X-axis and goes to a circle of radius $R = 80\,\mathrm{m}$ at center C (100 m, 100 m)

which at $\kappa = 0.0125$ reaches

$$X_0 = 177.5691613 \qquad Y_0 = 82.38074640 \tag{4.64}$$

at angle

$$\theta = \frac{1}{2\pi}a^2\kappa^2 = 1.3236\,\mathrm{rad} \approx 75.84\,\mathrm{deg} \tag{4.65}$$

Example 97 Using the design chart.

Assume we are asked to find a clothoid transition road to connect a straight road to a circle of radius $R = 58.824\,\mathrm{m}$. Having R is equivalent to have the destination curvature $\kappa = 1/R = 0.017$. The desired circle must be tangent to a clothoid with a given a at the point that the clothoid is intersecting the curve of $\kappa = 0.017$.

The clothoid for $a = 250\,\mathrm{m}$ hits the curve of $\kappa = 0.017$ at a point for which we have

$$X = 147.3884878\,\mathrm{m} \qquad Y = 176.4421850\,\mathrm{m} \tag{4.66}$$

$$\theta = 164.7102491\,\mathrm{deg} \qquad s = 338.2042540\,\mathrm{m} \tag{4.67}$$

The clothoid for $a = 210\,\mathrm{m}$ hits the curve of $\kappa = 0.017$ at a point for which we have

$$X = 157.4739501\,\mathrm{m} \qquad Y = 119.7133227\,\mathrm{m} \tag{4.68}$$

$$\theta = 116.2195518\,\mathrm{deg} \qquad s = 238.6369216\,\mathrm{m} \tag{4.69}$$

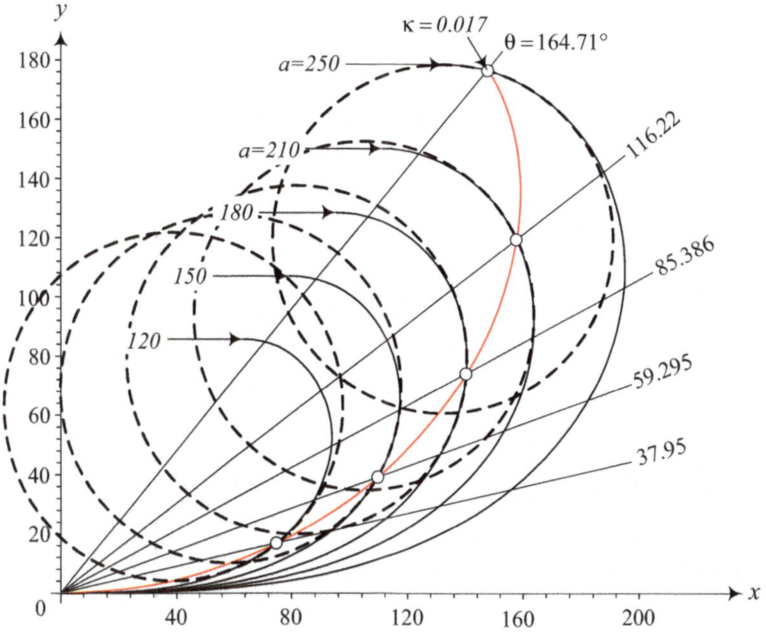

Fig. 4.8 A few clothoid transition road to connect a straight road to a circle of radius $R = 58.824\,\text{m}$

The clothoid for $a = 180\,\text{m}$ hits the curve of $\kappa = 0.017$ at a point for which we have

$$X = 140.1918463\,\text{m} \qquad Y = 74.21681673\,\text{m} \tag{4.70}$$

$$\theta = 85.38579313\,\text{deg} \qquad s = 175.3250853\,\text{m} \tag{4.71}$$

The clothoid for $a = 150\,\text{m}$ hits the curve of $\kappa = 0.017$ at a point for which we have

$$X = 109.3442240\,\text{m} \qquad Y = 38.89541829\,\text{m} \tag{4.72}$$

$$\theta = 59.29568967\,\text{deg} \qquad s = 121.7535314\,\text{m} \tag{4.73}$$

The clothoid for $a = 120\,\text{m}$ hits the curve of $\kappa = 0.017$ at a point for which we have

$$X = 74.57259185\,\text{m} \qquad Y = 16.67204291\,\text{m} \tag{4.74}$$

$$\theta = 37.94924139\,\text{deg} \qquad s = 77.92226012\,\text{m} \tag{4.75}$$

Figure 4.8 illustrates these solutions. The number of solutions is practically infinite and the best solution depends on safety, cost, and physical constraints of the field.

Example 98 ★ Clothoid spiral as an optimal curve problem.

Clothoid spiral is the shortest curve connecting two given points with given initial and final tangent angles and curvatures. The angle and curvature are varying. The problem can be formulated as follows:

Given two points (X_1, Y_1) and (X_2, Y_2) and two angles θ_1 and θ_2, find a curve (clothoid segment) which satisfies:

$$X(0) = X_1 \qquad Y(0) = Y_1 \qquad \tan\theta_1 = \frac{dY(0)}{dX(0)} = \frac{dY(0)/dt}{dX(0)/dt} \tag{4.76}$$

$$X(s) = X_2 \qquad Y(s) = Y_2 \qquad \tan\theta_2 = \frac{dY(s)}{dX(s)} = \frac{dY(s)/dt}{dX(s)/dt} \tag{4.77}$$

with minimal arc length s.

Example 99 Clothoid shift to meet a given circle.

It is not generally possible to design a clothoid starting at the origin and meet a given circle at an arbitrary center and radius. However, it is possible to start the clothoid from other points on the x-axis to meet the given circle.

Assume we need to design a clothoid starting on the x-axis to meet a given circle

$$(x - x_C)^2 + (y - y_C)^2 = R^2 \tag{4.78}$$

where (x_C, y_C) indicates the coordinates of the center of the circle, and R is the radius of the circle. Substituting t in terms of a and R

$$t = \frac{a}{\pi R} = \frac{a\kappa}{\pi} \tag{4.79}$$

we define the equation of clothoid to have the same radius of curvature as the circle at the end point.

$$\begin{aligned} x(\kappa) &= a \int_0^{a/(R\pi)} \cos\left(\frac{\pi}{2}u^2\right) du \\ y(\kappa) &= a \int_0^{a/(R\pi)} \sin\left(\frac{\pi}{2}u^2\right) du \end{aligned} \tag{4.80}$$

Let us assume there is a y at which the slope of the clothoid

$$\tan\theta = \tan\left(\frac{\pi t^2}{2}\right) = \tan\left(\frac{a^2}{2\pi R^2}\right) \tag{4.81}$$

and the circle

$$\tan\theta = -\frac{x - x_C}{y - y_C} \tag{4.82}$$

are equal.

$$\tan\left(\frac{a^2}{2\pi R^2}\right) = -\frac{x - x_C}{y - y_C} \tag{4.83}$$

Searching for a match point in the right half circle

$$y - y_C = \sqrt{R^2 - (x - x_C)^2} \tag{4.84}$$

makes the slope equation to be a function of a

$$\tan\left(\frac{a^2}{2\pi R^2}\right)\sqrt{R^2 - (x - x_C)^2} + \frac{x - x_C}{y - y_C} = 0 \tag{4.85}$$

or

$$\tan\left(\frac{a^2}{2\pi R^2}\right)\sqrt{R^2 - \left(a\int_0^{a/(R\pi)} \cos\left(\frac{\pi}{2}u^2\right) du - x_C\right)^2}$$
$$+a\int_0^{a/(R\pi)} \cos\left(\frac{\pi}{2}u^2\right) du - x_C = 0 \tag{4.86}$$

Solution of this equation provides us with an a for which the clothoid ends at a point with the same curvature as the circle. At the same y of the end point, the slope of the clothoid is also equal to the slope of the circle. A proper shift of the clothoid on the x-axis will match the clothoid and the circle.

As an example, let us assume that the circle is

$$(x - 60)^2 + (y - 60)^2 = 50^2 \tag{4.87}$$

and therefore, the slope equation will be

$$\tan\left(\frac{a^2}{2\pi 50^2}\right)\sqrt{50^2 - \left(a\int_0^{a/(50\pi)} \cos\left(\frac{\pi}{2}u^2\right) du - 60\right)^2}$$
$$+a\int_0^{a/(50\pi)} \cos\left(\frac{\pi}{2}u^2\right) du - 60 = 0 \tag{4.88}$$

Numerical solution of the equation is

$$a = 132.6477323 \tag{4.89}$$

The plot of the clothoid and the circle at this moment are shown in Fig. 4.9.

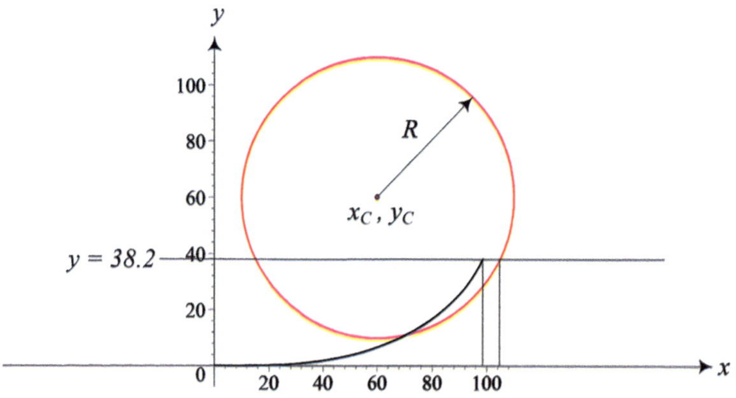

Fig. 4.9 A clothoid starting at origin ends at a point with the same slope and curvature as the given circle, at the same y

For the calculated a, the value of t at the end point of the clothoid is

$$t = \frac{a}{\pi R} = \frac{2.652954646}{\pi} = 0.84446 \tag{4.90}$$

and therefore, the coordinates of the end point are

$$x(\kappa) = 132.6 \int_0^{0.84} \cos\left(\frac{\pi}{2}u^2\right) du = 98.75389126$$
$$y(\kappa) = 132.6 \int_0^{0.84} \sin\left(\frac{\pi}{2}u^2\right) du = 38.22304651 \tag{4.91}$$

At the point, the radius of curvature of the clothoid is

$$R = \frac{1}{\kappa} = \frac{a}{\pi t} = \frac{132.6477323}{0.84446\pi} = 50 \tag{4.92}$$

and the slope is

$$\theta = \frac{\pi}{2}t^2 = \frac{\pi}{2}0.84446^2 = 1.1202 \text{ rad} \tag{4.93}$$

The x-coordinate of the circle at the same $y = 38.22304651$

$$38.22304651 - 60 = \sqrt{50^2 - (x-60)^2} \tag{4.94}$$

is

$$x_{circle} = 105.0084914 \tag{4.95}$$

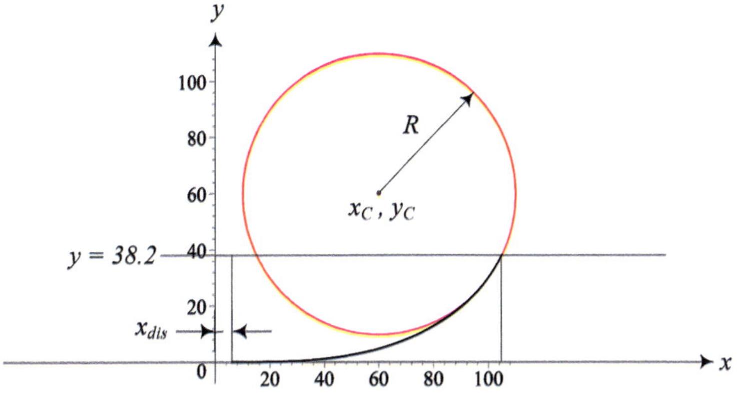

Fig. 4.10 A shifted clothoid starting on a point on the x-axis ends at a point on the given circle with the same slope and curvature

If we shift the clothoid by the difference between x_{circle} and $x_{clothoid}$

$$\begin{aligned} x_{dis} &= x_{circle} - x_{clothoid} \\ &= 105.0084914 - 98.75389126 = 6.2546 \end{aligned} \tag{4.96}$$

then the clothoid and circle meet at a point on the circle with all requirements to have a smooth transition. Figure 4.10 illustrates the result.

Example 100 Complex expression and proof of curvature.

Let us define a curve in complex plane as

$$C(s) = a \int_0^{s/a} e^{i\pi u^2/2} du \tag{4.97}$$

The derivative of the curve is an equation with absolute value of a.

$$\frac{dC(s)}{ds} = e^{i\pi s^2/(2a^2)} \tag{4.98}$$

The curvature of the curve is

$$\kappa = \left| \frac{d^2 C(s)}{ds^2} \right| = \left| i \frac{\pi s}{a^2} e^{i\pi s^2/(2a^2)} \right| = \frac{\pi s}{a^2} \tag{4.99}$$

Example 101 Parametric form of a straight road.

The equation of a straight road that connects two points $P_1(x_1, y_1, z_1)$ and $P_2(x_2, y_2, z_3)$ is

$$\frac{x - x_1}{x_2 - x_1} = \frac{y - y_1}{y_2 - y_1} = \frac{z - z_1}{z_2 - z_1} \tag{4.100}$$

This line may also be expressed by the following parametric equations.

$$\begin{aligned} x &= x_1 + (x_2 - x_1)\, t \\ y &= y_1 + (y_2 - y_1)\, t \\ z &= z_1 + (z_2 - z_1)\, t \end{aligned} \tag{4.101}$$

Example 102 Arc length of a planar road.

A planar road in the (x, y)-plane

$$y = f(x) \tag{4.102}$$

can be expressed vectorially by

$$\mathbf{r} = x\hat{\imath} + y\,(x)\,\hat{\jmath} \tag{4.103}$$

The displacement element on the curve

$$\frac{d\mathbf{r}}{dx} = \hat{\imath} + \frac{dy}{dx}\hat{\jmath} \tag{4.104}$$

provides us with

$$\left(\frac{ds}{dx}\right)^2 = \frac{d\mathbf{r}}{dx} \cdot \frac{d\mathbf{r}}{dx} = 1 + \left(\frac{dy}{dx}\right)^2 \tag{4.105}$$

Therefore, the arc length of the curve between $x = x_1$ and $x = x_2$ is

$$s = \int_{x_1}^{x_2} \sqrt{1 + \left(\frac{dy}{dx}\right)^2}\, dx \tag{4.106}$$

In case the curve is given parametrically,

$$x = x(t) \qquad y = y(t) \tag{4.107}$$

we have

$$\left(\frac{ds}{dt}\right)^2 = \frac{d\mathbf{r}}{dt} \cdot \frac{d\mathbf{r}}{dt} = \left(\frac{dx}{dt}\right)^2 + \left(\frac{dy}{dt}\right)^2 \tag{4.108}$$

and hence,

$$s = \int_{t_1}^{t_2} \left|\frac{d\mathbf{r}}{dt}\right| = \int_{t_1}^{t2} \sqrt{\left(\frac{dx}{dt}\right)^2 + \left(\frac{dy}{dt}\right)^2}\, dt \tag{4.109}$$

Let us show a circle with radius R by its polar expression using the angle θ as a parameter, as an example:

$$x = R\cos\theta \qquad y = R\sin\theta \tag{4.110}$$

The arc length between $\theta = 0$ and $\theta = \pi/2$ would then be one-fourth the perimeter of the circle. Therefore, the equation for calculating the perimeter of a circle with radius R is

$$\begin{aligned} s &= 4\int_0^{\pi/2}\sqrt{\left(\frac{dx}{d\theta}\right)^2+\left(\frac{dy}{d\theta}\right)^2}\,d\theta = R\int_0^{\pi/2}\sqrt{\sin^2\theta+\cos^2\theta}\,d\theta \\ &= 4R\int_0^{\pi/2} d\theta = 2\pi R \end{aligned} \tag{4.111}$$

Using Eq. (4.106), we can define the equation of the road as

$$y = \int_{x_1}^{x_2}\sqrt{\left(\frac{ds}{dx}\right)^2 - 1}\,dx \tag{4.112}$$

Example 103 A figure 8 as an approximately correct road.

Sometimes, matching slopes, instead of matching curvatures, can be used to design an approximately correct road. Let us make a closed road in the shape of a symmetric figure 8 with two 180 deg circular paths. Assuming

$$a = 200 \tag{4.113}$$

the equations of the clothoid road starting from the origin are:

$$X(t) = 200\int_0^t \cos\left(\frac{\pi}{2}u^2\right)du \tag{4.114}$$

$$Y(t) = 200\int_0^t \sin\left(\frac{\pi}{2}u^2\right)du \tag{4.115}$$

The slope (4.6) of the curve would be parallel to the symmetric line $Y = X$ when

$$\theta = \frac{\pi}{4} \qquad t = \sqrt{\frac{2\theta}{\pi}} = \frac{\sqrt{2}}{2} = t_0 \tag{4.116}$$

At $t = t_0$ the clothoid is at

$$X_0 = 200\int_0^{\sqrt{2}/2} \cos\left(\frac{\pi}{2}u^2\right)du = 132.943 \tag{4.117}$$

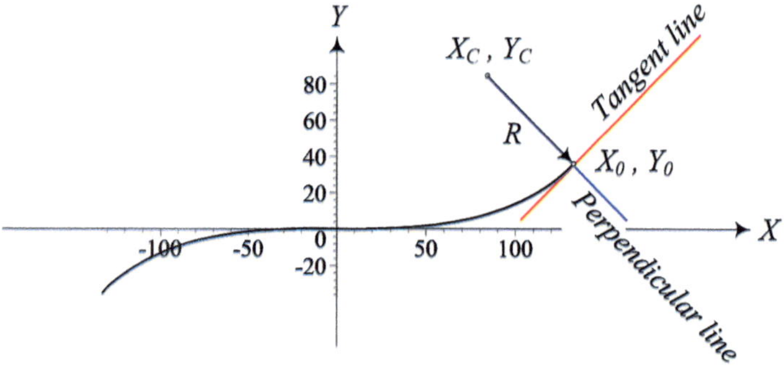

Fig. 4.11 A clothoid between two point at which the tangent lines are parallel to $Y = X$

$$Y_0 = 200 \int_0^{\sqrt{2}/2} \sin\left(\frac{\pi}{2}u^2\right) du = 35.424 \tag{4.118}$$

where the tangent and perpendicular lines respectively are

$$Y = 35.424 + (X - 132.943)\tan\theta = -97.519 + X \tag{4.119}$$

$$Y = 35.424 - (X - 132.943) / \tan\theta = 168.37 - X \tag{4.120}$$

as are shown in Fig. 4.11 for the clothoid from $t = -t_0$ to $t = t_0$.

The perpendicular line hits the symmetric line $Y = X$ at

$$X_C = Y_C = 84.184 \tag{4.121}$$

which would be the center of a circular path with

$$R = \sqrt{(X_0 - X_C)^2 + (Y_0 - Y_C)^2} = 68.956 \tag{4.122}$$

to connect (X_0, Y_0) to its mirror point with respect to $Y = X$ at

$$X_1 = 35.424 \qquad Y_1 = 132.943 \tag{4.123}$$

The mirror clothoid

$$X = 200 \int_{-\sqrt{2}/2}^{\sqrt{2}/2} \sin\left(\frac{\pi}{2}u^2\right) du \tag{4.124}$$

$$Y = 200 \int_{-\sqrt{2}/2}^{\sqrt{2}/2} \cos\left(\frac{\pi}{2}u^2\right) du \tag{4.125}$$

will complete the figure 8 road as is shown in Fig. 4.12.

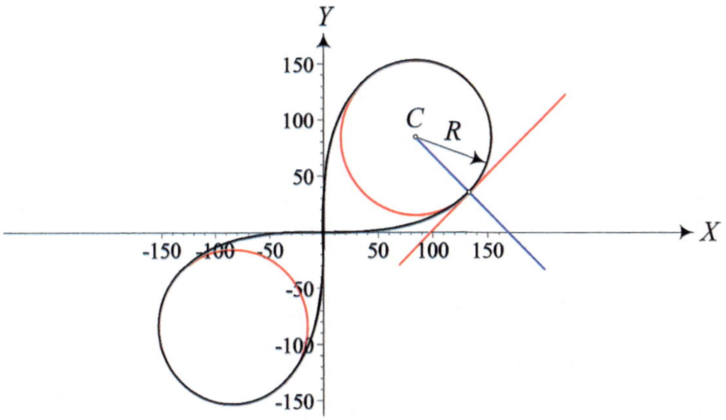

Fig. 4.12 A slope match symmetric figure 8 road based on two clothoid and two circular parts

Therefore, the parametric equations of the road beginning from the origin and moving in the X-direction are as follows. The parameter t is not continuous and not with the same dimension in all equations.

$$X(t) = 200 \int_0^t \cos\left(\frac{\pi}{2}u^2\right) du \qquad 0 \le t \le \frac{\sqrt{2}}{2} \tag{4.126}$$

$$Y(t) = 200 \int_0^t \sin\left(\frac{\pi}{2}u^2\right) du \qquad 0 \le t \le \frac{\sqrt{2}}{2} \tag{4.127}$$

$$X(t) = X_C + R\cos t \qquad -\frac{\pi}{4} \le t \le 2.3562 \tag{4.128}$$

$$Y(t) = Y_C + R\sin t \qquad -\frac{\pi}{4} \le t \le 2.3562 \tag{4.129}$$

$$X(t) = 200 \int_0^t \sin\left(\frac{\pi}{2}u^2\right) du \qquad \frac{\sqrt{2}}{2} \le t \le -\frac{\sqrt{2}}{2} \tag{4.130}$$

$$Y(t) = 200 \int_0^t \cos\left(\frac{\pi}{2}u^2\right) du \qquad \frac{\sqrt{2}}{2} \le t \le -\frac{\sqrt{2}}{2} \tag{4.131}$$

$$X(t) = X_C + R\cos t \qquad 2.3562 \le t \le 5.4978 \tag{4.132}$$

$$Y(t) = Y_C + R\sin t \qquad 2.3562 \le t \le 5.4978 \tag{4.133}$$

$$X(t) = 200 \int_0^t \cos\left(\frac{\pi}{2}u^2\right) du \qquad -\frac{\sqrt{2}}{2} \le t \le 0 \tag{4.134}$$

$$Y(t) = 200 \int_0^t \sin\left(\frac{\pi}{2}u^2\right) du \qquad -\frac{\sqrt{2}}{2} \le t \le 0 \tag{4.135}$$

Using s as the road length, we may define the equations with a smooth and continuous parameter.

$$X(t) = 200 \int_0^{s/200} \cos\left(\frac{\pi}{2}u^2\right) du \tag{4.136}$$

$$Y(t) = 200 \int_0^{s/200} \sin\left(\frac{\pi}{2}u^2\right) du \tag{4.137}$$

$$0 \le s \le 100\sqrt{2} \tag{4.138}$$

$$X(t) = X_C + R\cos\left(\frac{s - 100\sqrt{2}}{R} - \frac{\pi}{4}\right) \tag{4.139}$$

$$Y(t) = Y_C + R\sin\left(\frac{s - 100\sqrt{2}}{R} - \frac{\pi}{4}\right) \tag{4.140}$$

$$100\sqrt{2} \le s \le 358.05 \tag{4.141}$$

$$X(t) = 200 \int_0^{(499.47-s)/200} \sin\left(\frac{\pi}{2}u^2\right) du \tag{4.142}$$

$$Y(t) = 200 \int_0^{(499.47-s)/200} \cos\left(\frac{\pi}{2}u^2\right) du \tag{4.143}$$

$$358.05 \le s \le 640.89 \tag{4.144}$$

$$X(t) = -X_C + R\cos\left(\frac{640.89 - s}{R} - \frac{\pi}{4}\right) \tag{4.145}$$

$$Y(t) = -Y_C + R\sin\left(\frac{640.89 - s}{R} - \frac{\pi}{4}\right) \tag{4.146}$$

$$640.89 \le s \le 857.52 \tag{4.147}$$

$$X(t) = 200 \int_0^{\left(s-857.52-100\sqrt{2}\right)/200} \cos\left(\frac{\pi}{2}u^2\right) du \tag{4.148}$$

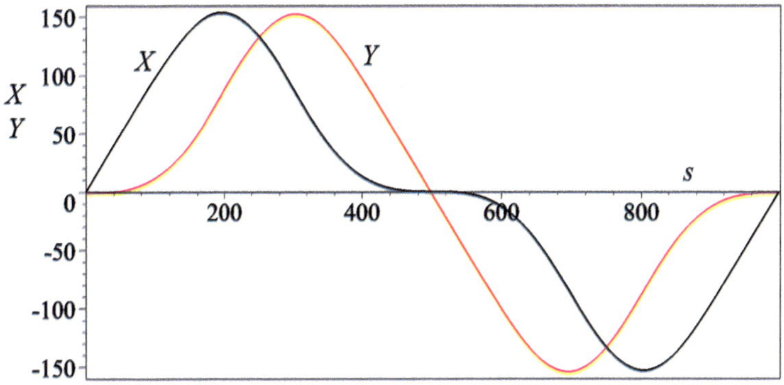

Fig. 4.13 The variation of X and Y for $0 \leq s \leq 998.94$ of the figure 8 road

$$Y(t) = 200 \int_0^{\left(s-857.52-100\sqrt{2}\right)/200} \sin\left(\frac{\pi}{2}u^2\right) du \tag{4.149}$$

$$857.52 \leq s \leq 998.94 \tag{4.150}$$

The variation of X and Y for $0 \leq s \leq 998.94$ is depicted in Fig. 4.13.

Example 104 A figure 8 correct road.

Let us design a closed road in the shape of a symmetric figure 8 with a curvature transition between the clothoids and the circular paths. Assuming

$$a = 200 \tag{4.151}$$

the equations of the clothoid road starting from the origin are:

$$X(t) = 200 \int_0^t \cos\left(\frac{\pi}{2}u^2\right) du \tag{4.152}$$

$$Y(t) = 200 \int_0^t \sin\left(\frac{\pi}{2}u^2\right) du \tag{4.153}$$

The slope (4.6) at t of the curve is

$$\theta = \frac{\pi}{2}t^2 \tag{4.154}$$

where the tangent and perpendicular lines respectively are:

$$Y = Y(t) + (X - X(t)) \tan\theta \tag{4.155}$$

$$Y = Y(t) - (X - X(t)) / \tan\theta \tag{4.156}$$

The radius of curvature of the clothoid at t is

$$R = \frac{1}{\kappa} = \frac{a}{\pi t} \tag{4.157}$$

If a clothoid point (X, Y) exists at which the radius of curvature is equal to the distance of the point from the line $Y = X$ on the perpendicular line, then we can have a circular path starting at the point. The intersection of the clothoid point (X, Y) and the symmetric line $Y = X$ is the point $Y_C = X_C$ where

$$Y_C = X_C = \frac{X(t) + Y(t)\tan\dfrac{\pi t^2}{2}}{\tan\dfrac{\pi t^2}{2} + 1} \tag{4.158}$$

The distance of the clothoid and the point (X_C, Y_C) on the perpendicular line is

$$\begin{aligned} d &= \sqrt{(Y(t) - Y_C)^2 + (X(t) - X_C)^2} \\ &= \sqrt{\left(\frac{Y(t) - X(t)}{\tan\dfrac{\pi t^2}{2} + 1}\right)^2 + \left(\frac{(X(t) - Y(t))\tan\dfrac{\pi t^2}{2}}{\tan\dfrac{\pi t^2}{2} + 1}\right)^2} \end{aligned} \tag{4.159}$$

Equating d and R provides us with an equation to be solved for the t at which the clothoid terminates and a circle with the same curvature starts.

$$d - \frac{a}{\pi t} = 0 \tag{4.160}$$

As is shown in Fig. 4.14, the equation has multiple solutions, and the first solution is at

$$t = t_0 = 0.9371211755 \tag{4.161}$$

At $t = t_0$ the clothoid and its kinematics are

$$X_0 = 200 \int_0^{0.9371211755} \cos\left(\frac{\pi}{2}u^2\right) du = 154.77 \tag{4.162}$$

$$Y_0 = 200 \int_0^{0.9371211755} \sin\left(\frac{\pi}{2}u^2\right) du = 75.154 \tag{4.163}$$

$$R = \frac{1}{\kappa} = 67.93355959 \tag{4.164}$$

$$\theta = 1.379467204 \text{ rad} = 79.038 \text{ deg} \tag{4.165}$$

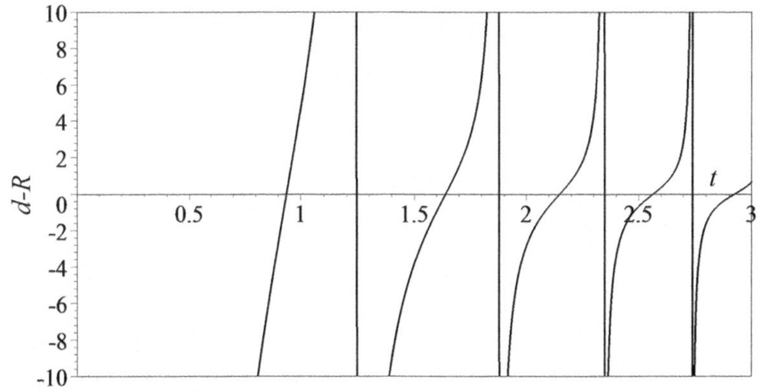

Fig. 4.14 Plot of the equation $d - R = d - \frac{a}{\pi t} = 0$ to find a t at which the curvature of the clothoid matches with a symmetric circle

The intersection of the perpendicular to the clothoid at (X_0, Y_0) with $Y = X$ is at

$$X_C = 88.0724138 < X_0 \tag{4.166}$$

$$Y_C = 88.0724138 > Y_0 \tag{4.167}$$

We can check the distance of (X_0, Y_0) and (X_C, Y_C) to be equal to R.

Therefore, using the parametric s as the road length, the equations of the road beginning from the origin and moving in the X-direction can be expressed as

$$X(t) = 200 \int_0^{s/a} \cos\left(\frac{\pi}{2}u^2\right) du \qquad 0 \le s \le s_0 \tag{4.168}$$

$$Y(t) = 200 \int_0^{s/a} \sin\left(\frac{\pi}{2}u^2\right) du \qquad 0 \le s \le s_0 \tag{4.169}$$

$$s_0 = a t_0 = 187.42 \tag{4.170}$$

$$X(t) = X_C + R\cos\left(\frac{s - s_0}{R} - \theta_0\right) \qquad s_0 \le s \le s_1 \tag{4.171}$$

$$Y(t) = Y_C + R\sin\left(\frac{s - s_0}{R} - \theta_0\right) \qquad s_0 \le s \le s_1 \tag{4.172}$$

$$\theta_0 = \arctan\frac{Y_C - Y_0}{X_0 - X_C} = 0.19132\,\mathrm{rad} = 10.962\,\mathrm{deg} \tag{4.173}$$

$$s_1 = s_0 + R\left(\frac{\pi}{2} + 2\theta_0\right) = 320.12 \tag{4.174}$$

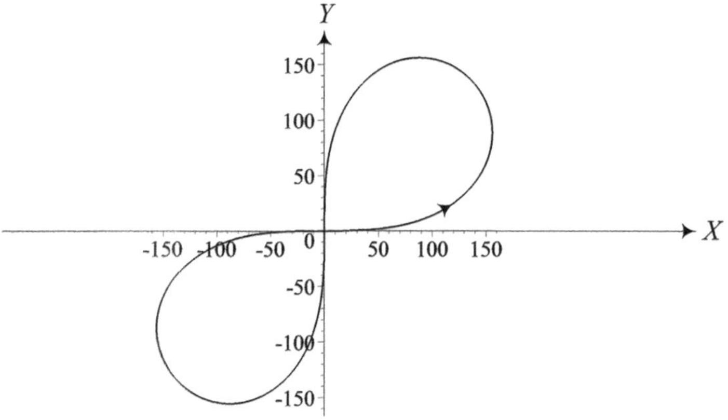

Fig. 4.15 A curvature match symmetric figure 8 road based on two clothoid and two circular parts

$$X(t) = 200 \int_0^{(s_1+s_0-s)/a} \sin\left(\frac{\pi}{2}u^2\right) du \qquad s_1 \le s \le s_2 \tag{4.175}$$

$$Y(t) = 200 \int_0^{(s_1+s_0-s)/a} \cos\left(\frac{\pi}{2}u^2\right) du \qquad s_1 \le s \le s_2 \tag{4.176}$$

$$s_2 = s_1 + 2s_0 = 694.96 \tag{4.177}$$

$$X(t) = -X_C + R\cos\left(\frac{s_2 - s}{R} - \left(\frac{\pi}{2} - \theta_0\right)\right) \qquad s_2 \le s \le s_3 \tag{4.178}$$

$$Y(t) = -Y_C + R\sin\left(\frac{s_2 - s}{R} - \left(\frac{\pi}{2} - \theta_0\right)\right) \qquad s_2 \le s \le s_3 \tag{4.179}$$

$$s_3 = s_2 + R\left(\frac{\pi}{2} + 2\theta_0\right) = 827.66 \tag{4.180}$$

$$X(t) = 200 \int_0^{(s-s_3-s_0)/200} \cos\left(\frac{\pi}{2}u^2\right) du \qquad s_3 \le s \le s_4 \tag{4.181}$$

$$Y(t) = 200 \int_0^{(s-s_3-s_0)/200} \sin\left(\frac{\pi}{2}u^2\right) du \qquad s_3 \le s \le s_4 \tag{4.182}$$

$$s_4 = s_3 + s_0 = 1015.1 \tag{4.183}$$

The road is shown in Fig. 4.15.

Example 105 ★ Spatial road.

If the position vector ${}^{G}\mathbf{r}_{P}$ of a moving car is such that each component is a function of a variable t,

$$ {}^{G}\mathbf{r} = {}^{G}\mathbf{r}(t) = x(t)\,\hat{\imath} + y(t)\,\hat{\jmath} + z(t)\,\hat{k} \tag{4.184} $$

then the end point of the position vector indicates a curve C in G. The curve ${}^{G}\mathbf{r} = {}^{G}\mathbf{r}(t)$ reduces to a point on C if we fix the parameter t. The functions

$$ x = x(t) \qquad y = y(t) \qquad z = z(t) \tag{4.185} $$

are the parametric equations of the curve. When the parameter t is the arc length s, the infinitesimal arc distance ds on the curve is

$$ ds^2 = d\mathbf{r} \cdot d\mathbf{r} \tag{4.186} $$

The arc length s of a curve is defined as the limit of the diagonal of a rectangular box as the length of the sides uniformly approaches zero.

When the space curve is a straight line that passes through point $P(x_0, y_0, z_0)$ where $x_0 = x(t_0)$, $y_0 = y(t_0)$, $z_0 = z(t_0)$, its equation can be shown by

$$ \frac{x - x_0}{\alpha} = \frac{y - y_0}{\beta} = \frac{z - z_0}{\gamma} \tag{4.187} $$

$$ \alpha^2 + \beta^2 + \gamma^2 = 1 \tag{4.188} $$

where α, β, and γ are the directional cosines of the line. The equation of the tangent line to the space curve (4.185) at a point $P(x_0, y_0, z_0)$ is

$$ \frac{x - x_0}{dx/dq} = \frac{y - y_0}{dy/dq} = \frac{z - z_0}{dz/dq} \tag{4.189} $$

$$ \left(\frac{dx}{dq}\right)^2 + \left(\frac{dy}{dq}\right)^2 + \left(\frac{dz}{dq}\right)^2 = 1 \tag{4.190} $$

To show this, let us consider a position vector ${}^{G}\mathbf{r} = {}^{G}\mathbf{r}(s)$ that describes a space curve using the length parameter s:

$$ {}^{G}\mathbf{r} = {}^{G}\mathbf{r}(s) = x(s)\,\hat{\imath} + y(s)\,\hat{\jmath} + z(s)\,\hat{k} \tag{4.191} $$

The arc length s is measured from a fixed point on the curve. By a very small change ds, the position vector will move to a very close point such that the increment in the position vector would be

$$ d\mathbf{r} = dx(s)\,\hat{\imath} + dy(s)\,\hat{\jmath} + dz(s)\,\hat{k} \tag{4.192} $$

The lengths of $d\mathbf{r}$ and ds are equal for infinitesimal displacement:

$$ds = \sqrt{dx^2 + dy^2 + dz^2} \tag{4.193}$$

The arc length has a better expression in the square form:

$$ds^2 = dx^2 + dy^2 + dz^2 = d\mathbf{r} \cdot d\mathbf{r} \tag{4.194}$$

If the parameter of the space curve is q instead of s, the increment arc length would be

$$\left(\frac{ds}{dt}\right)^2 = \frac{d\mathbf{r}}{dt} \cdot \frac{d\mathbf{r}}{dt} \tag{4.195}$$

Therefore, the arc length between two points on the curve can be found by integration:

$$s = \int_{t_1}^{t_2} \sqrt{\frac{d\mathbf{r}}{dt} \cdot \frac{d\mathbf{r}}{dt}}\, dt = \int_{t_1}^{t_2} \sqrt{\left(\frac{dx}{dt}\right)^2 + \left(\frac{dy}{dt}\right)^2 + \left(\frac{dz}{dt}\right)^2}\, dt \tag{4.196}$$

Let us expand the parametric equations of the curve (4.185) at a point $P(x_0, y_0, z_0)$,

$$\begin{aligned} x &= x_0 + \frac{dx}{dt}\Delta t + \frac{1}{2}\frac{d^2x}{dt^2}\Delta t^2 + \cdots \\ y &= y_0 + \frac{dy}{dt}\Delta t + \frac{1}{2}\frac{d^2y}{dt^2}\Delta t^2 + \cdots \\ z &= z_0 + \frac{dz}{dt}\Delta t + \frac{1}{2}\frac{d^2z}{dt^2}\Delta t^2 + \cdots \end{aligned} \tag{4.197}$$

and ignore the nonlinear terms to find the tangent line to the curve at the point:

$$\frac{x - x_0}{dx/dt} = \frac{y - y_0}{dy/dt} = \frac{z - z_0}{dz/dt} = \Delta t \tag{4.198}$$

Example 106 Length of a spatial road.

Consider a spatial closed road with the following parametric equations:

$$\begin{aligned} x &= (a + b\sin\theta)\cos\theta \\ y &= (a + b\sin\theta)\sin\theta \\ z &= b + b\cos\theta \end{aligned} \tag{4.199}$$

The total length of the road can be found by the integral of ds for θ from 0 to 2π:

$$s = \int_{\theta_1}^{\theta_2} \sqrt{\frac{d\mathbf{r}}{d\theta} \cdot \frac{d\mathbf{r}}{d\theta}} d\theta = \int_{\theta_1}^{\theta_2} \sqrt{\left(\frac{\partial x}{\partial \theta}\right)^2 + \left(\frac{\partial x}{\partial \theta}\right)^2 + \left(\frac{\partial x}{\partial \theta}\right)^2} d\theta$$

$$= \int_0^{2\pi} \frac{\sqrt{2}}{2} \sqrt{2a^2 + 3b^2 - b^2 \cos 2\theta + 4ab \sin\theta d\theta} \tag{4.200}$$

Example 107 History of clothoid.

The clothoid spiral is also called *Cornu spiral*, referring to *Alfred Cornu* (1841–1902), a French physicist who rediscovered the clothoid spiral. It may also be called *Euler spiral*, as *Leonard Euler* (1707–1783) was the first codiscoverer of the curve with *Jacques Bernoulli* (1654–1705) who formulated the clothoid spiral on deformations of elastic members. It is also called *Fresnel spiral* credited to *Augustin-Jean Fresnel spiral* (1788–1827) who independently rediscovered the curve in his work on the fringes of diffraction of light through a slot. In vehicle dynamics and road design industry it may also be called the *transition spiral* to refer to the road connections corners.

In the 19th century it became clear that we need a track shape with gradually varying curvature. Although circles were being used for most of the path, a correct transition curve was needed to gradually change the curvature from one path to the other. *Arthur Talbot* (1857–1942) in 1880 derived the same integrals as Bernoulli and Fresnel and introduced the railway transition spirals. Because of this contribution in railroad practical design, the clothoid spiral is also called *Talbot curve*. Talbot curve has been used in railways and road construction since.

It is said that Clotho was one of the three Fates who spun the thread of human life, by winding it around the spindle. At the beginning of the 20th century, the Italian mathematician *Ernesto Cesàro* (1859–1906), from this poetic reference gave the name "clothoid" to the curve with a double spiral shape.

4.2 Static Steering

Figure 4.16 illustrates a front-wheel-steering (FWS) vehicle that is turning to the left. There is a kinematic condition between the inner and outer wheels that allows them to turn slip-free at very low speed. The kinematic condition is called the *Ackerman condition* and is expressed by

$$\cot \delta_2 - \cot \delta_1 = \frac{w_f}{l} \tag{4.201}$$

where δ_1 and δ_2 are the steer angles of the wheel number 1 of the front left and the wheel number 2 of the front right wheel. In this equation the steer angles are measured from the x-axis and is positive if it is about positive z-axis (Jazar 2017).

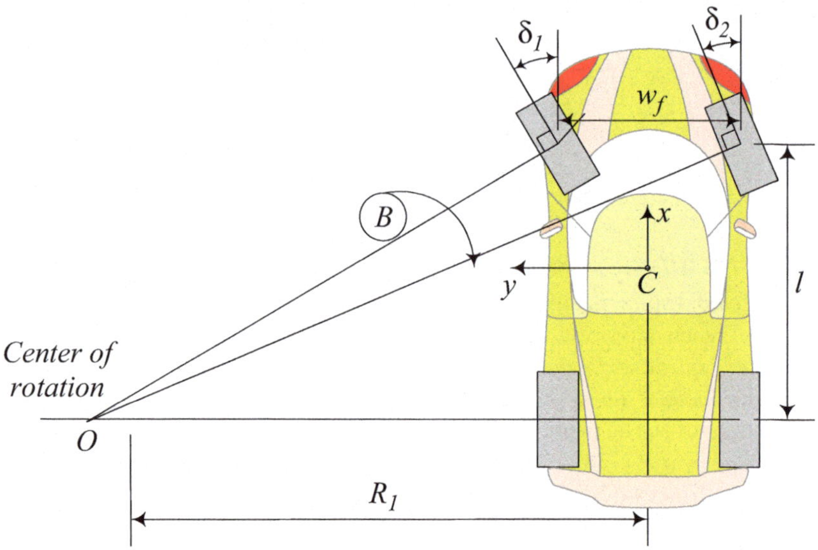

Fig. 4.16 A front-wheel-steering vehicle and the static steering condition

Ideally, perpendicular lines to all four wheels of a vehicle intersect at a single point called the kinematic center of rotation. When a vehicle is moving very slowly, we may assume the velocity vector of each wheel is in their tire plane. Therefore, the perpendicular lines to the tire planes intersect at the kinematic center of rotation of the vehicle, somewhere on the rear axis.

The distance between the tire planes of the left and right wheels is called the *track* and is shown by w. The distance between the front and real axles is called the *wheelbase* and is shown by l. Track w and wheelbase l are the kinematic width and length of the vehicle.

It is common to use one single steer angle command δ and calculate all other required steer angles based on that. When we employ the bicycle vehicle model the single steer angle is the cot-average of the inner and outer steer angles.

$$\cot\delta = \frac{\cot\delta_2 + \cot\delta_1}{2} \tag{4.202}$$

The angle δ is the equivalent steer angle of a bicycle having the same wheelbase l and radius of rotation ρ.

Proof To have all wheels turning freely on a curved road at very low speed, all the tire axes must intersect at a point. This criteria is the static steering condition. The tire axis is the perpendicular line to the tire-plane at the center of the tire.

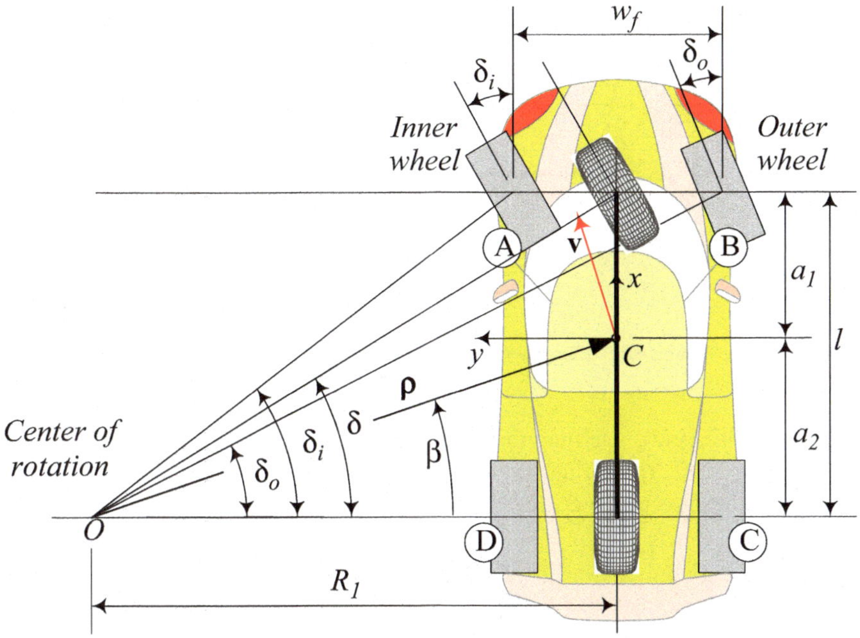

Fig. 4.17 A front-wheel-steering vehicle and steer angles of the inner and outer wheels, and equivalent bicycle model

Figure 4.17 illustrates a vehicle turning left about the turning center O. The inner and outer steer angles δ_1 and δ_2 may be calculated from the triangles $\triangle OAD$ and $\triangle OBC$ as:

$$\tan\delta_1 = \frac{l}{R_1 - \frac{w}{2}} \tag{4.203}$$

$$\tan\delta_2 = \frac{l}{R_1 + \frac{w}{2}} \tag{4.204}$$

Eliminating R_1

$$R_1 = \frac{1}{2}w + \frac{l}{\tan\delta_1} = -\frac{1}{2}w + \frac{l}{\tan\delta_2} \tag{4.205}$$

provides us with the static steering condition (4.201). The static steering condition is better defined by inner and outer steer angles.

$$\cot\delta_o - \cot\delta_i = \frac{w}{l} \tag{4.206}$$

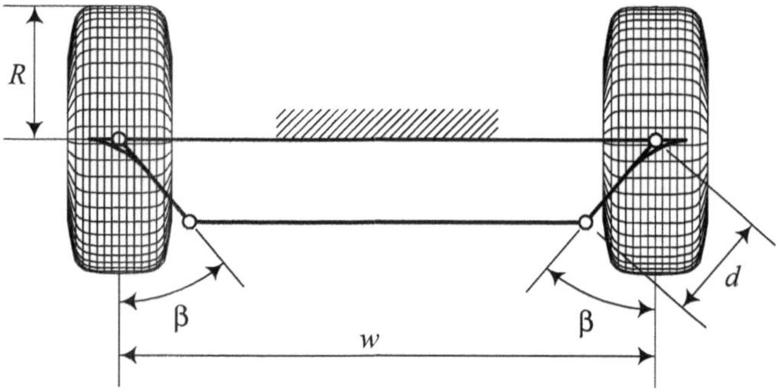

Fig. 4.18 Illustration of a trapezoidal steering mechanism

To find the vehicle's turning radius ρ, we look at the equivalent bicycle model, shown in Fig. 4.17. The radius of rotation ρ is perpendicular to the vehicle's velocity vector $\mathbf{v}$ at the mass center C.

$$\rho^2 = a_2^2 + R_1^2 \tag{4.207}$$

$$\cot \delta = \frac{R_1}{l} = \frac{1}{2} (\cot \delta_i + \cot \delta_o) \tag{4.208}$$

and therefore,

$$\rho = \sqrt{a_2^2 + l^2 \cot^2 \delta} \tag{4.209}$$

A device that provides steering according to the static condition (4.201) is called *static steering mechanism*, *Ackerman mechanism*, or *Ackerman geometry*. There is no practical mechanical linkage steering mechanism that can provide the static steering condition perfectly for every angle. However, we may design a multi-bar linkages to work close to the static condition and be exact at few angles. The ideal solution, however, would be steering by wires and controlling every wheel's steer angle independently. ■

Example 108 Trapezoidal steering mechanism.

The simplest practical steering mechanism that provides more steer angle to the inner wheel than the outer wheel is the trapezoidal steering mechanism as shown in Fig. 4.18. The trapezoidal mechanism is a symmetric four-bar linkage that has been used for more than 100 years as a steering connection. The mechanism is indicated by two parameters: angle β and offset arm length d. A steered position of the trapezoidal mechanism is shown in Fig. 4.19 to illustrate the inner and outer steer angles δ_i and δ_o (Genta 2007; Soni 1974; Hunt 1978).

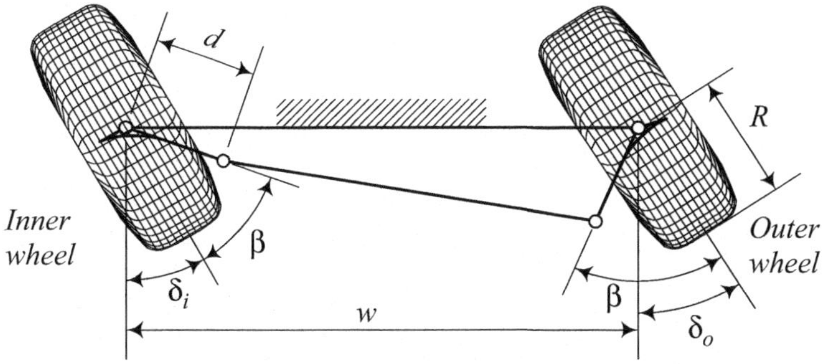

Fig. 4.19 A trapezoidal steering mechanism in steered position

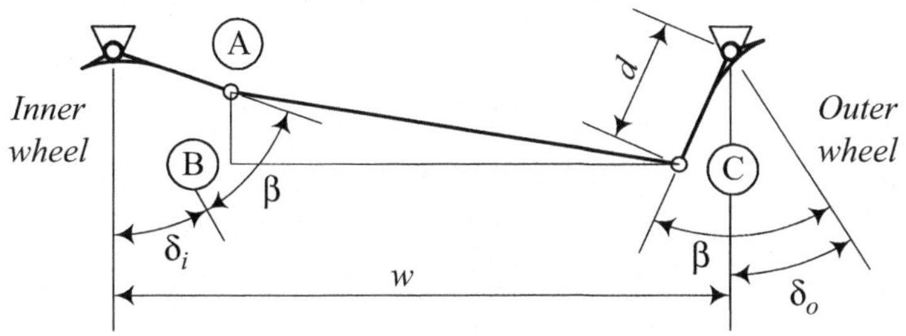

Fig. 4.20 Trapezoidal steering triangle ABC

The relationship between the inner and outer steer angles of a trapezoidal steering mechanism is given by:

$$\sin(\beta+\delta_i)+\sin(\beta-\delta_o) = \frac{w}{d}-\sqrt{\left(\frac{w}{d}-2\sin\beta\right)^2-(\cos(\beta-\delta_o)-\cos(\beta+\delta_i))^2} \tag{4.210}$$

To prove this equation, we examine Fig. 4.20. In the triangle $\triangle ABC$ we can write

$$(w-2d\sin\beta)^2 = (w-d\sin(\beta-\delta_o)-d\sin(\beta+\delta_i))^2 + (d\cos(\beta-\delta_o)-d\cos(\beta+\delta_i))^2 \tag{4.211}$$

and derive Eq. (4.210) with some manipulation.

Usually the functionality of a steering mechanism is tested by comparing the mechanism with the static steering condition (4.201). Figure 4.21 illustrates the inner-outer relationship of Eq. (4.210) for $l = 2.93\,\text{m} \approx 9.61\,\text{ft}$, $w = 1.66\,\text{m} \approx 5.45\,\text{ft}$ and respectively for $d = 0.4\,\text{m} \approx 1.3\,\text{ft}$ and $d = 0.2\,\text{m} \approx 0.65\,\text{ft}$. The horizontal axis shows the inner steer angle and the vertical axis shows the outer steer angle. It shows that for given l and w, a mechanism with $18\,\text{deg} \lessapprox \beta \lessapprox 22\,\text{deg}$ is the best simulator of the Ackerman mechanism if $\delta_i < 50\,\text{deg}$.

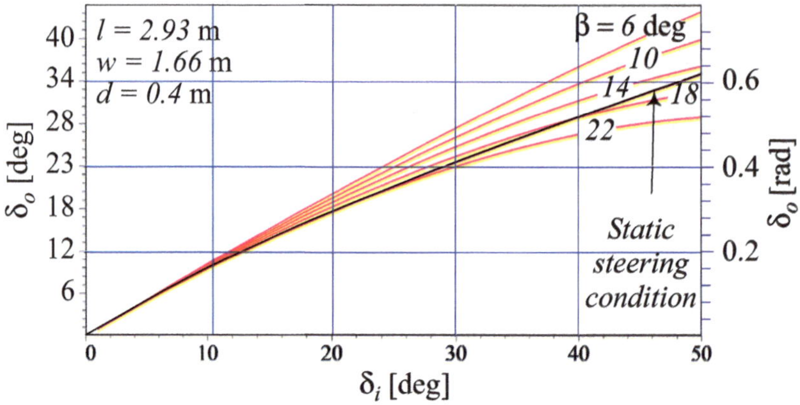

Fig. 4.21 Behavior of a trapezoidal steering mechanism, compared to the associated Ackerman mechanism $d = 0.4\,\mathrm{m}$

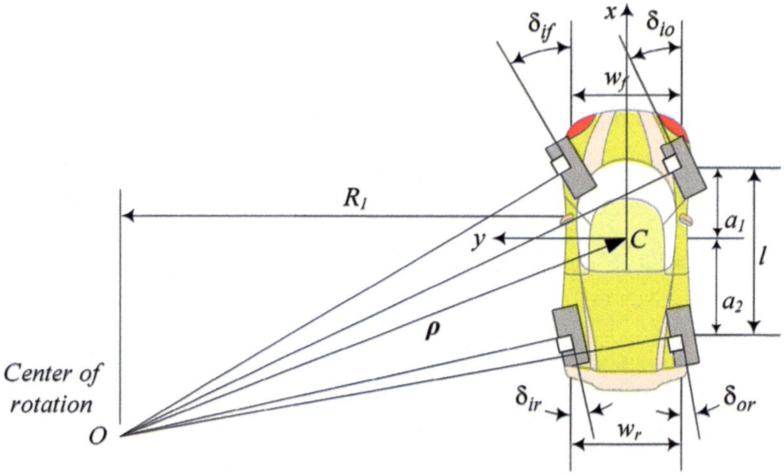

Fig. 4.22 A positive four-wheel steering vehicle

4.3 Four-Wheel Steering

At very low speeds, the kinematic steering condition that the perpendicular lines to each tire meet at one point must be applied. The intersection point is the *turning center* of the vehicle (Jazar et al. 2012).

Figure 4.22 illustrates a *positive* four-wheel steering vehicle, and Fig. 4.23 illustrates a *negative* $4WS$ vehicle. In a *positive* $4WS$ configuration the front and rear wheels steer in the same direction, and in a *negative* $4WS$ configuration the

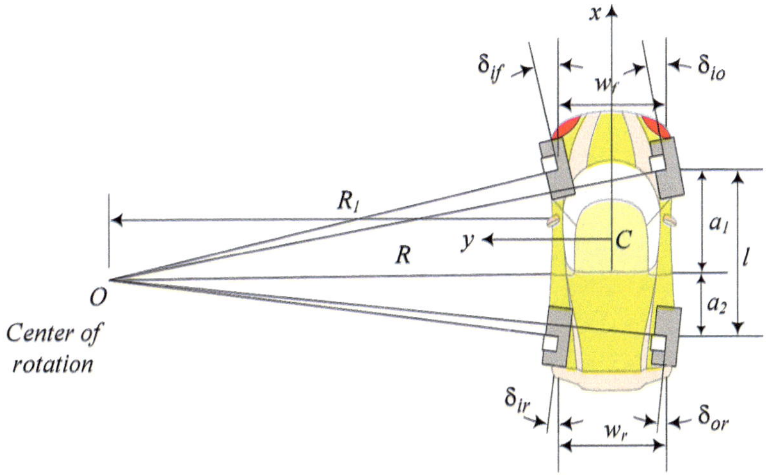

Fig. 4.23 A negative four-wheel steering vehicle

front and rear wheels steer opposite to each other. The kinematic condition between the steer angles of a $4WS$ vehicle is

$$\cot\delta_{of} - \cot\delta_{if} = \frac{w_f}{l} - \frac{w_r}{l}\frac{\cot\delta_{of} - \cot\delta_{if}}{\cot\delta_{or} - \cot\delta_{ir}} \tag{4.212}$$

where w_f and w_r are the front and rear tracks, δ_{if} and δ_{of} are the steer angles of the front inner and outer wheels, δ_{ir} and δ_{or} are the steer angles of the rear inner and outer wheels, and l is the wheelbase of the vehicle. We may also use the following more general equation for the kinematic condition between the steer angles of a $4WS$ vehicle

$$\cot\delta_{fr} - \cot\delta_{fl} = \frac{w_f}{l} - \frac{w_r}{l}\frac{\cot\delta_{fr} - \cot\delta_{fl}}{\cot\delta_{rr} - \cot\delta_{rl}} \tag{4.213}$$

where δ_{fl} and δ_{fr} are the steer angles of the front left and front right wheels, and δ_{rl} and δ_{rr} are the steer angles of the rear left and rear right wheels.

If we define the steer angles according to the sign convention shown in Fig. 4.24, then Eq. (4.213) expresses the kinematic condition for both positive and negative $4WS$ systems. Employing the wheel coordinate frame (x_w, y_w, z_w), we define the steer angle as the angle between the vehicle x-axis and the wheel x_w-axis, measured about the z-axis. Therefore, a steer angle is positive when the wheel is turned to the left, and it is negative when the wheel is turned to the right.

Proof The slip-free condition for wheels of a $4WS$ in a turn requires that the normal lines to the center of each tire-plane intersect at a common point. This is the kinematic steering condition.

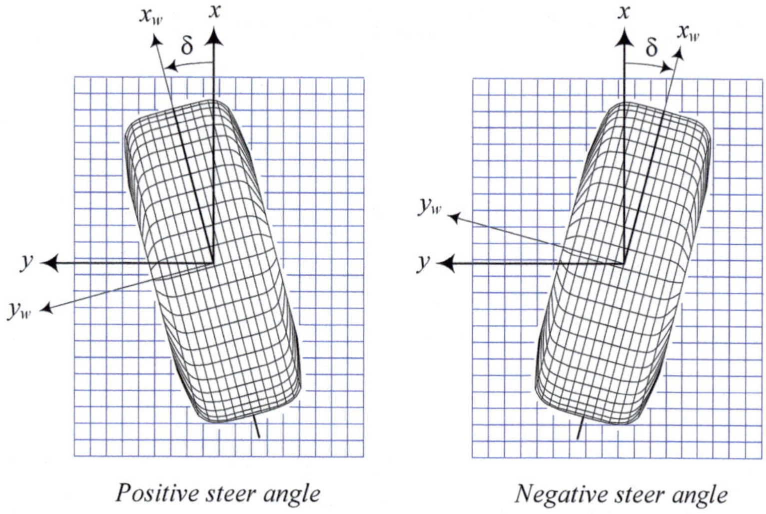

Fig. 4.24 Sign convention for steer angles

Figure 4.25 illustrates a positive $4WS$ vehicle in a left turn. The inner wheels are the left wheels that are closer to the turning center O. The longitudinal distance between point O and the axles of the car are indicated by c_1 and c_2 measured in the body coordinate frame.

The front inner and outer steer angles δ_{if}, δ_{of} may be calculated from the triangles $\triangle OAE$ and $\triangle OBF$, while the rear inner and outer steer angles δ_{ir}, δ_{or} may be calculated from the triangles $\triangle ODG$ and $\triangle OCH$ as follows.

$$\tan \delta_{if} = \frac{c_1}{R_1 - \frac{w_f}{2}} \tag{4.214}$$

$$\tan \delta_{of} = \frac{c_1}{R_1 + \frac{w_f}{2}} \tag{4.215}$$

$$\tan \delta_{ir} = \frac{c_2}{R_1 - \frac{w_r}{2}} \tag{4.216}$$

$$\tan \delta_{or} = \frac{c_2}{R_1 + \frac{w_r}{2}} \tag{4.217}$$

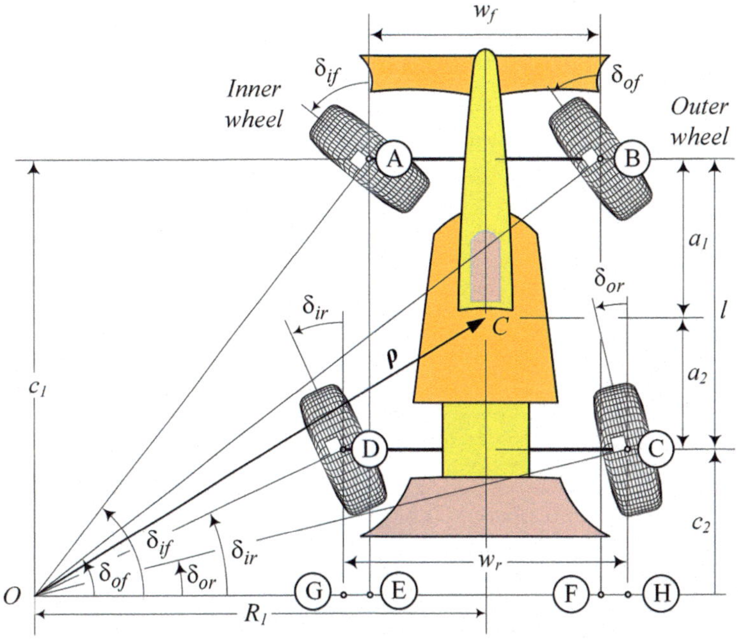

Fig. 4.25 Illustration of a positive four-wheel steering vehicle in a left turn

Eliminating R_1

$$R_1 = \frac{1}{2}w_f + \frac{c_1}{\tan\delta_{if}} = -\frac{1}{2}w_f + \frac{c_1}{\tan\delta_{of}} \tag{4.218}$$

between (4.214) and (4.215) provides the kinematic condition between the front steering angles δ_{if} and δ_{of}.

$$\cot\delta_{of} - \cot\delta_{if} = \frac{w_f}{c_1} \tag{4.219}$$

Similarly, we may eliminate R_1

$$R_1 = \frac{1}{2}w_r + \frac{c_2}{\tan\delta_{ir}} = -\frac{1}{2}w_r + \frac{c_2}{\tan\delta_{or}} \tag{4.220}$$

between (4.216) and (4.217) to provide the kinematic condition between the rear steering angles δ_{ir} and δ_{or}.

$$\cot\delta_{or} - \cot\delta_{ir} = \frac{w_r}{c_2} \tag{4.221}$$

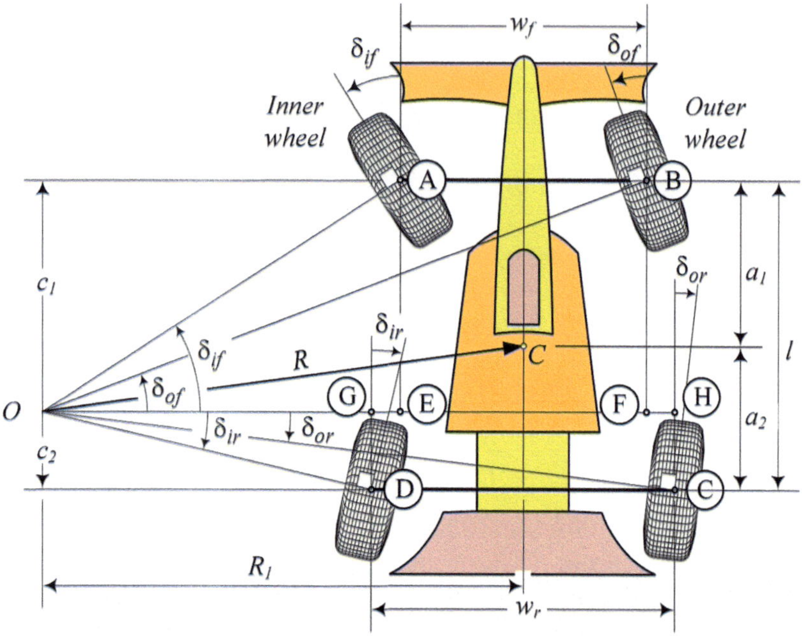

Fig. 4.26 Illustration of a negative four-wheel steering vehicle in a left turn

Using the constraint

$$c_1 - c_2 = l \tag{4.222}$$

we may combine Eqs. (4.219) and (4.221)

$$\frac{w_f}{\cot \delta_{of} - \cot \delta_{if}} - \frac{w_r}{\cot \delta_{or} - \cot \delta_{ir}} = l \tag{4.223}$$

to find the kinematic condition (4.212) between the steer angles of the front and rear wheels for a positive $4WS$ vehicle.

Figure 4.26 illustrates a negative $4WS$ vehicle in a left turn. The inner wheels are the left wheels that are closer to the turning center O. The front inner and outer steer angles δ_{if}, δ_{of} can be calculated from the triangles $\triangle OAE$ and $\triangle OBF$, while the rear inner and outer steer angles δ_{ir}, δ_{or} may be calculated from the triangles $\triangle ODG$ and $\triangle OCH$ as follows.

$$\tan \delta_{if} = \frac{c_1}{R_1 - \dfrac{w_f}{2}} \tag{4.224}$$

$$\tan \delta_{of} = \frac{c_1}{R_1 + \dfrac{w_f}{2}} \tag{4.225}$$

$$-\tan\delta_{ir} = \frac{-c_2}{R_1 - \dfrac{w_r}{2}} \tag{4.226}$$

$$-\tan\delta_{or} = \frac{-c_2}{R_1 + \dfrac{w_r}{2}} \tag{4.227}$$

Eliminating R_1

$$R_1 = \frac{1}{2}w_f + \frac{c_1}{\tan\delta_{if}} = -\frac{1}{2}w_f + \frac{c_1}{\tan\delta_{of}} \tag{4.228}$$

between (4.224) and (4.225) provides us with the kinematic condition between the front steering angles δ_{if} and δ_{of}.

$$\cot\delta_{of} - \cot\delta_{if} = \frac{w_f}{c_1} \tag{4.229}$$

Similarly, we may eliminate R_1

$$R_1 = \frac{1}{2}w_r + \frac{c_2}{\tan\delta_{ir}} = -\frac{1}{2}w_r + \frac{c_2}{\tan\delta_{or}} \tag{4.230}$$

between (4.226) and (4.227) to provide the kinematic condition between the rear steering angles δ_{ir} and δ_{or}.

$$\cot\delta_{or} - \cot\delta_{ir} = \frac{w_r}{c_2} \tag{4.231}$$

Using the constraint

$$c_1 - c_2 = l \tag{4.232}$$

we may combine Eqs. (4.229) and (4.231)

$$\frac{w_f}{\cot\delta_{of} - \cot\delta_{if}} - \frac{w_r}{\cot\delta_{or} - \cot\delta_{ir}} = l \tag{4.233}$$

to find the kinematic condition (4.212) between the steer angles of the front and rear wheels for a negative $4WS$ vehicle.

Using the sign convention of Fig. 4.24, we may re-examine Figs. 4.25 and 4.26. When the steer angle of the front wheels are positive, then the steer angle of the rear wheels are negative in a negative $4WS$ system, and are positive in a positive $4WS$ system. Therefore, Eq. (4.213)

$$\cot\delta_{fr} - \cot\delta_{fl} = \frac{w_f}{l} - \frac{w_r}{l}\frac{\cot\delta_{fr} - \cot\delta_{fl}}{\cot\delta_{rr} - \cot\delta_{rl}} \tag{4.234}$$

expresses the kinematic condition for both positive and negative $4WS$ systems. Similarly, the following equations can uniquely determine c_1 and c_2 regardless of the positive or negative $4WS$ system.

$$c_1 = \frac{w_f}{\cot\delta_{fr} - \cot\delta_{fl}} \tag{4.235}$$

$$c_2 = \frac{w_r}{\cot\delta_{rr} - \cot\delta_{rl}} \tag{4.236}$$

Four-wheel steering or all wheel steering AWS may be applied on vehicles to improve steering response, increase the stability at high speed maneuvering, or decrease turning radius at low speeds. A negative $4WS$ has shorter turning radius R than a front-wheel steering (FWS) vehicle.

For a FWS vehicle, the perpendicular to the front wheels meet at a point on the extension of the rear axle. However, for a $4WS$ vehicle, the intersection point can be any point in the xy plane. The point is the *turning center* of the car and its position depends on the steer angles of the wheels. Positive steering is also called *same steer*, and a negative steering is also called *counter steer*. ■

Example 109 Steering angles relationship.

Consider a car with the following dimensions.

$$l = 2.8\,\text{m} \qquad w_f = 1.35\,\text{m} \qquad w_r = 1.4\,\text{m} \tag{4.237}$$

The set of Eqs. (4.214)–(4.217) which are the same as (4.224)–(4.227) must be used to find the kinematic steer angles of the wheels. Assume one of the angles, such as

$$\delta_{if} = 15\,\text{deg} \tag{4.238}$$

is known as an input steer angle. To find the other steer angles, we need to know the position of the turning center O. The position of the turning center can be determined if we have one of the three parameters c_1, c_2, R_1. To clarify this fact, let us assume that the car is turning left and we know the value of δ_{if}. Therefore, the perpendicular line to the front left wheel is known. The turning center can be any point on this line. When we pick a point, the other wheels can be adjusted accordingly.

The steer angles for a $4WS$ system is a set of four equations, each with two variables.

$$\delta_{if} = \delta_{if}\,(c_1, R_1) \qquad \delta_{of} = \delta_{of}\,(c_1, R_1) \tag{4.239}$$

$$\delta_{ir} = \delta_{ir}\,(c_2, R_1) \qquad \delta_{or} = \delta_{or}\,(c_2, R_1) \tag{4.240}$$

If c_1 and R_1 are known, we will be able to determine the steer angles δ_{if}, δ_{of}, δ_{ir}, and δ_{or} uniquely. However, a practical situation is when we have one of the steer angles, such as δ_{if}, and we need to determine the required steer angle of the other wheels, δ_{of}, δ_{ir}, δ_{or}. It can be done if we know c_1 or R_1.

The turning center is the curvature center of the path of motion. If the path of motion is known, then at any point of the road, the turning center can be found in the vehicle coordinate frame.

In this example, let us assume

$$R_1 = 50\,\mathrm{m} \tag{4.241}$$

therefore, from Eq. (4.214), we have

$$c_1 = \left(R_1 - \frac{w_f}{2}\right)\tan\delta_{if} = \left(50 - \frac{1.35}{2}\right)\tan\frac{\pi}{12} = 13.217\,\mathrm{m} \tag{4.242}$$

Because $c_1 > l$ and $\delta_{if} > 0$ the vehicle is in a positive $4WS$ configuration and the turning center is behind the rear axle of the car.

$$c_2 = c_1 - l = 13.217 - 2.8 = 10.417\,\mathrm{m} \tag{4.243}$$

Now, employing Eqs. (4.215)–(4.217) provides us the other steer angles.

$$\begin{aligned}\delta_{of} &= \tan^{-1}\frac{c_1}{R_1 + \frac{w_f}{2}} = \tan^{-1}\frac{13.217}{50 + \frac{1.35}{2}} \\ &= 0.25513\,\mathrm{rad} \approx 14.618\,\mathrm{deg}\end{aligned} \tag{4.244}$$

$$\begin{aligned}\delta_{ir} &= \tan^{-1}\frac{c_2}{R_1 - \frac{w_r}{2}} = \tan^{-1}\frac{10.417}{50 - \frac{1.4}{2}} \\ &= 0.20824\,\mathrm{rad} \approx 11.931\,\mathrm{deg}\end{aligned} \tag{4.245}$$

$$\begin{aligned}\delta_{or} &= \tan^{-1}\frac{c_2}{R_1 + \frac{w_r}{2}} = \tan^{-1}\frac{10.417}{50 + \frac{1.4}{2}} \\ &= 0.20264\,\mathrm{rad} \approx 11.61\,\mathrm{deg}\end{aligned} \tag{4.246}$$

Example 110 Position of the turning center.

The turning center of a vehicle, in the vehicle body coordinate frame, is at a point with coordinates (x_O, y_O). The coordinates of the turning center are

$$x_O = -a_2 - c_2 = -a_2 - \frac{w_r}{\cot\delta_{or} - \cot\delta_{ir}} \tag{4.247}$$

$$y_O = R_1 = \frac{l + \frac{1}{2}\left(w_f \tan\delta_{if} - w_r \tan\delta_{ir}\right)}{\tan\delta_{if} - \tan\delta_{ir}} \tag{4.248}$$

Equation (4.248) is found by substituting c_1 and c_2 from (4.228) and (4.230) in (4.232), and defining y_O in terms of δ_{if} and δ_{ir}. It is also possible to define y_O in terms of δ_{of} and δ_{or}.

Equations (4.247) and (4.248) can be used to define the coordinates of the turning center for both positive and negative $4WS$ systems.

As an example, let us examine a car with

$$l = 2.8\,\text{m} \qquad w_f = 1.35\,\text{m} \qquad w_r = 1.4\,\text{m} \qquad a_1 = a_2 \tag{4.249}$$

$$\begin{aligned} \delta_{if} &= 0.26180\,\text{rad} \approx 15\,\text{deg} \\ \delta_{of} &= 0.25513\,\text{rad} \approx 14.618\,\text{deg} \\ \delta_{ir} &= 0.20824\,\text{rad} \approx 11.931\,\text{deg} \\ \delta_{or} &= 0.20264\,\text{rad} \approx 11.61\,\text{deg} \end{aligned} \tag{4.250}$$

and find the position of the turning center.

$$\begin{aligned} x_O &= -a_2 - \frac{w_r}{\cot\delta_{or} - \cot\delta_{ir}} \\ &= -\frac{2.8}{2} - \frac{1.4}{\cot 0.20264 - \cot 0.20824} = -11.802\,\text{m} \end{aligned} \tag{4.251}$$

$$\begin{aligned} y_O &= \frac{l + \frac{1}{2}\left(w_f \tan\delta_{if} - w_r \tan\delta_{ir}\right)}{\tan\delta_{if} - \tan\delta_{ir}} \\ &= \frac{2.8 + \frac{1}{2}(1.35 \tan 0.26180 - 1.4 \tan 0.20824)}{\tan 0.26180 - \tan 0.20824} = 50.011\,\text{m} \end{aligned} \tag{4.252}$$

The position of turning center for a FWS vehicle is at

$$x_O = -a_2 \qquad y_O = \frac{1}{2}w_f + \frac{l}{\tan\delta_{if}} \tag{4.253}$$

and for a RWS vehicle is at

$$x_O = a_1 \qquad y_O = \frac{1}{2}w_r + \frac{l}{\tan\delta_{ir}} \tag{4.254}$$

Example 111 Curvature radius.

Consider a road as a path of motion that is expressed mathematically by a function $Y = f(X)$, in a global coordinate frame. The radius of curvature R_κ of such a road at point X is

$$R_\kappa = \frac{\left(1 + Y'^2\right)^{3/2}}{Y''} \tag{4.255}$$

where

$$Y' = \frac{dY}{dX} \qquad Y'' = \frac{d^2Y}{dX^2} \tag{4.256}$$

Consider a road with a given equation

$$Y = \frac{X^2}{200} \qquad Y' = \frac{X}{100} \qquad Y'' = \frac{1}{100} \tag{4.257}$$

where both X and Y are measured in meter [m]. The curvature radius of the road is.

$$R_\kappa = \frac{\left(1 + Y'^2\right)^{3/2}}{Y''} = 100\left(\frac{1}{10{,}000}X^2 + 1\right)^{3/2} \tag{4.258}$$

At $X = 30\,\mathrm{m}$, we have

$$Y = \frac{9}{2}\,\mathrm{m} \qquad Y' = \frac{3}{10} \qquad Y'' = \frac{1}{100}\,\mathrm{m}^{-1} \tag{4.259}$$

and therefore,

$$R_\kappa = 113.80\,\mathrm{m} \tag{4.260}$$

Example 112 Autodriver.

Consider a car at the global position (X, Y) that is moving on a road, as shown in Fig. 4.27. Point C indicates the center of curvature of the road at the car's position. The center of curvature of the road is supposed to be the turning center of the car at the instant of consideration (Jazar 2010b; Bourmistrova et al. 2011).

There is a global coordinate frame G attached to the ground, and a vehicle coordinate frame B attached to the car at its mass center C. The z and Z axes are parallel and the angle ψ indicates the angle between the X and x axes. If (X_C, Y_C) are the coordinates of C in the global coordinate frame G, then the coordinates of C in B would be

$$^{B}\mathbf{r}_C = R_{z,\psi}\left(^{G}\mathbf{r}_C - {}^{G}\mathbf{d}\right) \tag{4.261}$$

$$\begin{bmatrix} x_C \\ y_C \\ 0 \end{bmatrix} = \begin{bmatrix} \cos\psi & \sin\psi & 0 \\ -\sin\psi & \cos\psi & 0 \\ 0 & 0 & 1 \end{bmatrix} \left(\begin{bmatrix} X_C \\ Y_C \\ 0 \end{bmatrix} - \begin{bmatrix} X \\ Y \\ 0 \end{bmatrix} \right)$$

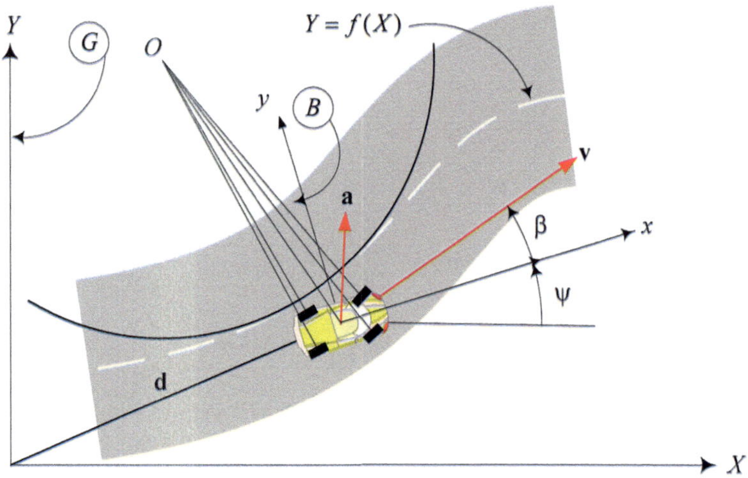

Fig. 4.27 Illustration of a car that is moving on a road at the point that O is the center of curvature

$$= \begin{bmatrix} (X_C - X)\cos\psi + (Y_C - Y)\sin\psi \\ (Y_C - Y)\cos\psi - (X_C - X)\sin\psi \\ 0 \end{bmatrix} \tag{4.262}$$

Having coordinates of C in the vehicle coordinate frame is enough to determine R_1, c_1, and c_2.

$$R_1 = y_C = (Y_C - Y)\cos\psi - (X_C - X)\sin\psi \tag{4.263}$$

$$c_2 = -a_2 - x_C = -(X_C - X)\cos\psi - (Y_C - Y)\sin\psi - a_2 \tag{4.264}$$

$$c_1 = c_2 + l = -(X_C - X)\cos\psi - (Y_C - Y)\sin\psi + a_1 \tag{4.265}$$

Then, the required steer angles of the wheels can be uniquely determined by Eqs. (4.214)–(4.217).

It is possible to define a road by a mathematical function $Y = f(X)$ in a global coordinate frame. At any point X of the road, the position of the vehicle and the position of the turning center in the vehicle coordinate frame can be determined. The required steer angles can accordingly be set to keep the vehicle on the road and run the vehicle in the correct direction. This principle may be used to design an **autodriver**.

As an example, let us consider a car that is moving tangent to a road with a given equation

$$Y = \frac{X^2}{200} \tag{4.266}$$

where both X and Y are measured in meter [m]. At $X = 30$ m, we have $Y = 4.5$ m and $Y' = 0.3$, $Y'' = 0.01$, and therefore

$$\psi = \arctan\frac{dY}{dX} = \arctan 0.3 = 0.29146\,\mathrm{rad} \approx 16.7\,\mathrm{deg} \tag{4.267}$$

The curvature radius at (30, 4.5), from Example 111, is

$$R_\kappa = 113.80\,\mathrm{m} \tag{4.268}$$

The tangent line to the road is

$$Y - 4.5 = 0.3\,(X - 30) \tag{4.269}$$

and therefore the perpendicular line to the road is

$$Y - 4.5 = -\frac{10}{3}\,(X - 30) \tag{4.270}$$

Having $R_\kappa = 113.80$ m,

$$R_\kappa = \frac{\left(1 + Y'^2\right)^{3/2}}{Y''} = \frac{\left(1 + 0.3^2\right)^{3/2}}{0.01} = 113.8\,\mathrm{m} \tag{4.271}$$

we have

$$(X - X_C)^2 + (Y - Y_C)^2 = R_\kappa^2 \tag{4.272}$$

and we can find the global coordinates of the curvature center (X_C, Y_C) at the proper intersection of the line (4.270) and circle (4.272).

$$X_C = -2.7002\,\mathrm{m} \qquad Y_C = 113.5\,\mathrm{m} \tag{4.273}$$

The coordinates of the turning center in the body frame would then be

$$\begin{bmatrix} x_C \\ y_C \\ 0 \end{bmatrix} = \begin{bmatrix} \cos\psi & \sin\psi & 0 \\ -\sin\psi & \cos\psi & 0 \\ 0 & 0 & 1 \end{bmatrix} \left(\begin{bmatrix} X_C \\ Y_C \\ 0 \end{bmatrix} - \begin{bmatrix} X \\ Y \\ 0 \end{bmatrix} \right) = \begin{bmatrix} 0 \\ 113.8 \\ 0 \end{bmatrix} \tag{4.274}$$

Example 113 Curvature equation.

Consider a vehicle that is moving on a path $Y = f(X)$ with velocity $\mathbf{v}$ and acceleration $\mathbf{a}$. The curvature $\kappa = 1/R$ of the path that the vehicle is moving on is

$$\kappa = \frac{1}{R} = \frac{a_n}{v^2} \tag{4.275}$$

where a_n is the normal component of the acceleration **a**. The normal component a_n is toward the rotation center and is equal to

$$a_n = \left| \frac{\mathbf{v}}{v} \times \mathbf{a} \right| = \frac{1}{v} \left| \mathbf{v} \times \mathbf{a} \right|$$
$$= \frac{1}{v} \left(a_Y v_X - a_X v_Y \right) = \frac{\ddot{Y}\dot{X} - \ddot{X}\dot{Y}}{\sqrt{\dot{X}^2 + \dot{Y}^2}} \tag{4.276}$$

and therefore,

$$\kappa = \frac{\ddot{Y}\dot{X} - \ddot{X}\dot{Y}}{\left(\dot{X}^2 + \dot{Y}^2\right)^{3/2}} = \frac{\ddot{Y}\dot{X} - \ddot{X}\dot{Y}}{\dot{X}^3} \frac{1}{\left(1 + \frac{\dot{Y}^2}{\dot{X}^2}\right)^{3/2}} \tag{4.277}$$

However,

$$Y' = \frac{dY}{dX} = \frac{\dot{Y}}{\dot{X}} \tag{4.278}$$

$$Y'' = \frac{d^2Y}{dX^2} = \frac{d}{dx}\left(\frac{\dot{Y}}{\dot{X}}\right) = \frac{d}{dt}\left(\frac{\dot{Y}}{\dot{X}}\right)\frac{1}{\dot{X}} = \frac{\ddot{Y}\dot{X} - \ddot{X}\dot{Y}}{\dot{X}^3} \tag{4.279}$$

and we find the equation for the curvature of the path and radius of the curvature based on the equation of the path (Jazar 2011).

$$\kappa = \frac{Y''}{\left(1 + Y'^2\right)^{3/2}} \tag{4.280}$$

$$R_\kappa = \frac{1}{\kappa} = \frac{\left(1 + Y'^2\right)^{3/2}}{Y''} = \frac{\left(\dot{X}^2 + \dot{Y}^2\right)^{3/2}}{\ddot{Y}\dot{X} - \ddot{X}\dot{Y}} \tag{4.281}$$

As an example, let us consider a road with a given equation

$$Y = \frac{X^2}{200} \tag{4.282}$$

At a point with $X = 30$ m, we have

$$Y = \frac{9}{2}\,\mathrm{m} \qquad Y' = \frac{3}{10} \qquad Y'' = \frac{1}{100}\,\mathrm{m}^{-1} \tag{4.283}$$

and therefore,

$$\kappa = 8.7874 \times 10^{-3}\,\mathrm{m}^{-1} \qquad R_\kappa = 113.80\,\mathrm{m} \tag{4.284}$$

Example 114 Center of curvature in global coordinate frame.

Assume a planar road being expressed by a parametric equation in a global frame as

$$X = X(t) \qquad Y = Y(t) \tag{4.285}$$

The perpendicular line to the road at a point (X_0, Y_0) is:

$$Y = Y_0 - (X - X_0)/\tan\theta \tag{4.286}$$

$$\tan\theta = \frac{dY}{dX} = \frac{dY/dt}{dX/dt} = \frac{\dot{Y}}{\dot{X}} \tag{4.287}$$

The curvature center (X_C, Y_C) is on the perpendicular line and is at a distance R_κ from the point $(X, Y) = (X_0, Y_0)$. Therefore,

$$X_C = X - \frac{\dot{Y}\left(\dot{X}^2 + \dot{Y}^2\right)}{\ddot{Y}\dot{X} - \ddot{X}\dot{Y}} \qquad Y_C = Y + \frac{\dot{X}\left(\dot{X}^2 + \dot{Y}^2\right)}{\ddot{Y}\dot{X} - \ddot{X}\dot{Y}} \tag{4.288}$$

Employing Eq. (4.274), the curvature center in the body coordinate frame of a moving vehicle that its x-axis makes the angle ψ with the global X-axis is

$$\begin{aligned} x_C &= (X_C - X)\cos\psi + (Y_C - Y)\sin\psi \\ &= \left(\dot{X}\sin\psi - \dot{Y}\cos\psi\right)\frac{\dot{X}^2 + \dot{Y}^2}{\ddot{Y}\dot{X} - \ddot{X}\dot{Y}} \end{aligned} \tag{4.289}$$

$$\begin{aligned} y_C &= (Y_C - Y)\cos\psi - (X_C - X)\sin\psi \\ &= \left(\dot{X}\cos\psi + \dot{Y}\sin\psi\right)\frac{\dot{X}^2 + \dot{Y}^2}{\ddot{Y}\dot{X} - \ddot{X}\dot{Y}} \end{aligned} \tag{4.290}$$

As an example, the global coordinate frame of a parabolic path

$$X = t \qquad Y = \frac{t^2}{200} \tag{4.291}$$

would be

$$X_C = X - \frac{\dot{X}^2 + \dot{Y}^2}{\ddot{Y}\dot{X} - \ddot{X}\dot{Y}}\dot{Y} = 2.5244 \times 10^{-29}t - 0.0001t^3 \tag{4.292}$$

$$Y_C = Y + \frac{\dot{X}^2 + \dot{Y}^2}{\ddot{Y}\dot{X} - \ddot{X}\dot{Y}}\dot{X} = 0.015\,t^2 + 100 \tag{4.293}$$

Example 115 An elliptic path and curvature center.

Consider an elliptic path with equations

$$X = a\cos t \qquad Y = b\sin t \tag{4.294}$$

$$a = 100\,\mathrm{m} \qquad b = 65\,\mathrm{m} \tag{4.295}$$

The curvature center of the road in the global coordinate frame is at

$$X_C = X - \frac{\dot{Y}\left(\dot{X}^2 + \dot{Y}^2\right)}{\ddot{Y}\dot{X} - \ddot{X}\dot{Y}} = \frac{a^2 - b^2}{a}\cos^3 t = \frac{231}{4}\cos^3 t \tag{4.296}$$

$$Y_C = Y + \frac{\dot{X}\left(\dot{X}^2 + \dot{Y}^2\right)}{\ddot{Y}\dot{X} - \ddot{X}\dot{Y}} = -\frac{a^2 - b^2}{b}\sin^3 t = \frac{-1155}{13}\sin^3 t \tag{4.297}$$

Therefore, the curvature center in the vehicle coordinate frame would be

$$\begin{aligned} x_C &= \left(\dot{X}\sin\psi - \dot{Y}\cos\psi\right)\frac{\dot{X}^2 + \dot{Y}^2}{\ddot{Y}\dot{X} - \ddot{X}\dot{Y}} \\ &= \left(\frac{a^2 - b^2}{a}\cos^3 t - a\cos t\right)\cos\psi \\ &\quad - \left(\frac{a^2 - b^2}{b}\sin^3 t + b\sin t\right)\sin\psi \end{aligned} \tag{4.298}$$

$$\begin{aligned} y_C &= \left(\dot{X}\cos\psi + \dot{Y}\sin\psi\right)\frac{\dot{X}^2 + \dot{Y}^2}{\ddot{Y}\dot{X} - \ddot{X}\dot{Y}} \\ &= -\left(\frac{a^2 - b^2}{a}\sin^3 t + b\sin t\right)\cos\psi \\ &\quad - \left(\frac{a^2 - b^2}{a}\cos^3 t - a\cos t\right)\sin\psi \end{aligned} \tag{4.299}$$

Figure 4.28 illustrates the elliptic path and its curvature center.

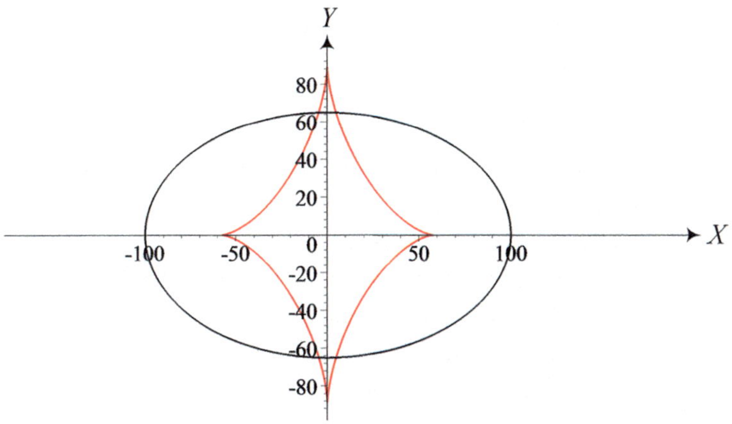

Fig. 4.28 An elliptic path and its curvature center

4.4 Chapter Summary

Passenger cars are developed to move on smooth paved pre-designed roads. To keep vehicles on road, we need a steering mechanism to provide steer angle as an input to the vehicle dynamic system. Ideally, all wheels of a vehicle should be able to steer independently such that the vehicle follows the desired path at the given speed. In this chapter we review steer and road dynamics.

Roads are made by continuously connecting straight and circular paths by proper transition turning sections. Having a continuous and well-behaved curvature is a necessary criterion in road design. The *clothoid spiral* is the best smooth transition connecting curve in road design which is expressed by parametric equations called *Fresnel Integrals*. The curvature of the clothoid curve varies linearly with arc length and this linearity makes clothoid the smoothest driving transition curve. Having a road with linearly increasing curvature is equivalent to entering the path with a steering wheel at the neutral position and turning the steering wheel with a constant angular velocity. This is a desirable and natural driving action.

Ideally, perpendicular lines to all wheels of a vehicle intersect at a single point called the kinematic center of rotation. When a vehicle is moving very slowly, we may assume the velocity vector of each wheel is in their tire plane. Therefore, the perpendicular lines to the tire planes intersect at the kinematic center of rotation of the vehicle, somewhere on the rear axis. However, when the vehicle moves faster, the actual center of rotation will move away from the kinematic center of rotation. Steering mechanism relates the left and right steerable wheels and provide a mathematical relationship to calculate all steer angles based on the angle of the steering wheel or the steer angle one of the wheels.

4.5 Key Symbols

$4WS$	Four-wheel-steering
a, b, c, d	Lengths of the links of a four-bar linkage
a_i	Distance of the axle number i from the mass center
A, B, C	Input angle parameters of a four-bar linkage
AWS	All-wheel-steering
b_1	Distance of the hinge point from rear axle
b_2	Distance of trailer axle from the hinge point
c	Stability index of a trailer motion
c_1	Longitudinal distance of turn center and front axle
c_2	Longitudinal distance of turn center and rear axle
c_s	$4WS$ factor
C	Mass center, curvature center
$C_1, C_2, \cdots$	Constants of integration
d	Arm length in trapezoidal steering mechanism
e	Error
e	Length of the offset arm
FWS	Front-wheel-steering
g	Overhang distance
J	Link parameters of a four-bar linkage
l	Wheelbase
l_s	Steering length
n	Number of increments
O	Center of rotation in a turn, curvature center
p	Perturbation in u
q	Perturbation in v
r	Yaw velocity of a turning vehicle
$\mathbf{r}$	Position vector of a car at the hinge
R	Radius of rotation at mass center
R_1	Radius of rotation at the center of the rear axle for FWS
R_1	Horizontal distance of O and the center of axles
R_c	Trailer's radius of rotation
R_t	Radius of rotation at the center of the trailer axle
R_w	Radius of the rear wheel
R_κ	Curvature radius

RWS	Rear wheel steering
$\mathbf{s}$	Position vector of a trailer at the axle center
t	Time
u	Temporary variable in car-trailer analysis
u_R	Steering rack translation
$\hat{u}$	Unit vector
$v \equiv \dot{x},\ \mathbf{v}$	Vehicle velocity, temporary variable in car-trailer analysis
v_{ri}	Speed of the inner rear wheel
v_{ro}	Speed of the outer rear wheel
w	Track
w_f	Front track
w_r	Rear track
$x, y, z,\ \mathbf{x}$	Displacement
$\mathbf{z} = \mathbf{r} - \mathbf{s}$	Position vector of a trailer relative to the car
β	Arm angle in trapezoidal steering mechanism
δ	cot-average of the inner and outer steer angles
$\delta_1 = \delta_{fl}$	Front left wheel steer angle
$\delta_2 = \delta_{fr}$	Front right wheel steer angle
δ_{Ac}	Steer angle based on Ackerman condition
δ_{fl}	Front left wheel steer angle
δ_{fr}	Front right wheel steer angle
δ_i	Inner wheel
δ_{rl}	Rear left wheel steer angle
δ_{rr}	Rear right wheel steer angle
δ_o	Outer wheel
δ_S	Steer command
$\Delta = \delta_2 - \delta_{Ac}$	Steer angle difference
θ	Angle between trailer and vehicle longitudinal axes
κ	Curvature of a road
λ	Eigenvalue
ω	Angular velocity
$\omega_i = \omega_{ri}$	Angular velocity of the rear inner wheel
$\omega_o = \omega_{ro}$	Angular velocity of the rear outer wheel

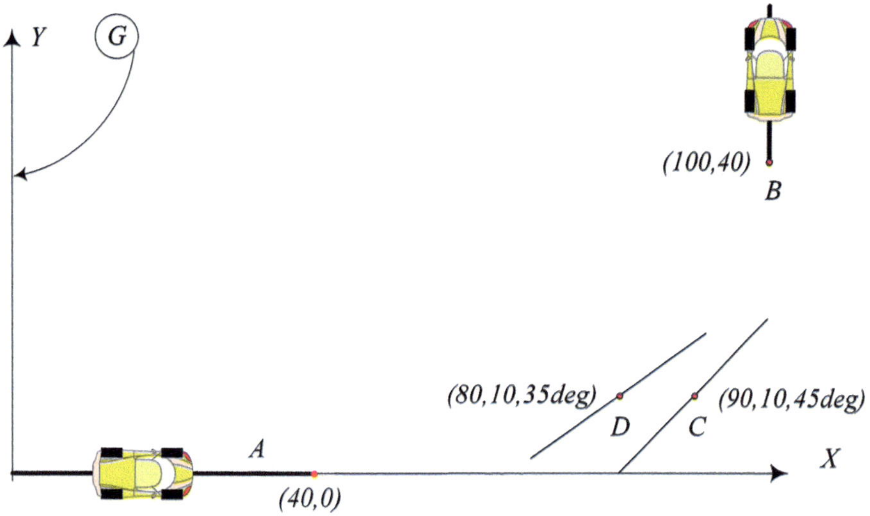

Fig. 4.29 Two straight roads need to be connected. The connecting clothoids must meet at a given middle point

Exercises

1. Radius of rotation.
   In Fig. 4.16 show that the mass center of the steered vehicle will turn on a circle with radius ρ,

$$\rho = \sqrt{a_2^2 + l^2 \cot^2 \delta} \tag{4.300}$$

2. 90 deg connection to straight roads.
   In Fig. 4.29, connect road A at point (40, 0) m to road B at point (100, 40) m. The connecting road will be made by two pieces of clothoids. The first one begins from road A to a designed point at a certain slop angle, and the second piece starts from the middle point and ends to the road B. Assume the middle point is:
   (a) point C at (90, 10) m and at slope 45 deg.
   (b) point D at (80, 10) m and at slope 35 deg.

3. Radius of rotation.
   Consider a two-axle truck that is offered in different wheelbases.

$$l = 109\,\text{in} \qquad l = 132.5\,\text{in} \qquad l = 150.0\,\text{in} \qquad l = 176.0\,\text{in} \tag{4.301}$$

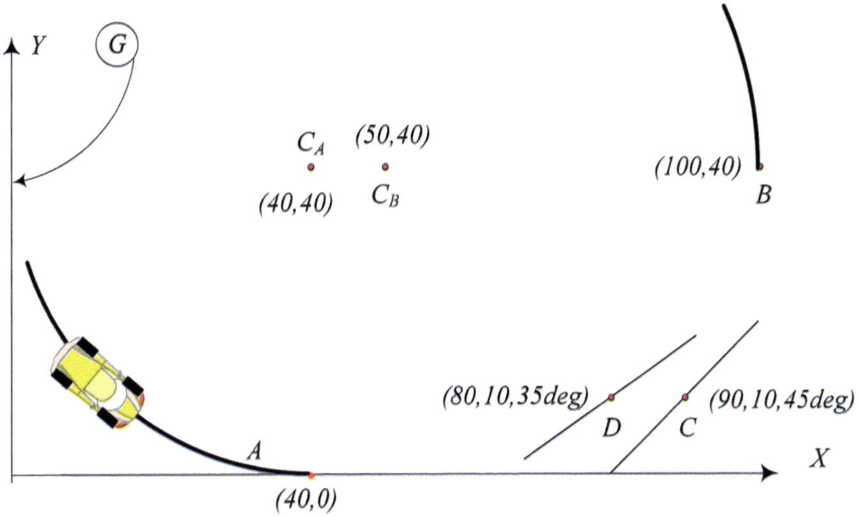

Fig. 4.30 Two circular roads need to be connected. The connecting clothoids must meet at a given middle point

If the front track of the vehicles is

$$w = 70\,\text{in} \tag{4.302}$$

and $a_1 = a_2$, calculate the radius of rotations for $\delta = 30$ deg.

4. Connecting road with given a and κ.
Assume $a = 200$. determine the clothoid connecting road from point $(0, 0)$ m to a point at which the slope is as below. Determine the coordinate of the end point, slope, and coordinate of the curvature center of the road at the end point.
   (a) $\kappa = 0.05$
   (b) $\kappa = 0.01$
   (c) $\kappa = 0.02$

5. 90 deg connection to circular roads.
In Fig. 4.30, assume the road A and B to be circular path with given center of rotation. Connect road A at point $(40, 0)$ m to road B at point $(100, 40)$ m. The connecting road will be made by two pieces of clothoids. The first begins from road A to a designed point at a certain slop angle, and the second piece starts from the middle point and ends to the road B. Assume the middle point is:
   (a) point C at $(90, 10)$ m and at slope 45 deg.
   (b) point D at $(80, 10)$ m and at slope 35 deg.

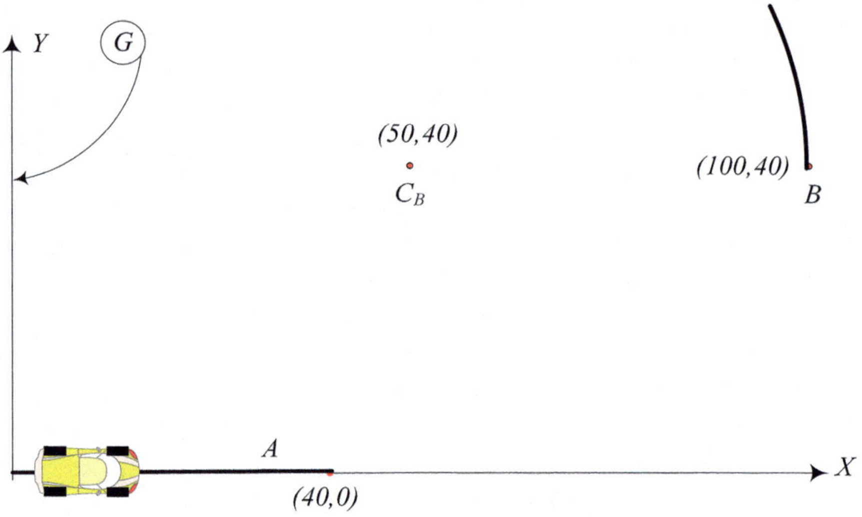

Fig. 4.31 A straight load needs to be connected to a circular road

6. Using design chart.
   Use the design chart to connect a straight road to a circular path with
   (a) radius $R = 100\,\mathrm{m}$ for $a = 200$.
   (b) radius $R = 100\,\mathrm{m}$ for $a = 100$.
   (c) radius $R = 200\,\mathrm{m}$ for $a = 100$.
   (d) radius $R = 200\,\mathrm{m}$ for $a = 200$.
7. Bank angle of the road.
   The straight road A in Fig. 4.31 will be connected to road B using clothoid connecting curve. Determine the required clothoid road.
8. Trapezoidal steering mechanism.
   Derive the steering equations for the trapezoidal steering mechanism of Fig. 4.19 if the d link is unequal for left and right, $d_f = 0.9d_r$.

Appendix A
Trigonometric Formulas

Definitions in Terms of Exponentials

$$\cos z = \frac{e^{iz} + e^{-iz}}{2} \tag{A.1}$$

$$\sin z = \frac{e^{iz} - e^{-iz}}{2i} \tag{A.2}$$

$$\tan z = \frac{e^{iz} - e^{-iz}}{i\left(e^{iz} + e^{-iz}\right)} \tag{A.3}$$

$$e^{iz} = \cos z + i \sin z \tag{A.4}$$

$$e^{-iz} = \cos z - i \sin z \tag{A.5}$$

Angle Sum and Difference

$$\sin(\alpha \pm \beta) = \sin\alpha\cos\beta \pm \cos\alpha\sin\beta \tag{A.6}$$

$$\cos(\alpha \pm \beta) = \cos\alpha\cos\beta \mp \sin\alpha\sin\beta \tag{A.7}$$

$$\tan(\alpha \pm \beta) = \frac{\tan\alpha \pm \tan\beta}{1 \mp \tan\alpha\tan\beta} \tag{A.8}$$

$$\cot(\alpha \pm \beta) = \frac{\cot\alpha\cot\beta \mp 1}{\cot\beta \pm \cot\alpha} \tag{A.9}$$

R. N. Jazar, *Advanced Vehicle Dynamics*,
https://doi.org/10.1007/978-3-030-13062-6

Symmetry

$$\sin(-\alpha) = -\sin\alpha \tag{A.10}$$

$$\cos(-\alpha) = \cos\alpha \tag{A.11}$$

$$\tan(-\alpha) = -\tan\alpha \tag{A.12}$$

Multiple Angles

$$\sin(2\alpha) = 2\sin\alpha\cos\alpha = \frac{2\tan\alpha}{1+\tan^2\alpha} \tag{A.13}$$

$$\cos(2\alpha) = 2\cos^2\alpha - 1 = 1 - 2\sin^2\alpha = \cos^2\alpha - \sin^2\alpha \tag{A.14}$$

$$\tan(2\alpha) = \frac{2\tan\alpha}{1-\tan^2\alpha} \tag{A.15}$$

$$\cot(2\alpha) = \frac{\cot^2\alpha - 1}{2\cot\alpha} \tag{A.16}$$

$$\sin(3\alpha) = -4\sin^3\alpha + 3\sin\alpha \tag{A.17}$$

$$\cos(3\alpha) = 4\cos^3\alpha - 3\cos\alpha \tag{A.18}$$

$$\tan(3\alpha) = \frac{-\tan^3\alpha + 3\tan\alpha}{-3\tan^2\alpha + 1} \tag{A.19}$$

$$\sin(4\alpha) = -8\sin^3\alpha\cos\alpha + 4\sin\alpha\cos\alpha \tag{A.20}$$

$$\cos(4\alpha) = 8\cos^4\alpha - 8\cos^2\alpha + 1 \tag{A.21}$$

$$\tan(4\alpha) = \frac{-4\tan^3\alpha + 4\tan\alpha}{\tan^4\alpha - 6\tan^2\alpha + 1} \tag{A.22}$$

$$\sin(5\alpha) = 16\sin^5\alpha - 20\sin^3\alpha + 5\sin\alpha \tag{A.23}$$

$$\cos(5\alpha) = 16\cos^5\alpha - 20\cos^3\alpha + 5\cos\alpha \tag{A.24}$$

$$\sin(n\alpha) = 2\sin((n-1)\alpha)\cos\alpha - \sin((n-2)\alpha) \tag{A.25}$$

$$\cos(n\alpha) = 2\cos((n-1)\alpha)\cos\alpha - \cos((n-2)\alpha) \tag{A.26}$$

$$\tan(n\alpha) = \frac{\tan((n-1)\alpha) + \tan\alpha}{1 - \tan((n-1)\alpha)\tan\alpha} \tag{A.27}$$

Half Angle

$$\cos\left(\frac{\alpha}{2}\right) = \pm\sqrt{\frac{1+\cos\alpha}{2}} \tag{A.28}$$

$$\sin\left(\frac{\alpha}{2}\right) = \pm\sqrt{\frac{1-\cos\alpha}{2}} \tag{A.29}$$

$$\tan\left(\frac{\alpha}{2}\right) = \frac{1-\cos\alpha}{\sin\alpha} = \frac{\sin\alpha}{1+\cos\alpha} = \pm\sqrt{\frac{1-\cos\alpha}{1+\cos\alpha}} \tag{A.30}$$

$$\sin\alpha = \frac{2\tan\frac{\alpha}{2}}{1+\tan^2\frac{\alpha}{2}} \tag{A.31}$$

$$\cos\alpha = \frac{1-\tan^2\frac{\alpha}{2}}{1+\tan^2\frac{\alpha}{2}} \tag{A.32}$$

Powers of Functions

$$\sin^2\alpha = \frac{1}{2}(1-\cos(2\alpha)) \tag{A.33}$$

$$\sin\alpha\cos\alpha = \frac{1}{2}\sin(2\alpha) \tag{A.34}$$

$$\cos^2\alpha = \frac{1}{2}(1+\cos(2\alpha)) \tag{A.35}$$

$$\sin^3\alpha = \frac{1}{4}(3\sin(\alpha) - \sin(3\alpha)) \tag{A.36}$$

$$\sin^2\alpha\cos\alpha = \frac{1}{4}(\cos\alpha - 3\cos(3\alpha)) \tag{A.37}$$

$$\sin\alpha\cos^2\alpha = \frac{1}{4}(\sin\alpha + \sin(3\alpha)) \tag{A.38}$$

$$\cos^3\alpha = \frac{1}{4}\left(\cos(3\alpha) + 3\cos\alpha\right) \tag{A.39}$$

$$\sin^4\alpha = \frac{1}{8}\left(3 - 4\cos(2\alpha) + \cos(4\alpha)\right) \tag{A.40}$$

$$\sin^3\alpha\cos\alpha = \frac{1}{8}\left(2\sin(2\alpha) - \sin(4\alpha)\right) \tag{A.41}$$

$$\sin^2\alpha\cos^2\alpha = \frac{1}{8}\left(1 - \cos(4\alpha)\right) \tag{A.42}$$

$$\sin\alpha\cos^3\alpha = \frac{1}{8}\left(2\sin(2\alpha) + \sin(4\alpha)\right) \tag{A.43}$$

$$\cos^4\alpha = \frac{1}{8}\left(3 + 4\cos(2\alpha) + \cos(4\alpha)\right) \tag{A.44}$$

$$\sin^5\alpha = \frac{1}{16}\left(10\sin\alpha - 5\sin(3\alpha) + \sin(5\alpha)\right) \tag{A.45}$$

$$\sin^4\alpha\cos\alpha = \frac{1}{16}\left(2\cos\alpha - 3\cos(3\alpha) + \cos(5\alpha)\right) \tag{A.46}$$

$$\sin^3\alpha\cos^2\alpha = \frac{1}{16}\left(2\sin\alpha + \sin(3\alpha) - \sin(5\alpha)\right) \tag{A.47}$$

$$\sin^2\alpha\cos^3\alpha = \frac{1}{16}\left(2\cos\alpha - 3\cos(3\alpha) - 5\cos(5\alpha)\right) \tag{A.48}$$

$$\sin\alpha\cos^4\alpha = \frac{1}{16}\left(2\sin\alpha + 3\sin(3\alpha) + \sin(5\alpha)\right) \tag{A.49}$$

$$\cos^5\alpha = \frac{1}{16}\left(10\cos\alpha + 5\cos(3\alpha) + \cos(5\alpha)\right) \tag{A.50}$$

$$\tan^2\alpha = \frac{1 - \cos(2\alpha)}{1 + \cos(2\alpha)} \tag{A.51}$$

Products of sin and cos

$$\cos\alpha\cos\beta = \frac{1}{2}\cos(\alpha - \beta) + \frac{1}{2}\cos(\alpha + \beta) \tag{A.52}$$

$$\sin\alpha\sin\beta = \frac{1}{2}\cos(\alpha - \beta) - \frac{1}{2}\cos(\alpha + \beta) \tag{A.53}$$

$$\sin\alpha\cos\beta = \frac{1}{2}\sin(\alpha - \beta) + \frac{1}{2}\sin(\alpha + \beta) \tag{A.54}$$

$$\cos\alpha\sin\beta = \frac{1}{2}\sin(\alpha + \beta) - \frac{1}{2}\sin(\alpha - \beta) \tag{A.55}$$

$$\sin(\alpha + \beta)\sin(\alpha - \beta) = \cos^2\beta - \cos^2\alpha = \sin^2\alpha - \sin^2\beta \tag{A.56}$$

$$\cos(\alpha + \beta)\cos(\alpha - \beta) = \cos^2\beta + \sin^2\alpha \tag{A.57}$$

Sum of Functions

$$\sin\alpha \pm \sin\beta = 2\sin\frac{\alpha \pm \beta}{2}\cos\frac{\alpha \pm \beta}{2} \tag{A.58}$$

$$\cos\alpha + \cos\beta = 2\cos\frac{\alpha + \beta}{2}\cos\frac{\alpha - \beta}{2} \tag{A.59}$$

$$\cos\alpha - \cos\beta = -2\sin\frac{\alpha + \beta}{2}\sin\frac{\alpha - \beta}{2} \tag{A.60}$$

$$\tan\alpha \pm \tan\beta = \frac{\sin(\alpha \pm \beta)}{\cos\alpha\cos\beta} \tag{A.61}$$

$$\cot\alpha \pm \cot\beta = \frac{\sin(\beta \pm \alpha)}{\sin\alpha\sin\beta} \tag{A.62}$$

$$\frac{\sin\alpha + \sin\beta}{\sin\alpha - \sin\beta} = \frac{\tan\frac{\alpha+\beta}{2}}{\tan\frac{\alpha-+\beta}{2}} \tag{A.63}$$

$$\frac{\sin\alpha + \sin\beta}{\cos\alpha - \cos\beta} = \cot\frac{-\alpha + \beta}{2} \tag{A.64}$$

$$\frac{\sin\alpha + \sin\beta}{\cos\alpha + \cos\beta} = \tan\frac{\alpha + \beta}{2} \tag{A.65}$$

$$\frac{\sin\alpha - \sin\beta}{\cos\alpha + \cos\beta} = \tan\frac{\alpha - \beta}{2} \tag{A.66}$$

Trigonometric Relations

$$\sin^2\alpha - \sin^2\beta = \sin(\alpha + \beta)\sin(\alpha - \beta) \tag{A.67}$$

$$\cos^2\alpha - \cos^2\beta = -\sin(\alpha + \beta)\sin(\alpha - \beta) \tag{A.68}$$

Appendix B
Unit Conversions

General Conversion Formulas

$$
\begin{aligned}
\mathrm{N}^a\,\mathrm{m}^b\,\mathrm{s}^c &\approx 4.448^a \times 0.3048^b \times \mathrm{lb}^a\,\mathrm{ft}^b\,\mathrm{s}^c \\
&\approx 4.448^a \times 0.0254^b \times \mathrm{lb}^a\,\mathrm{in}^b\,\mathrm{s}^c \\
\mathrm{lb}^a\,\mathrm{ft}^b\,\mathrm{s}^c &\approx 0.2248^a \times 3.2808^b \times \mathrm{N}^a\,\mathrm{m}^b\,\mathrm{s}^c \\
\mathrm{lb}^a\,\mathrm{in}^b\,\mathrm{s}^c &\approx 0.2248^a \times 39.37^b \times \mathrm{N}^a\,\mathrm{m}^b\,\mathrm{s}^c
\end{aligned}
$$

Conversion Factors

Acceleration

$$1\,\mathrm{ft/s^2} \approx 0.3048\,\mathrm{m/s^2} \quad 1\,\mathrm{m/\,s^2} \approx 3.2808\,\mathrm{ft/s^2}$$

Angle

$$1\,\mathrm{deg} \approx 0.01745\,\mathrm{rad} \quad 1\,\mathrm{rad} \approx 57.307\,\mathrm{deg}$$

R. N. Jazar, *Advanced Vehicle Dynamics*,
https://doi.org/10.1007/978-3-030-13062-6

Area

$$1\,\text{in}^2 \approx 6.4516\,\text{cm}^2 \quad 1\,\text{cm}^2 \approx 0.155\,\text{in}^2$$
$$1\,\text{ft}^2 \approx 0.09290304\,\text{m}^2 \quad 1\,\text{m}^2 \approx 10.764\,\text{ft}^2$$
$$1\,\text{acre} \approx 4046.86\,\text{m}^2 \quad 1\,\text{m}^2 \approx 2.471 \times 10^{-4}\,\text{acre}$$
$$1\,\text{acre} \approx 0.4047\,\text{ha} \quad 1\,\text{ha} \approx 2.471\,\text{acre}$$

Damping

$$1\,\text{N s/m} \approx 6.85218 \times 10^{-2}\,\text{lb s/ft} \quad 1\,\text{lb s/ft} \approx 14.594\,\text{N s/m}$$
$$1\,\text{N s/m} \approx 5.71015 \times 10^{-3}\,\text{lb s/in} \quad 1\,\text{lb s/in} \approx 175.13\,\text{N s/m}$$

Energy and Heat

$$1\,\text{Btu} \approx 1055.056\,\text{J} \quad 1\,\text{J} \approx 9.4782 \times 10^{-4}\,\text{Btu}$$
$$1\,\text{cal} \approx 4.1868\,\text{J} \quad 1\,\text{J} \approx 0.23885\,\text{cal}$$
$$1\,\text{kW h} \approx 3600\,\text{kJ} \quad 1\,\text{MJ} \approx 0.27778\,\text{kW h}$$
$$1\,\text{ft lbf} \approx 1.355818\,\text{J} \quad 1\,\text{J} \approx 0.737562\,\text{ft lbf}$$

Force

$$1\,\text{lb} \approx 4.448222\,\text{N} \quad 1\,\text{N} \approx 0.22481\,\text{lb}$$

Fuel Consumption

$$1\ \text{l/100 km} \approx 235.214583\,\text{mi/gal} \quad 1\,\text{mi/gal} \approx 235.214583\,\text{l/100 km}$$
$$1\ \text{l/100 km} = 100\,\text{km/l} \quad 1\,\text{km/l} = 100\ \text{l/100 km}$$
$$1\,\text{mi/gal} \approx 0.425144\,\text{km/l} \quad 1\,\text{km/l} \approx 2.352146\,\text{mi/gal}$$

Length

$$1\,\text{in} \approx 25.4\,\text{mm} \quad 1\,\text{cm} \approx 0.3937\,\text{in}$$
$$1\,\text{ft} \approx 30.48\,\text{cm} \quad 1\,\text{m} \approx 3.28084\,\text{ft}$$
$$1\,\text{mi} \approx 1.609347\,\text{km} \quad 1\,\text{km} \approx 0.62137\,\text{mi}$$

Mass

$$1\,\text{lb} \approx 0.45359\,\text{kg} \quad 1\,\text{kg} \approx 2.204623\,\text{lb}$$
$$1\,\text{slug} \approx 14.5939\,\text{kg} \quad 1\,\text{kg} \approx 0.068522\,\text{slug}$$
$$1\,\text{slug} \approx 32.174\,\text{lb} \quad 1\,\text{lb} \approx 0.03.1081\,\text{slug}$$

Moment and Torque

$$1\,\text{lb ft} \approx 1.35582\,\text{N m} \quad 1\,\text{N m} \approx 0.73746\,\text{lb ft}$$
$$1\,\text{lb in} \approx 8.85075\,\text{N m} \quad 1\,\text{N m} \approx 0.11298\,\text{lb in}$$

Mass Moment

$$1\,\text{lb ft}^2 \approx 0.04214\,\text{kg m}^2 \quad 1\,\text{kg m}^2 \approx 23.73\,\text{lb ft}^2$$

Power

$$1\,\text{Btu/h} \approx 0.2930711\,\text{W} \quad 1\,\text{W} \approx 3.4121\,\text{Btu/h}$$
$$1\,\text{hp} \approx 745.6999\,\text{W} \quad 1\,\text{kW} \approx 1.341\,\text{hp}$$
$$1\,\text{hp} \approx 550\,\text{lb ft/s} \quad 1\,\text{lb ft/ s} \approx 1.8182 \times 10^{-3}\,\text{hp}$$
$$1\,\text{lb ft/h} \approx 3.76616 \times 10^{-4}\,\text{W} \quad 1\,\text{W} \approx 2655.2\,\text{lb ft/h}$$
$$1\,\text{lb ft/min} \approx 2.2597 \times 10^{-2}\,\text{W} \quad 1\,\text{W} \approx 44.254\,\text{lb ft/min}$$

Pressure and Stress

$$1\,\text{lb/in}^2 \approx 6894.757\,\text{Pa} \quad 1\,\text{MPa} \approx 145.04\,\text{lb/in}^2$$
$$1\,\text{lb/ft}^2 \approx 47.88\,\text{Pa} \quad 1\,\text{Pa} \approx 2.0886 \times 10^{-2}\,\text{lb/ft}^2$$
$$1\,\text{Pa} \approx 0.00001\,\text{atm} \quad 1\,\text{atm} \approx 101{,}325\,\text{Pa}$$

Stiffness

$$1\,\text{N/m} \approx 6.85218 \times 10^{-2}\,\text{lb/ft} \quad 1\,\text{lb/ft} \approx 14.594\,\text{N/m}$$
$$1\,\text{N/m} \approx 5.71015 \times 10^{-3}\,\text{lb/in} \quad 1\,\text{lb/in} \approx 175.13\,\text{N/m}$$

Temperature

$$^\circ\text{C} = (^\circ\text{F} - 32)/1.8$$
$$^\circ\text{F} = 1.8\,^\circ\text{C} + 32$$

Velocity

$$1\,\text{mi/h} \approx 1.60934\,\text{km/h} \quad 1\,\text{km/h} \approx 0.62137\,\text{mi/h}$$
$$1\,\text{mi/h} \approx 0.44704\,\text{m/s} \quad 1\,\text{m/s} \approx 2.2369\,\text{mi/h}$$
$$1\,\text{ft/s} \approx 0.3048\,\text{m/s} \quad 1\,\text{m/s} \approx 3.2808\,\text{ft/s}$$
$$1\,\text{ft/min} \approx 5.08 \times 10^{-3}\,\text{m/s} \quad 1\,\text{m/s} \approx 196.85\,\text{ft/min}$$

Volume

$$1\,\text{in}^3 \approx 16.39\,\text{cm}^3 \quad 1\,\text{cm}^3 \approx 0.0061013\,\text{in}^3$$
$$1\,\text{ft}^3 \approx 0.02831685\,\text{m}^3 \quad 1\,\text{m}^3 \approx 35.315\,\text{ft}^3$$
$$1\,\text{gal} \approx 3.785\,\text{l} \quad 1\,\text{l} \approx 0.2642\,\text{gal}$$
$$1\,\text{gal} \approx 3785.41\,\text{cm}^3 \quad 1\,\text{l} \approx 1000\,\text{cm}^3$$

References

Abe, M. (2009). *Vehicle handling dynamics: Theory and application*. Oxford, UK: Butterworth Heinemann.

Andrzejewski, R., & Awrejcewicz, J. (2005). *Nonlinear dynamics of a wheeled vehicle*. New York: Springer.

Beatty, M. F. (1986). *Principles of engineering mechanics: Kinematics—the geometry of motion* (Vol. 1). New York: Plenum Press.

Bourmistrova, A., Simic, M., Hoseinnezhad, R., & Jazar, R. N. (2011). Autodriver algorithm. *Journal of Systemics, Cybernetics and Informatics, 9*(1), 56–66.

Bottema, O., & Roth, B. (1979). *Theoretical kinematics*. Amsterdam: North-Holland Publication.

Clark, S. K. (1971). *Mechanics of pneumatic tires*. Washington, DC: US Government Printing Office.

Cossalter, V. (2002). *Motorcycle dynamics*. Greendale, WI: Race Dynamic Publishing.

Dai, Q. V., Marzbani H., Fard M., & Jazar, R. N. (2016). Caster–Camber relationship in vehicles. In R. N. Jazar & L. Dai (Eds.), *Nonlinear approaches in engineering applications* (Vol. 4). New York: Springer.

Dai, Q. V., Jazar, R. N., & Fard M. (2017a). Kinematics of a steering tyre with adjustable caster. *International Journal of Vehicle Design, 24*(2), 1741–5314. https://doi.org/10.1504/IJVD.2017.085448

Dai, Q. V., Marzbani H., Fard M., & Jazar, R. N. (2017b). Variable caster steering in vehicle dynamics. *Proceedings of the Institution of Mechanical Engineers, Part D: Journal of Automobile Engineering*. https://doi.org/10.1177/0954407017728650

Dieter, S., Hiller, M., & Bardini, R. (2018). *Vehicle dynamics, modeling and simulation* (2nd ed.). Berlin: Springer.

Dixon, J. C. (1996). *Tire, suspension and handling*. Warrendale, PA: SAE.

Dukkipati, R. V., Pang, J. Qatu, M. S., Sheng, G., & Shuguang, Z. (2008). *Road vehicle dynamics*. Warrendale, PA: SAE.

Ellis, J. R. (1994). *Vehicle handling kinematics*. London: Mechanical Engineering Publications.

Fenton, J. (1996). *Handbook of vehicle design analysis*. Warrendale, PA: Society of Automotive Engineers International.

Genta, G. (2007). *Motor vehicle dynamics, modeling and simulation*. Singapore: World Scientific.

Genta, G., & Morello, L. (2009a). *The automotive chassis: Components design* (Vol. 1). New York: Springer.

Genta, G., & Morello, L. (2009b). *The automotive chassis: System design* (Vol. 2). New York: Springer.

R. N. Jazar, *Advanced Vehicle Dynamics*,
https://doi.org/10.1007/978-3-030-13062-6

Goldstein, H., Poole, C., & Safko, J. (2002). *Classical mechanics* (3rd ed.). New York: Addison Wesley.

Haney, P. (2003). *The racing and high-performance tire*. Warrendale, PA: SAE.

Hartman, J. C., Marzbani, H., Alam, F., Fard, M., & Jazar, R. N. (2018). Friction coefficient of pneumatic tires and Bitumen roads. In L. Dai & R. N. Jazar (Eds.), *Nonlinear approaches in engineering applications energy: Vibrations, and modern applications*. New York: Springer.

Hunt, K. H. (1978). *Kinematic geometry of mechanisms*. London: Oxford University Press.

Jazar, R. N. (2010a). *Theory of applied robotics: Kinematics, dynamics, and control* (2nd ed.). New York: Springer.

Jazar, R. N. (2010b). Mathematical theory of autodriver for autonomous vehicles. *Journal of Vibration and Control, 16*(2), 253–279.

Jazar, R. N. (2011). *Advanced dynamics: Rigid body, multibody, and aerospace applications*. New York: Wiley.

Jazar, R. N. (2012). Derivative and coordinate frames. *Journal of Nonlinear Engineering, 1*(1), 25–34. https://doi.org/10.1515/nleng-2012-0001

Jazar, R. N. (2013). *Advanced vibrations: A modern approach*. New York: Springer.

Jazar, R. N., Subic A., & Zhong N. (2012). Kinematics of a smart variable caster mechanism for a vehicle steerable wheel. *Vehicle System Dynamics, 50*(12), 1861–1875.

Jazar, R. N. (2017). *Vehicle dynamics: Theory and application* (3rd ed.). New York: Springer.

Karnopp, D. (2013). *Vehicle dynamics, stability, and control* (2nd ed.). London, UK: CRC Press.

MacMillan, W. D. (1936). *Dynamics of rigid bodies*. New York: McGraw-Hill.

Marzbani, H., & Jazar, R. N. (2015). Steady-state vehicle dynamics. In L. Dai & R. N. Jazar (Eds.), *Nonlinear approaches in engineering applications* (Vol. 3). New York: Springer.

Marzbani, H., Simic, M., Fard, M., & Jazar, R.N. (2015). Better road design for autonomous vehicles using clothoids. In E. Damiani, R. Howlett, L. Jain, L. Gallo & G. De Pietro (Eds.), *Intelligent interactive multimedia systems and services. Smart innovation, systems and technologies* (Vol. 40). Cham: Springer.

Mason, M. T. (2001). *Mechanics of robotic manipulation*. Cambridge, MA: MIT Press.

Milliken, W. F., & Milliken, D. L. (1995). *Race car vehicle dynamics*. Warrendale, PA: SAE.

Milliken, W. F., & Milliken, D. L. (2002). *Chassis design*. Warrendale, PA: SAE.

Pacejka, H. (2012). *Tire and vehicle dynamics* (3rd ed.). Oxford, UK: Butterworth-Heinemann.

Popp, K., & Schiehlen, W. (2010). *Ground vehicle dynamics*. Berlin: Springer.

Schiehlen, W. O. (1982). *Dynamics of high-speed vehicles*. Wien-New York: Springer.

Soni, A. H. (1974). *Mechanism synthesis and analysis*. New York: McGraw-Hill.

Yang, S., Chen, L., & Li, S. (2015). *Dynamics of vehicle-road coupled system*. Berlin: Springer.

Index

R. N. Jazar, *Advanced Vehicle Dynamics*,
https://doi.org/10.1007/978-3-030-13062-6

Printed by Printforce, the Netherlands